Lecture Notes in Physics

The Editorial Policy for Proceedings

The series Lecture Notes in Physics reports new developments in physical research and teaching – quickly, informally, and at a high level. The proceedings to be considered for publication in this series should be limited to only a few areas of research, and these should be closely related to each other. The contributions should be of a high standard and should avoid lengthy redraftings of papers already published or about to be published elsewhere. As a whole, the proceedings should aim for a balanced presentation of the theme of the conference including a description of the techniques used and enough motivation for a broad readership. It should not be assumed that the published proceedings must reflect the conference in its entirety. (A listing or abstracts of papers presented at the meeting but not included in the proceedings could be added as an appendix.)

When applying for publication in the series Lecture Notes in Physics the volume's editor(s) should submit sufficient material to enable the series editors and their referees to make a fairly accurate evaluation (e.g. a complete list of speakers and titles of papers to be presented and abstracts). If, based on this information, the proceedings are (tentatively) accepted, the volume's editor(s), whose name(s) will appear on the title pages, should select the papers suitable for publication and have them refereed (as for a journal) when appropriate. As a rule discussions will not be accepted. The series editors and Springer-Verlag will normally not interfere with the detailed editing except in fairly obvious cases or on technical matters.

Final acceptance is expressed by the series editor in charge, in consultation with Springer-Verlag only after receiving the complete manuscript. It might help to send a copy of the authors' manuscripts in advance to the editor in charge to discuss possible revisions with him. As a general rule, the series editor will confirm his tentative acceptance if the final manuscript corresponds to the original concept discussed, if the quality of the contribution meets the requirements of the series, and if the final size of the manuscript does not greatly exceed the number of pages originally agreed upon. The manuscript should be forwarded to Springer-Verlag shortly after the meeting. In cases of extreme delay (more than six months after the conference) the series editors will check once more the timeliness of the papers. Therefore, the volume's editor(s) should establish strict deadlines, or collect the articles during the conference and have them revised on the spot. If a delay is unavoidable, one should encourage the authors to update their contributions if appropriate. The editors of proceedings are strongly advised to inform contributors about these points at an early stage.

The final manuscript should contain a table of contents and an informative introduction accessible also to readers not particularly familiar with the topic of the conference. The contributions should be in English. The volume's editor(s) should check the contributions for the correct use of language. At Springer-Verlag only the prefaces will be checked by a copy-editor for language and style. Grave linguistic or technical shortcomings may lead to the rejection of contributions by the series editors. A conference report should not exceed a total of 500 pages. Keeping the size within this bound should be achieved by a stricter selection of articles and not by imposing an upper limit to the length of the individual papers. Editors receive jointly 30 complimentary copies of their book. They are entitled to purchase further copies of their book at a reduced rate. As a rule no reprints of individual contributions can be supplied. No royalty is paid on Lecture Notes in Physics volumes. Commitment to publish is made by letter of interest rather than by signing a formal contract. Springer-Verlag secures the copyright for each volume.

The Production Process

The books are hardbound, and the publisher will select quality paper appropriate to the needs of the author(s). Publication time is about ten weeks. More than twenty years of experience guarantee authors the best possible service. To reach the goal of rapid publication at a low price the technique of photographic reproduction from a camera-ready manuscript was chosen. This process shifts the main responsibility for the technical quality considerably from the publisher to the authors. We therefore urge all authors and editors of proceedings to observe very carefully the essentials for the preparation of camera-ready manuscripts, which we will supply on request. This applies especially to the quality of figures and halftones submitted for publication. In addition, it might be useful to look at some of the volumes already published. As a special service, we offer free of charge LATEX and TEX macro packages to format the text according to Springer-Verlag's quality requirements. We strongly recommend that you make use of this offer, since the result will be a book of considerably improved technical quality. To avoid mistakes and time-consuming correspondence during the production period the conference editors should request special instructions from the publisher well before the beginning of the conference. Manuscripts not meeting the technical standard of the series will have to be returned for improvement.

For further information please contact Springer-Verlag, Physics Editorial Department V, Tiergartenstrasse 17, D-69121 Heidelberg, FRG

F. J. Chinea   L. M. González-Romero (Eds.)

Rotating Objects and Relativistic Physics

Proceedings of the El Escorial Summer School
on Gravitation and General Relativity 1992:
Rotating Objects and Other Topics
Held at El Escorial, Spain, 24-28 August 1992

Springer-Verlag
Berlin Heidelberg GmbH

Editors

F. J. Chinea
L. M. González-Romero
Departamento de Física Teórica II
Facultad de Ciencias Físicas
Universidad Complutense
E-28040 Madrid, Spain

ISBN 978-3-662-13934-9   ISBN 978-3-540-48087-7 (eBook)
DOI 10.1007/978-3-540-48087-7

Originally published by Springer-Verlag Berlin Heidelberg New York in 1993
Softcover reprint of the hardcover 1st edition 1993

58/3140-543210 - Printed on acid-free paper

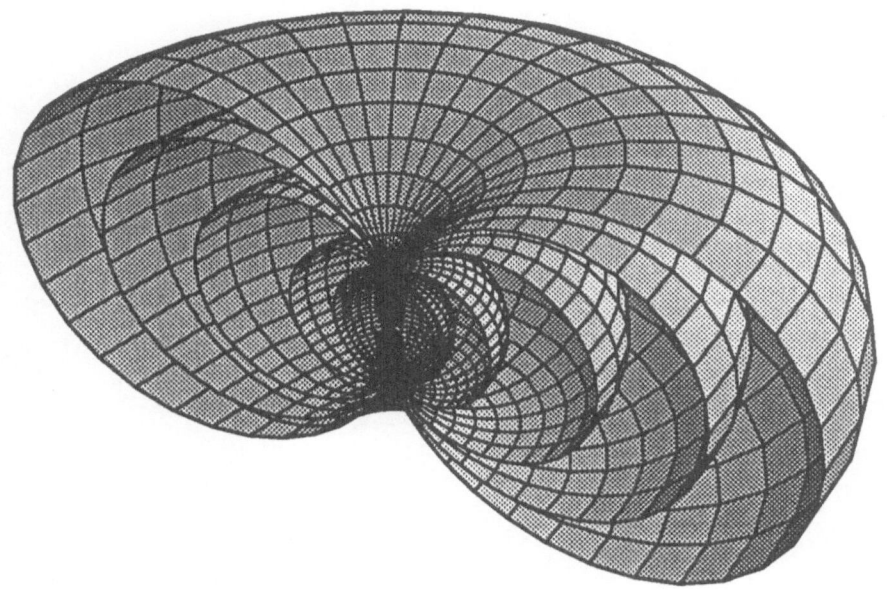
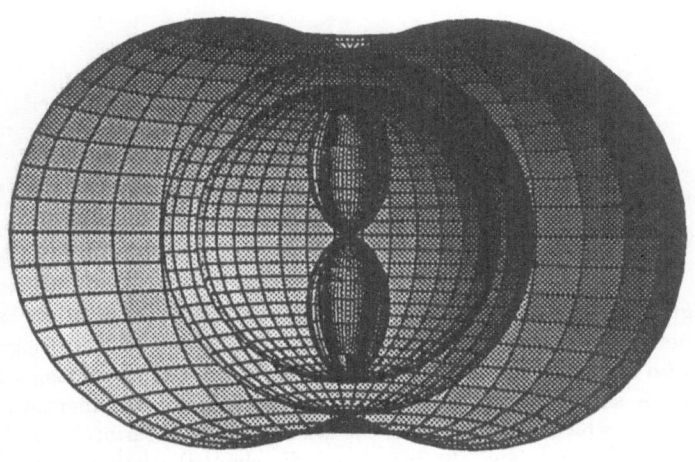

The upper figure in the previous page shows a section of vacuum vorticity surfaces in a Kerr gravitational field; the surfaces of constant twist potential are orthogonal to them. The figure also includes some other physically relevant surfaces: A magnification of the central region is shown below; proceeding inwards, we find the innermost of the outer vorticity surfaces (which coincides with the stationary limit surface), the horizon $(r = r_+)$, the surface $r = r_-$, and the inner (spindle shaped) vorticity surfaces. In order to enhance some of the features, the value 0.99 has been used for the square of the spin-to-mass ratio.

Dedication

To the memory of Basilis C. Xanthopoulos, colleague and friend

Preface

This volume contains invited lectures and contributed talks presented at the Summer School on Gravitation and General Relativity held in El Escorial, Spain, in August 1992. Traditionally, scientists working on general relativity in Spain meet on a yearly basis (*Encuentros Relativistas Españoles*), hosted by the different active groups in succession. When our turn came, we decided to use the permanent facilities owned by Universidad Complutense de Madrid in El Escorial for the purpose of holding Summer Courses in different fields, thus merging the Summer School with the relativists' meeting.

The School centered on the study of the gravitational fields corresponding to rotating objects of astrophysical interest, from the various viewpoints: Theoretical, numerical and observational. Special emphasis was put on the analysis of interior and exterior fields of stationary axisymmetric systems. Lectures and contributions ranged from basic surveys, useful for the younger participant, to technical points pertaining to current research in this area; they are collected in Part I of this volume.

Part II contains lectures and contributions on other topics of gravitation theory. Some of the talks are summarized in the form of abstracts.

F.J. Chinea
L.M. González-Romero

Madrid
July 1993

Acknowledgements

The El Escorial Summer Courses are run by Universidad Complutense de Madrid; funds for the organization of the courses are provided by Banco Central Hispano. We would like to thank both institutions, as well as Prof. M.A. Alario, who coordinated the scientific courses and provided much help in ironing out the problems in the day-to-day operation of the School. Partial financial support from Grupo Interuniversitario de Física Teórica is gratefully acknowledged.

Finally, our thanks go also to the other co-organizers of the School, Dr. J.A. Ruiz-Martín and Mr. L. Fernández-Jambrina.

Contents

Contributed Talks

I

Rotating Objects

Equilibrium Configurations of General Relativistic Rotating Stars

Y. Eriguchi

Department of Earth Science and Astronomy, College of Arts and
Sciences, University of Tokyo, Komaba, Meguro, Tokyo 153, Japan

Abstract: The present status of studies of equilibrium structures of rotating stars is
reviewed. First we briefly summarize the results of equilibrium states of rotating stars
in Newtonian gravity. Second the problem of equilibrium of rotating stars in general
relativity is formulated. After we explain several approximate methods such as the
post-Newtonian scheme and the slow rotation approximation, we discuss two powerful
schemes to solve rapidly rotating general relativistic stars. These schemes have been
applied to construct rapidly rotating neutron stars. However there seems a discrepancy
between results obtained by these two different schemes. As another application of
the powerful numerical scheme, we discuss the structures of toroids around compact
objects such as black holes and the relation to the axisymmetric instability of mass
overflow from the Roche lobe of the toroids.

1 Introduction

Equilibrium configurations of self-gravitating rotating bodies have been investigated since Newton discovered the theory of gravity and applied it to the shape
of the earth. Shapes of rotating *fluids* were mainly studied because configurations
with a constant density could be easily handled.

In 1742 Maclaurin discovered that spheroidal axisymmetric configurations
can be in equilibrium when they rotate uniformly (Maclaurin spheroids). Equilibrium models of this Maclaurin sequence can be characterized by one parameter
such as the ellipticity of the meridional cross section or the angular momentum.
It should be noted that the angular velocity is not a good parameter to specify
the Maclaurin spheroid because there are two different equilibrium states for
one value of the angular velocity. About 100 years later (1834) Jacobi discovered
that non-axisymmetric ellipsoids can be equilibrium states of uniformly rotating fluids (Jacobi ellipsoids). This Jacobi sequence exists only when the angular
momentum of the fluid exceeds a certain critical value. It implies that uniformly
and slowly rotating fluids must be axisymmetric but that rapidly rotating fluids

can become non-axisymmetric as well as axisymmetric. Furthermore since the Jacobi sequence bifurcates from the Maclaurin sequence, this change can occur continuously by increasing the amount of the angular momentum if the mass and the density of the fluid are kept fixed.

In 1860 Riemann introduced uniform vorticity as well as uniform rotation and found that, as far as the velocity field is expressed by a linear combination of the Cartesian coordinates, ellipsoidal configurations are allowed to be in equilibrium states (Riemann ellipsoids). In other words fluids with a constant density can be in equilibrium when there is a special kind of internal motion together with uniform rotation. As a special family of Riemann ellipsoids there are ellipsoidal configurations whose non-axisymmetric shapes are maintained by the internal motion alone. These configurations are called Dedekind ellipsoids.

As far as rotating configurations with spheroidal or ellipsoidal shapes are concerned, many important effects of rotation on equilibrium configurations have been understood. However configurations with a constant density seem to be far from realistic configurations of compressible gases, i.e., real stars.

Therefore many efforts have been devoted to the study of the effect of rotation on compressible stars. It has been so difficult to solve the governing equations analytically that various approximate schemes have been developed. When rotation is considered to be slow, the problem can be treated perturbationally with respect to a small parameter which characterizes the slow rotation. Since the perturbational treatments are not powerful enough to clarify the nature of rotating gases, one needs other methods to get a full understanding of rapidly rotating stars.

The advent of high speed computers seemed to give us useful tools to obtain equilibrium models for rapidly rotating stars. However computers with moderate computational power could not solve equilibrium structures of compressible gaseous stars. Mathematically speaking we have to solve boundary value problems for elliptical partial differential equations with free boundary. In an ordinary situation elliptical type equations are approximately handled by solving a set of linear equations, i.e., by inverting a matrix. Even for axisymmetric configurations the size of the matrix becomes so large that a huge amount of computational time is required to solve it. Moreover since the surface of the star is unknown, it is difficult to apply boundary conditions at the surface.

Under such a situation, equilibrium configurations of rapidly rotating compressible Newtonian stars were first solved by James [1] in 1964. However his scheme was too complicated to be applied to general problems. Moreover his formulation did not work for models with high compressibility. Around 1970 Ostriker and his colleagues [2]-[9] developed the Self-Consistent-Field (SCF) method and applied it to rotating Newtonian barotropes. However, their SCF method could not be applied to highly deformed and/or very rapidly rotating configurations. About 10 years later Eriguchi and Hachisu [10]-[28] succeeded in developing simple and powerful numerical schemes for obtaining equilibrium structures of rotating stars. Among several approaches developed by them two different schemes are powerful and practical. One is the Hachisu's Self-

Consistent-Field (HSCF) scheme developed by Hachisu [26,28] and the other is the StraightForward-Newton-Raphson (SFNR) scheme developed by Eriguchi and Müller [22]. They computed many kinds of configurations and found many new equilibrium sequences. Thus as far as the Newtonian gravity is concerned, we are in a state where structures of rotating stars can be easily obtained numerically by using the HSCF scheme or the SFNR scheme.

Concerning the strong gravity, since no exact solutions have been found, the effect of rotation has been estimated by extrapolating the results of Newtonian configurations. It was in 1967 that the problem for slowly rotating general relativistic configurations was formulated perturbationally by Hartle [29]. Using this formulation many authors have computed slowly rotating general relativistic configurations.

As for rapidly rotating configurations a thin disk was solved by Bardeen and Wagoner [30]. The thin disk can, however, be treated as a boundary condition to the Einstein equations. Therefore it would be precise to say that they solved vacuum solutions with special boundary conditions on the equator.

The internal structure of general relativistic rotating stars was, therefore, first solved by Wilson [31,32]. He found that an ergo-region appeared inside the matter of rapidly rotating models. However, one of his boundary conditions was Newtonian-like so that his solutions could not be accepted as accurate for highly relativistic cases. Almost at the same time Bonazzola and Schneider [33] used integral representations of the Einstein equations and computed relativistic incompressible fluids and compressible models of ideal Fermi gases. In their treatment, however, there was a very severe restriction that the metric component g_{tt} could never become positive (in this paper the sign convention is taken as $(-,+,+,+)$). Thus they could not solve models with very strong gravity.

Butterworth and Ipser [34,35] succeeded in computing equilibrium structures of incompressible fluids even for very strong gravity. Their numerical scheme was an extension of the Newtonian scheme developed by Stoeckly [36]. The non-linear differential equations, i.e., the Einstein equations, were solved by a Newton-Raphson-like iteration scheme. In particular they treated the boundary conditions very carefully because in numerical computations only a finite region could be handled. However its scheme did not work for the compressible gas [37].

Although the reason of the failure of the application of their scheme to *compressible* gases was not clearly explained, it may be that their scheme could not be applied to significantly deformed configurations. It may sound strange that their scheme can give solutions for fluids with very rapid rotation and very strong gravity but not for gases with rapid rotation and strong gravity. This situation can be understood as follows. For incompressible fluids, the mass in the outer region is large enough to weaken the effect of rotation which is apt to deform the configuration. Therefore deformation of incompressible fluids cannot become so large. On the other hand for compressible gases the density of the outer region is so low that the rotational effect becomes significant in the surface region of the stars. Compared with models in Newtonian gravity, general relativistic configurations are much more centrally condensed. In other words the effect of general relativity can be considered to make the matter more compressible, which implies that rotation can deform shapes of general relativistic compressible stars considerably. Therefore only numerical schemes which will succeed in handling significantly deformed configurations can solve rapidly rotating general relativistic *compressible* gases.

Komatsu, Eriguchi and Hachisu [38,39] have succeeded in developing a computational scheme which satisfied the requirements mentioned above. They extended the Newtonian scheme developed by Hachisu [26,28] to general relativistic configurations. They can compute many kinds of configurations even for very soft equation of states. Thus in general relativistic regime, too, we have reached a stage where we can solve structures of rapidly rotating bodies in a practically short time as far as mechanical equilibrium is concerned.

2 Axisymmetric Equilibrium States of Rotating Stars

2.1 Assumptions

Equilibrium configurations of rotating stars are considered on the following assumptions.

In Newtonian gravity there can be rotating non-axisymmetric equilibrium states, while in the framework of general relativity rotating non-axisymmetric spacetime and configurations will not be in equilibrium states unless shapes of configurations are standing still in the inertial frame. Thus we assume that the spacetime and the matter are *axisymmetric* and *stationary*. It implies that there are two Killing vectors. The matter is assumed to be *perfect fluid* and be confined in a compact region in the space so that the spacetime is *asymptotically flat*. The energy momentum tensor can be expressed as

$$T^{ij} = (\varepsilon + p)u^i u^j + pg^{ij}, \tag{1}$$

where T^{ij}, ε, p, u^i, and g^{ij} are the energy momentum tensor, the energy density, the pressure, the four velocity and the metric, respectively. The equation of state is assumed *barotropic*:

$$p = p(\varepsilon). \tag{2}$$

2.2 Metric

The stationary and axisymmetric spacetime can be expressed by the following metric [40,41]:

$$ds^2 = g_{00}dt^2 + 2g_{03}dtd\varphi + g_{33}d\varphi^2 + g_{11}(dx^1)^2 + g_{22}(dx^2)^2, \tag{3}$$

where t and φ are the time and the azimuthal coordinates corresponding to two Killing vectors, respectively. When we use the spherical coordinates (r, θ, φ), this metric can be written as:

$$ds^2 = -e^{2\nu}dt^2 + e^{2\alpha}(dr^2 + r^2 d\theta^2) + r^2 \sin^2 \theta e^{2\beta}(d\varphi - \omega dt)^2, \tag{4}$$

where ν, α, β and ω are four gravitational potentials. In particular ω is the frame dragging potential [40,41].

When we are interested only in the vacuum region, the potential $\beta + \nu$ can be set to vanish and the metric can be reduced to the following form [42]:

$$ds^2 = f^{-1}[e^{2\bar{\gamma}}(dR^2 + dz^2) + R^2 d\varphi^2] - f(dt - \tilde{\omega}d\varphi)^2, \tag{5}$$

in the cylindrical coordinates. However we will not refer to this form of the metric in this review any more.

2.3 Boundary Conditions

Since the matter is confined in a compact region, there is a surface which divides the matter region and the vacuum region. On that surface the pressure vanishes. When a rotating star or a black hole is surrounded by axisymmetric configuration(s) such as toroid(s), we need to consider two or more surfaces corresponding to matter configurations, e.g., the surface of the central star and the surface of the toroid and so on.

The asymptotically flat conditions for the metric (4) are expressed as:

$$\nu \rightarrow -\frac{M}{r} + O(\frac{1}{r^2}), \tag{6a}$$

$$\beta \rightarrow \frac{M}{r} + \frac{\beta_0}{r^2} + O(\frac{1}{r^3}), \tag{6b}$$

$$\omega \rightarrow \frac{2J}{r^3} - \frac{6MJ}{r^4} + O(\frac{1}{r^5}), \tag{6c}$$

$$\alpha \rightarrow \beta + O(\frac{1}{r^4}), \tag{6d}$$

where M, J and β_0 are the gravitational mass, the total angular momentum and a certain constant, respectively.

When we consider a rotating star, the gravitational potential must be regular at its center. However the situation is different for a system consisting of a black hole and a surrounding axisymmetric matter. If we choose coordinates in which the horizon is located on the surface of $r = constant$ [40], the values of the metric on the horizon of the black hole should be:

$$e^\nu = 0, \tag{7a}$$

$$\omega = \omega_{\rm h}(= {\rm constant}), \tag{7b}$$

and

$$\beta = {\rm finite}, \tag{7c}$$

where $\omega_{\rm h}$ is the angular velocity of the horizon.

One more boundary condition is derived from the local flatness of the spacetime on the rotation axis, which reads

$$\alpha = \beta, \quad {\rm on\ the\ z\text{-axis}} \quad . \tag{8}$$

The problem is, therefore, to solve the Einstein equations derived from the metric (4) with boundary conditions mentioned above. However it is not easy to solve them directly. Historically speaking, as discussed in the introduction, models with weak gravity and/or for slow rotation have been investigated in the first stage.

3 Post-Newtonian Rotating Stars

Equilibrium configurations of Newtonian stars are not discussed in this paper but various schemes and results are found in references [1]-[28,36].

When the gravity becomes a little stronger, the problem can be handled in the post-Newtonian frame. Chandrasekhar [43] formulated the problem to the order of $1/c^2$ and Chandrasekhar and Nutku [44] extended it to the order of $1/c^4$. In this section the light velocity is explicitly written.

In the post-Newtonian treatment of *uniformly* rotating stars the metric can be expressed by using three potentials, U, Φ and D as follows:

$$g_{00} = -c^2 \left(1 + 2\frac{U}{c^2} + \frac{2U^2 + 4\Phi}{c^4} \right), \tag{9a}$$

$$g_{11} = g_{22} = g_{33} = 1 - \frac{2U}{c^2}, \tag{9b}$$

$$g_{01} = \frac{4\Omega x_2 D}{c^3}, \tag{9c}$$

and

$$g_{02} = -\frac{4\Omega x_1 D}{c^3}, \tag{9d}$$

where Ω and (x_1, x_2, x_3) are the angular velocity and the Cartesian coordinates, respectively. These potentials must satisfy the following differential equations:

$$\nabla^2 U = 4\pi G \tilde{\rho}, \tag{10a}$$

$$\nabla^2 D + \frac{1}{r}\frac{\partial D}{\partial r} - \frac{\mu}{r^2}\frac{\partial D}{\partial \mu} = 4\pi G \tilde{\rho}, \tag{10b}$$

and

$$\nabla^2 \Phi = 4\pi G \tilde{\rho} \left[\Omega^2 r^2 (1 - \mu^2) - U + \frac{1}{2}\Pi + \frac{3}{2}\frac{p}{\tilde{\rho}} \right], \tag{10c}$$

where $\mu = \cos\theta$, $\tilde{\rho}$ and $\tilde{\rho}\Pi$ are the mass density and the internal energy density which is defined by

$$\varepsilon = \tilde{\rho}(c^2 + \Pi). \tag{11}$$

The hydrostatic equilibrium equation can be written as

$$\left[1 - \frac{\Pi + p/\tilde{\rho}}{c^2} \right] \nabla p = \tilde{\rho} \nabla \left[-U + \frac{1}{2}\Omega^2 r^2 \sin^2\theta \right]$$

$$+\frac{1}{c^2}\left(\frac{1}{4}(\Omega r\sin\theta)^4-2U(\Omega r\sin\theta)^2-2\Phi+4(\Omega r\sin\theta)^2D\right)\right]. \qquad (12)$$

These equations were solved for slowly rotating polytropes by Fahlman and Anand [45] when $1.5 \le N \le 3$ where N is the polytropic index. For rapidly rotating cases Miketinac and Barton [46] applied the Stoeckly's scheme [36] and obtained equilibrium sequences for $N = 1.5$ and $N = 3$.

4 Slow Rotation Approximation

The rotation can be characterized by the following quantity:

$$\omega^2 \equiv \frac{\Omega^2}{4\pi G\varepsilon_c}, \qquad (13)$$

where ε_c is the maximum energy density. This parameter is roughly equal to the ratio of the centrifugal force to the gravitational force or the ratio of the rotational energy to the gravitational energy. Thus when stars rotate slowly, this parameter is so small that we can expand physical quantities in terms of this parameter.

The convenient formulation to treat slowly rotating stars was first derived by Hartle [29]. The metric for slowly rotating stars can be expressed as:

$$ds^2 = -e^{2\nu}dt^2 + e^{2\lambda}dr^2 + r^2F^2[d\theta^2 + \sin^2\theta(d\varphi - \omega\,dt)^2], \qquad (14)$$

where

$$e^{2\nu} = e^{2\nu_0}[1 + 2h_0(r) + 2h_2(r)P_2(\cos\theta)], \qquad (15a)$$

$$e^{2\lambda} = e^{2\lambda_0}\left(1 + \frac{e^{2\lambda_0}}{r}[2m_0(r) + 2m_2(r)P_2(\cos\theta)]\right), \qquad (15b)$$

$$F^2 = 1 + 2k_2(r)P_2(\cos\theta), \qquad (15c)$$

and ν_0 and λ_0 are those for spherical models. From the $t - \varphi$ component of the Einstein equations we can derive the following equation for the dragging potential ω:

$$\frac{1}{r^4}\frac{d}{dr}\left(r^4 j\frac{d\bar\omega}{dr}\right) + \frac{4}{r}\frac{dj}{dr}\bar\omega = 0, \qquad (16)$$

where

$$\bar\omega \equiv \Omega - \omega, \qquad (17)$$

and

$$j \equiv e^{-(\nu_0+\lambda_0)}. \qquad (18)$$

The boundary condition for $\bar\omega$ at the infinity is

$$\bar\omega \to \Omega - \frac{2J}{r^3}. \qquad (19)$$

It is important to note that we can solve (16) for the dragging potential without knowing other perturbed quantities. Other perturbed quantities are obtained after the dragging potential is solved.

These equations are numerically computed by Hartle and Thorne[47] for models with the Harrison-Wheeler equation of state, with the V_γ equation of state and for massive stars with $N = 3$ polytrope. They applied their results to estimate the increase of the gravitational mass due to rotation and found that rotation could increase the gravitational mass about 20% as far as uniform rotation is concerned. This formulation has been used by many authors for slowly rotating polytropic models [48,49] and for slowly rotating neutron stars [50,51].

5 Rapidly Rotating Relativistic Disks

Thin disks are not always interesting objects from the standpoint of the internal structure. Nonetheless the behavior of the spacetime deserves investigation because the rotational effect of the matter in a highly relativistic regime can be revealed. Bardeen and Wagoner [30] treated uniformly rotating infinitesimally thin disks. They used almost the same metric as (4) in the cylindrical coordinates (R, z, φ). Since the space is vacuum except on a part of the equatorial plane ($R \leq a$, where a is the coordinate radius of the disk), the metric component $\beta + \nu$ can be set to vanish, i.e.

$$\beta + \nu = 0. \tag{20}$$

Moreover the infinitesimally thin disk approximation results in $p = 0$ within the disk.

Two components of the Einstein equations for the vacuum become as follows:

$$\nabla^2 \nu = \frac{1}{2} R^2 e^{-4\nu} \nabla\omega\nabla\omega, \tag{21a}$$

and

$$R^{-2}\nabla(R^2\nabla\omega) = 4\nabla\nu\nabla\omega. \tag{21b}$$

Here it should be noted that in these equations the metric α does not appear. Thus we can obtain two metric components ν and ω first and then the metric α is obtained from a first order differential equation which we will not write here.

The boundary conditions can be obtained by integrating the Einstein equations with matter terms through the disk, which read:

$$\frac{\partial\nu}{\partial\xi} = 2\pi a|\eta|\sigma\frac{1+v^2}{1-v^2}, \tag{22a}$$

$$\frac{\partial\omega}{\partial\xi} = -8\pi a|\eta|\sigma\frac{\Omega-\omega}{1-v^2}, \tag{22b}$$

where ξ and η are the oblate spheroidal coordinates defined by

$$R = a(1+\xi)^{1/2}(1-\eta^2)^{1/2}, \qquad z = a\xi\eta, \tag{23}$$

and σ is the surface density

$$\sigma \equiv \int \varepsilon e^{2\alpha} dz, \tag{24}$$

and

$$v \equiv (\Omega - \omega) R e^{-2\nu}. \tag{25}$$

Furthermore the matter in the disk is in an equilibrium state so that the equation of hydrostatic equilibrium can be integrated to give:

$$e^\nu (1 - v^2)^{1/2} = \text{constant}, \tag{26}$$

inside the matter on the equator.

Structures of the disk and the spacetime can be obtained by solving (21a) and (21b) for ν, ω and σ with conditions (6a) and (6c) at infinity and conditions (22a), (22b) and (26) on the equator.

Bardeen and Wagoner [30] made use of the expansion of physical quantities in terms of the quantity δ which is defined by

$$\delta \equiv 1 - e^{2\nu_c}, \tag{27}$$

where ν_c is the value of the metric at the center of the disk. This quantity measures the strength of gravity or is related to the red shift factor at the center z_c as follows:

$$\delta = \frac{z_c}{1 + z_c}. \tag{28}$$

For example the metric ν is expanded as

$$\nu = \sum_{n=1}^{\infty} \nu_n(\xi, \eta) \delta^n. \tag{29}$$

In this way Bardeen and Wagoner [30] obtained the binding energy (E_b), the gravitational mass (M) and so on of the thin disks. They can be written as:

$$\frac{E_b}{M_0} = \frac{\delta}{5}[1 + \frac{2}{5}\delta + 0.1879\delta^2 + 0.1002\delta^3 + \cdots], \tag{30}$$

$$\frac{M^2}{J} = \frac{10}{3\pi}\delta^{1/2}[1 + 0.05\delta - 0.02362\delta^2 - 0.02399\delta^3 - \cdots], \tag{31}$$

where M_0 and J are the rest mass and the total angular momentum, respectively. The limiting values of these quantities for $\delta \to 1$ are

$$\frac{E_b}{M_0} \to 0.373, \tag{32}$$

$$\frac{M^2}{J} \to 1.00014. \tag{33}$$

These values are those of the extreme Kerr metric so that in this limit the spacetime will become the extreme Kerr spacetime. One special feature of the spacetime around the thin disk is the appearance of the ergo-region. In particular the topology of the ergo-region is not spheroidal but toroidal, i.e., no part of the rotation axis is included inside the ergo-region.

Salpeter and Wagoner [52] treated the effect of the pressure of the disks.

6 Rapidly Rotating Relativistic Stars

6.1 Various Attempts

Wilson [31] computed structures of massive stars with the polytropic index $N = 3$ and the spacetime around it in the framework of general relativity. He wrote down the Einstein equations and solved the differential equations iteratively and obtained solutions.

However his treatment had two shortcomings. First he fixed the matter distribution. Thus he obtained the gravitational field and the rotation law which are consistent with the prescribed matter distribution. By this kind of approach, unless we know the matter distribution rather precisely, we cannot obtain realistic models.

Second point is related to the boundary condition of the metric. Since in the numerical computation we can only handle a finite region, we need to be very careful about the *asymptotically flat* condition. Wilson applied Newtonian-like boundary condition to the metric ν. Therefore his computation would give results different from the real solutions especially when very strong gravity is treated.

Although his scheme had limitations as mentioned above, he found the ergo-toroidal region for the first time when the matter was taken into account.

After that Wilson [32] obtained models of rapidly rotating neutron stars by revising his code whose details, however, were not clearly described. He computed differentially rotating models and applied the stability criterion of the Newtonian analysis that rotating configurations may become unstable against the mode with $m = 2$ deformation when the ratio of the rotational energy to the gravitational energy exceeds 0.14. By using this criterion he concluded that the rotation could increase the gravitational mass of the neutron star by $50 - 70\%$ compared with that of spherical stars.

Almost at the same time Bonazzola and Schneider [33] developed a completely different numerical scheme to handle general relativistic rotating stars. Their idea was to use the integral representation of the potentials. If the non-linear terms with respect to the potentials are treated as source terms, the remaining linear differential operators can be transformed to integral forms by using appropriate Green functions corresponding to the operators. The most important point in this kind of ideas is that we can easily include the boundary conditions at infinity. Furthermore we need not fit or match the interior solutions to the outer solutions at the surface of the rotating stars.

Although their basic ideas were fine, their code was not appropriate. They used $\ln(-g_{tt})$ as one of the variables. It implies that they could not treat the spacetime in which a region with $g_{tt} > 0$ appeared, i.e., very strong gravity case. (In their paper [33] the sign convention is taken as $(+, -, -, -)$ but in this paper the sign convention is taken as $(-, +, +, +)$ as mentioned before.)

They computed incompressible fluids and completely degenerate Fermi gas models. For incompressible fluids their results were compared with results of Newtonian (Maclaurin sequence) and post-Newtonian treatments. The agreement was good as far as the strength of gravity was not so large. However in their results iso-density contours in the central region of the degenerate gases became prolate. It is very hard to deform the central region to prolate shape by rotation, while they interpreted that the effect of the dragging could make the shape prolate. The dragging potential can never exceed the value of the angular velocity at the same position so that it cannot act as "anti-centrifugal" force which might result in prolate shape.

6.2 Butterworth and Ipser Scheme

First satisfactory configurations for rapidly rotating and highly general relativistic stars were obtained by Butterworth and Ipser [34,35]. They solved the differential equations by applying a Newton-Raphson-like iteration scheme. The boundary conditions were carefully analyzed because only a finite region could be handled in numerical computations as mentioned before. In practice multipole expansions for the metric were done to the orders of $1/r^5$, $1/r^7$ and $1/r^6$ for ν, ω and β, respectively.

If a set of non-linear equations, i.e., the Einstein equations, were solved by a Newton-Raphson scheme, a matrix of a very large size would be inverted, which would be very time consuming and thus far from practical. Therefore they handled the problem as follows. When the second order derivative of a certain metric with respect to coordinates appears in a certain equation, its equation is treated as the equation for its metric by assuming that other components of the metric which are necessarily included in the equation are treated as given quantities. Then the equation is linearized with respect to the metric concerned and the "Newton-Raphson" iteration scheme is applied. After the iteration converges, the second metric and the second equation are chosen and the same procedure is followed. While the second metric is being solved, the newly converged values for the first metric is used in the equations for the second metric. The convergence of this scheme is not assured but as the uniformly rotating incompressible fluids are concerned, they have succeeded in computing many equilibrium configurations.

Their models for the incompressible fluids cover very wide range of parameters in both rotation and gravity. For equilibrium sequences with very strong gravity they found that an ergo-toroidal region appears when rotation becomes fast. Therefore their scheme can be said very satisfactory for investigation of general relativistic rotating stars.

However their scheme did not give solutions for *Newtonian* or *post-Newtonian* configurations if the rotation is very rapid or if deformation becomes significant. They did not obtain equilibrium configurations with the axis ratio less than 0.4, i.e., $r_p/r_e < 0.4$, where r_p and r_e are the polar radius and the equatorial radius of the star, respectively. As the strength of gravity decreases, the role of gravity becomes weak compared with that of rotation. Configurations with relatively weak gravity are apt to deform much more easily due to rotation. Therefore highly deformed configurations might not be solved by their code.

Butterworth [37] applied this scheme to the compressible stars. However his code could not give equilibrium solutions for rapidly rotating and highly relativistic models. He computed polytropic configurations with $0.5 \le N \le 3.0$. The strength of gravity can be characterized by the following quantity:

$$\kappa \equiv \frac{p_c}{\varepsilon_c}, \tag{34}$$

where subscript $_c$ denotes that the quantities are evaluated at the position where the energy density becomes maximum. He could compute models with $\kappa < 0.25$, which was not highly relativistic. This contrasted with the fact that for incompressible cases their code gave solutions even for $\kappa = 2 \sim 3$. Butterworth [37] discussed that the reason of the failure was unclear. Reasons may be that their scheme was not suitable for highly deformed configurations or that he could use too small numbers of mesh points to resolve the structure of compressible gases.

Butterworth and Ipser's scheme is, therefore, nicely applied to models with wide range of parameters both for rotation and for gravity but may not be powerful enough to obtain models with any strength of parameters for rotation and gravity. Numerical schemes with much wider applicability are desired for solving highly deformed configurations.

6.3 KEH Scheme

In the Newtonian framework Hachisu [26,28] succeeded in developing a very versatile scheme to treat any kind of rotation law and any kind of deformations so far as barotropes are concerned. The basic idea of Hachisu's scheme is the extension of the SCF method by Ostriker et al. [2]

In the SCF method the problem is divided into two parts: the potential part and the density part. In the potential part the assumed density distribution is used to compute the gravitational potential. After obtaining the gravitational potential, the density is calculated from the equation of the hydrostatic equilibrium by using the obtained gravitational potential. This iteration cycle is pursued until the density and the potential change no more. In this iteration scheme the most important point is the choice of model parameters whose values are fixed all through iteration cycles. If improper parameters are chosen, the iteration will diverge.

Komatsu et al. [38,39] have extended the Hachisu's scheme to general relativistic stars. The basic idea was to use the integral representation of the Einstein equations just as Bonazzola and Schneider [33] did and to apply the Hachisu's scheme for iteration to solve the non-linear integral equations.

We briefly summarize the Komatsu el al.'s scheme (hereafter KEH scheme). The Einstein equations for ρ, γ and ω can be written as follows:

$$\Delta \left[\rho e^{\gamma/2} \right] = S_\rho(r, \mu), \tag{35}$$

$$(\Delta + \frac{1}{r}\frac{\partial}{\partial r} - \frac{1}{r^2}\mu\frac{\partial}{\partial \mu})\gamma e^{\gamma/2} = S_\gamma(r, \mu), \tag{36}$$

$$(\Delta + \frac{2}{r}\frac{\partial}{\partial r} - \frac{2}{r^2}\mu\frac{\partial}{\partial \mu})\omega = S_\omega(r, \mu), \tag{37}$$

where

$$S_\rho = e^{\gamma/2}[8\pi e^{2\alpha}(\varepsilon + p)\frac{1+v^2}{1-v^2} + r^2(1-\mu^2)e^{-2\rho}\nabla\omega \cdot \nabla\omega + \frac{1}{r}\gamma_{'r} - \frac{\mu}{r^2}\gamma_{'\mu}$$

$$+ \frac{\rho}{2}(16\pi e^{2\alpha}p - \frac{1}{r}\gamma_{'r} + \frac{\mu}{r^2}\gamma_{'\mu} - \frac{1}{2}\nabla\gamma \cdot \nabla\gamma)], \tag{38}$$

$$S_\gamma = e^{\gamma/2}[16\pi e^{2\alpha}p + \frac{\gamma}{2}(16\pi e^{2\alpha}p - \frac{1}{2}\nabla\gamma \cdot \nabla\gamma)], \tag{39}$$

$$S_\omega = \nabla(4\nu - 3\gamma) \cdot \nabla\omega - 16\pi e^{2\alpha}(\varepsilon + p)\frac{\Omega - \omega}{1 - v^2}, \tag{40}$$

$$\gamma \equiv \nu + \beta, \tag{41}$$

and

$$\rho \equiv \nu - \beta. \tag{42}$$

Here Δ and ∇ are the Laplacian and the gradient operators in the flat 3-space, respectively, $\mu = \cos\theta$, and the subscript $_'$ denotes differentiation with respect to the following variable. By using the Einstein equations, the metric function α satisfies the following equation:

$$\alpha_{'\mu} = -\nu_{'\mu} - \{(1-\mu^2)(1+rB^{-1}B_{'r})^2 + [\mu - (1-\mu^2)B^{-1}B_{'\mu}]^2\}^{-1}$$

$$[\frac{1}{2}B^{-1}\{r^2B_{'rr} - [(1-\mu^2)B_{'\mu}]_{'\mu} - 2\mu B_{'\mu}\}[-\mu + (1-\mu^2)B^{-1}B_{'\mu}]$$

$$+ rB^{-1}B_{'r}[\frac{1}{2}\mu + \mu rB^{-1}B_{'r} + \frac{1}{2}(1-\mu^2)B^{-1}B_{'\mu}]$$

$$+ \frac{3}{2}B^{-1}B_{'\mu}[-\mu^2 + \mu(1-\mu^2)B^{-1}B_{'\mu}] - (1-\mu^2)rB^{-1}B_{'\mu r}(1+rB^{-1}B_{'r})$$

$$- \mu r^2 v_{'r}^2 - 2(1-\mu^2)r\nu_{'\mu}\nu_{'r} + \mu(1-\mu^2)v_{'\mu}^2 - 2(1-\mu^2)r^2B^{-1}B_{'r}$$

$$\times \nu_{'\mu}\nu_{'r} + (1-\mu^2)B^{-1}B_{'\mu}[r^2v_{'r}^2 - (1-\mu^2)v_{'\mu}^2] + (1-\mu^2)B^2e^{-4\nu}$$

$$\times \{\frac{1}{4}\mu r^4\omega_{'r}^2 + \frac{1}{2}(1-\mu^2)r^3\omega_{'\mu}\omega_{'r} - \frac{1}{4}\mu(1-\mu^2)r^2\omega_{'\mu}^2 + \frac{1}{2}(1-\mu^2)$$

$$\times r^4B^{-1}B_{'r}\omega_{'\mu}\omega_{'r} - \frac{1}{4}(1-\mu^2)r^2B^{-1}B_{'\mu}[r^2\omega_{'r}^2 - (1-\mu^2)\omega_{'\mu}^2]\}], \tag{43}$$

where

$$B \equiv e^\gamma. \tag{44}$$

The description of the system becomes complete by considering the conservation laws, which in the case of hydrostatic equilibrium can be written as:

$$\nabla p + (\varepsilon + p)[\nabla\nu + \frac{1}{1-v^2}(-v\nabla v + v^2\frac{\nabla\Omega}{\Omega - \omega})] = 0. \tag{45}$$

In addition to these basic equations we must specify a rotation law of the star. The integrability condition of the equation of motion allows one to choose the followings as rotation laws:(1) uniform rotation,

$$\Omega = \text{constant}, \tag{46a}$$

or (2) differential rotation,

$$A^2(\Omega_c - \Omega) = \frac{(\Omega - \omega)r^2 \sin^2 \theta e^{2(\gamma - 2\nu)}}{1 - (\Omega - \omega)^2 r^2 \sin^2 \theta e^{2(\gamma - 2\nu)}}, \tag{46b}$$

where Ω_c is a certain constant and A is a constant which is referred to as a rotation parameter hereafter. The smaller the rotation parameter becomes, the stronger the degree of differential rotation is. In the Newtonian limit, this relation becomes:

$$\Omega/\Omega_c = A^2/(A^2 + r^2 \sin^2 \theta). \tag{47}$$

The rotation law (47), with a small value of A, gives constant specific angular momentum distribution except near the rotation axis.

The boundary conditions (6a)-(6d) for the metric functions can be easily taken into account by using Green functions. Thus we rewrite (35)-(37) in the integral forms which read:

$$\rho = -\sum_{n=0}^{\infty} e^{-\gamma/2} \int_0^{\infty} dr' \int_0^1 d\mu' r'^2 f_{2n}^2(r, r') P_{2n}(\mu) P_{2n}(\mu') S_\rho(r', \mu'), \tag{48}$$

$$r \sin \theta \gamma = -\frac{2}{\pi} \sum_{n=1}^{\infty} e^{-\gamma/2} \int_0^{\infty} dr' \int_0^1 d\mu' r'^2 f_{2n-1}^1(r, r') \frac{1}{2n-1} \sin(2n-1)\theta$$
$$\times \sin(2n-1)\theta' S_\gamma(r', \mu'), \tag{49}$$

$$r \sin \theta \omega = -\sum_{n=1}^{\infty} \int_0^{\infty} dr' \int_0^1 d\mu' r'^3 \sin \theta' f_{2n-1}^2(r, r') \frac{1}{2n(2n-1)} P_{2n-1}^1(\mu)$$
$$\times P_{2n-1}^1(\mu') S_\omega(r', \mu'), \tag{50}$$

where

$$f_n^1(r, r') = (r'/r)^n, \text{ for } r'/r \le 1,$$
$$(r/r')^n, \text{ for } r'/r > 1, \tag{51}$$

and

$$f_n^2(r, r') = (1/r)(r'/r)^n, \text{ for } r'/r \le 1,$$
$$(1/r')(r/r')^n, \text{ for } r'/r > 1. \tag{52}$$

If we assume a polytropic relation

$$p = K\varepsilon^{1+1/N}, \tag{53}$$

where K and N are a certain constant and the polytropic index, respectively, the equation of motion for the star is integrated to give

$$(1 + N)\ln(K\varepsilon^{1/N} + 1) + \nu + \frac{1}{2}\ln(1 - v^2) - \frac{1}{2}A^2(\Omega - \Omega_c)^2 = C, \tag{54}$$

where C is a constant of integration.

In order to obtain an equilibrium state, we need to specify the values of four parameters: 1) the axis ratio of the star (r_p/r_e); 2) the ratio of the maximum pressure to the maximum energy density of the toroid (κ); 3) the polytropic index of the star (N); 4) the rotation parameter of the star (A). Basic equations are (43), (46), (48)-(50), (53) and (54). These seven equations are solved for the seven unknown functions, α, ρ, γ, ω, ε, p and Ω.

The numerical computation begins with specifying the four parameters mentioned above. Next, we need to prepare initial guesses for the variables ρ, γ, ω, α, ε, p and Ω. Substituting them into the right hand side of (48)-(50), we can obtain new values of ρ, γ and ω. Using these new values, we can integrate (43), starting from the rotation axis and ending at the equator, to obtain a new value for α. After that, we compute ε and Ω in order to keep consistency with the new potentials ρ, γ, ω and α by using (46) and (54).

We regard newly obtained values as an improved set of guesses and return to the next iteration step by keeping the model parameters fixed. We continue this iteration cycle until the differences of each physical quantity between two successive cycles become sufficiently small and then we regard those values as true values which satisfy the basic equations and the boundary conditions.

This scheme has worked almost perfectly as far as problems treated thus far and seems to have almost no limitations in its applicability.

In practice Komatsu et al. [38,39] computed rapidly rotating relativistic polytropes with uniform rotation without difficulty. They found that the critical states where the "centrifugal" force balanced the gravity at the equatorial surface were characterized by the same value of the axis ratio, $r_p/r_e \approx 0.6$, irrespective of the strength of gravity for $N = 1.5$ polytropes. They also applied their scheme to differentially rotating polytropes. For highly differentially rotating cases equilibrium configurations became ring-like. Obtained ring-like configurations were highly relativistic and considerably deformed.

Therefore if we adopt the scheme developed by Komatsu et al.[38,39], i.e., KEH scheme, we are now at the stage where any configurations in general relativity can be solved as far as barotropes are concerned (in the weak gravity cases Hachisu's scheme (HSCF-scheme) [26,28] or Eriguchi and Müller's scheme (SFNR-scheme) [22] can give solutions for almost every situation). We can apply this scheme to realistic models such as rapidly rotating neutron stars or extend their scheme to treat baroclinic states, which may be very difficult.

7 Rapidly Rotating Neutron Stars

7.1 Importance of Modeling of Rapidly Rotating Neutron Stars

Up to now it is still a hard task to determine the equation of state for the density region of the nuclear matter and higher with laboratory experiments on the earth and nuclear physics theories. On the other hand we have observed many neutron stars whose densities are in the range of the nuclear matter density and higher. This has motivated several attempts to find clues to the equation of state of neutron stars by constructing theoretical models of neutron stars [53]-[56]. For the softest equation of state, the maximum mass and the radius with a given mass of the neutron stars are smaller. Neutron stars with smaller radius can rotate more rapidly. On the other hand, a stiffer equation of state would result in the larger radius and the larger maximum mass. Then rapid rotation would more easily cause matter to shed from the equatorial surface by a large centrifugal force. Since the mass of the binary pulsar is known [57], the equation of state cannot be too soft to sustain the observed mass of the binary pulsar. As for the rotation, the observed period of pulsars can also set some constraints on the equation of state. Thus it is important to determine the maximum mass and the maximum angular velocity of rapidly rotating neutron stars for a range of stiffness of the equation of state.

In order to obtain structures of rotating neutron stars general relativity must be taken into account. As explained in Sect.6 two schemes [34,35,38,39] gave satisfactory results for the problem we are concerned with.

The formulation developed by Butterworth and Ipser [34,35] has been used to compute rapidly rotating neutron stars by Friedman et al. [53]-[56]. They developed two independent numerical codes and applied them to realistic neutron stars by using many kinds of equation of state for the high density region and obtained many equilibrium configurations. However they did not explain how their codes overcame difficulties encountered by Butterworth [37] as discussed in Sect.6.2.

Recently Lattimer et al. [58] and Wu et al. [59] computed structures of rapidly rotating neutron stars for various equations of state. The numerical code by Lattimer et al. [59] is the same as used by Butterworth and Ipser [34,35], so that their results are almost the same as obtained by Friedman et al. [53]-[56] for the same equations of state. Although Wu et al. [59] used the Neugebauer's minimal surface formalism [60], the performance and limitation, if exists, of their code have not been published yet.

Because the existence and the structure of realistic neutron stars have an important influence on the nuclear matter physics, it is desirable to construct the same equilibrium models as solved by Friedman et al. [53]-[56] with totally different and reliable numerical scheme to compare the results with theirs. Eriguchi et al. [61] extended the scheme developed by Komatsu el al. [38,39] to compute realistic neutron star models to compare results with those thus far obtained with different codes.

7.2 Rotating Neutron Star Models

Eriguchi el al. [61] treated *axisymmetric* and *stationary* states of rotating *perfect* fluid. The metric of the spacetime and the energy momentum were expressed by (4) and (1), respectively.

As for the rotation laws for neutron stars, two kinds of rotation laws, uniform rotation (46a) and differential rotation (46b), were employed. It is widely assumed that neutron stars are uniformly rotating because the angular momentum will be re-distributed to that of uniform rotation due to the viscosity. The time scale for this re-distribution can be estimated by using the shear viscosity coefficients calculated by Flowers and Itoh [62,63], whose values are analytically fitted by Cutler and Lindblom [64], as:

$$\tau = 18 \times \left(\frac{\rho(\text{gcm}^{-3})}{10^{15}}\right)^{-5/4} \left(\frac{T(\text{K})}{10^9}\right)^2 \left(\frac{R(\text{cm})}{10^6}\right) \quad \text{yr.,} \tag{55}$$

(see also Sawyer [65] for the bulk viscosity). It should be noticed that this viscosity is calculated for $\varepsilon \le 4 \times 10^{14}\,\text{gcm}^{-3}$. Therefore the above time scale has a large uncertainty for the neutron star whose density is higher than that value. However since there may be a chance to observe newly born neutron stars after 10 years or so from its birth (e.g., SN1987A) and thus rotation will not always settle down to completely uniform one, it is important to study slightly differentially rotating neutron stars as well. Although it is difficult to estimate the extent of differential rotation and the rotation law itself, the above rotation law with a rather large value for the parameter A is chosen by Eriguchi et al. [61] because the constant specific angular momentum distribution is an extreme case which satisfies the stability condition against the axisymmetric collapse or expansion.

In Eriguchi et al. [61] four different equations of state were applied: 1) Pandharipande's equation of state for the neutron matter [66] which was cited as the model A by Arnett and Bowers [67]; 2) Friedman-Pandharipande's equation of state [68]; 3) and 4) Bethe -Johnson's equations of state (I) and (V) [69] which were referred to as models C and D by Arnett and Bowers [67], respectively.

The Pandharipande's equation of state assumes composition above $\varepsilon > 6.97 \times 10^{14}\,\text{gcm}^{-3}$ to be neutron and uses the Reid soft core potential, thereby, being rather soft. The Bethe-Johnson's equations of state include hyperons together with the modified Reid soft core potential, thus being regarded rather stiff. Three-nucleon interactions are included in the Friedman-Pandharipande's equation of state, which is very stiff at the higher density. For the lower energy density region the equation of state of Baym-Pethick-Sutherland [70] was used.

7.2.1 Uniformly Rotating Neutron Stars

For uniformly rotating neutron stars the results of Friedman et al. [56] and Eriguchi et al. [61] are summarized in Fig.1. In Fig.1 the maximum angular velocity is drawn against the gravitational mass of the neutron star. As seen from this figure the results of Eriguchi et al. [61] are in good agreement with those of Friedman et al. [56] except at high energy densities. The difference in the maximum angular velocity is several percent at the highest density region especially for the Friedman-Pandharipande's equation of state.

This may stem from the difference in the formulation and numerical codes. In order to obtain the maximum angular velocity we have to compute the equatorial radius of the neutron star very accurately. If the radius determined by a certain numerical code is a little larger than the true value, the obtained angular velocity is smaller than the true value and vice versa. In particular the accuracy of the obtained model at very high energy densities is crucial to determine the maximum angular velocity. In fact the central energy density and the ratio of the central pressure to the central energy density for the critical state obtained by Eriguchi et al. [61] are:

$$\kappa = 0.67, \quad \varepsilon_c \sim 3.4 \times 10^{15}\,\text{gcm}^{-3}, \tag{56}$$

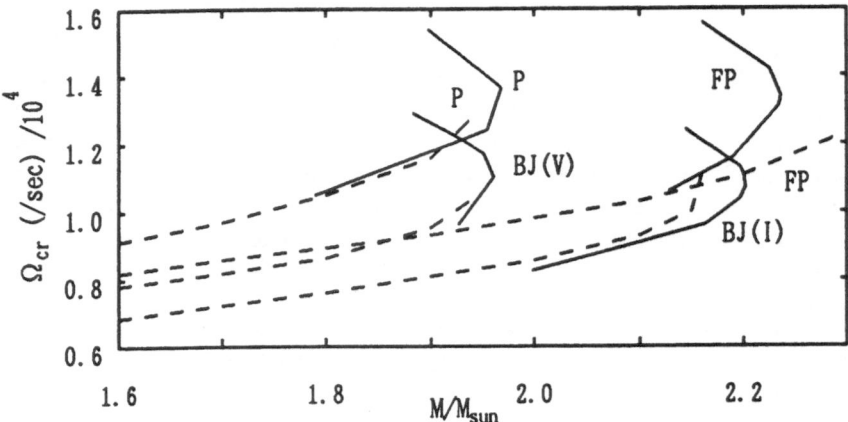

Fig. 1. The maximum angular velocities for uniformly rotating neutron stars are drawn against the gravitational mass of the rotating neutron star. Different curves correspond to different equations of state whose abbreviation are attached to curves. P : Pandhari-pande's neutron matter. FP: Friedman-Pandharipande's equation of state. BJ: Bethe-Johnson's equation of state. Full curves denote the results of Eriguchi et al. [61] and dashed curves denote Friedman et al.'s results [56].

whereas those for Friedman et al.'s results [56] are:

$$\kappa \sim 0.38, \quad \varepsilon_c \sim 2.5 \times 10^{15} \, \mathrm{g cm^{-3}}. \tag{57}$$

As mentioned before, the effective action of general relativity is to increase the compressibility. It implies that we need to use numerical codes which can be applied to highly deformed compressible stars.

The discrepancy shown in Fig.1 must be compared with the numerical accuracy. The accuracy of the radius and the angular velocity was discussed by Friedman et al. [53,54], which is 5 percent with their code. The accuracy of their quantities may stem from the problem how to find critical equilibrium states from numerical computations. Non-convergence of the iteration may be considered to be one approximate "signal" of non-existence of equilibrium states. In most situations this "criterion" can practically work. However since any numerical computational scheme is far from perfect, non-convergence of iterations does not always mean the non-existence of equilibrium states.

The mass shedding states or Kepler states can be found by investigating the behavior of the total (gravitational plus the centrifugal) potential. The surface of the equilibrium configuration is one of the constant surfaces of the total potential and its values inside the surface are lower than that value on the surface. If a state where matter begins to shed from the equator is reached, the value of the total potential at the surface will be the same as that of the local maximum above which no equilibrium configurations exist. Thus we can determine the equilibrium state very near to the critical states from numerical computations by examining the total potential. Furthermore the equatorial surface of the critical state becomes cusp-like so that we can make use of the shape of the configuration to see whether the obtained models are located near the critical state. As shown in Fig.2 the last model of equilibrium sequence of Eriguchi et al. [61] shows cusp-like shape near the equatorial surface.

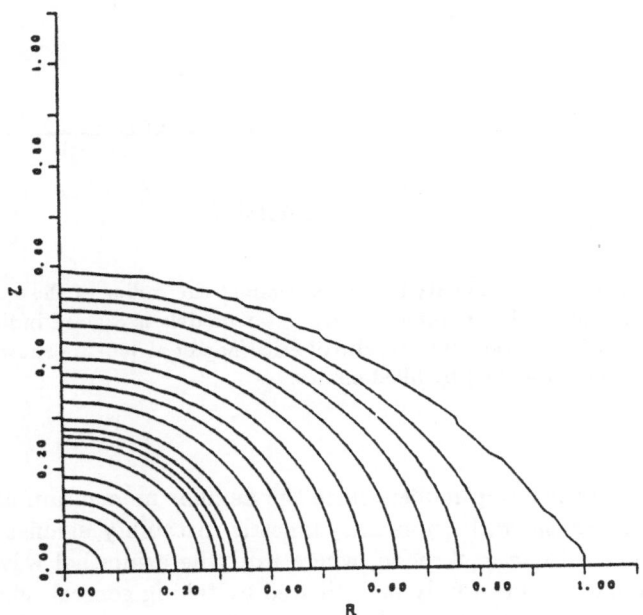

Fig. 2. The equi-density contours of rotating neutron star in a critical state obtained by Eriguchi et al. [61] are shown. The difference between two contours are 1/20 of the maximum energy density. The equation of state is that of Friedman-Pandharipande.

In Fig.3 the angular velocity for critical models is plotted against the radius of spherical neutron stars, which can be considered to represent the softness of the equation of state. From this figure, too, we can see that there is several percent difference between the results of Eriguchi et al. [61] and those of Friedman et al. [56].

7.2.2 Differentially Rotating Neutron Stars

Eriguchi et al. [61] also computed slightly differentially rotating models. The value of parameter A was chosen rather large, i.e., $A = 2R_e$, where R_e is the equatorial radius. They showed results only for the Friedman-Pandharipande's equation of state.

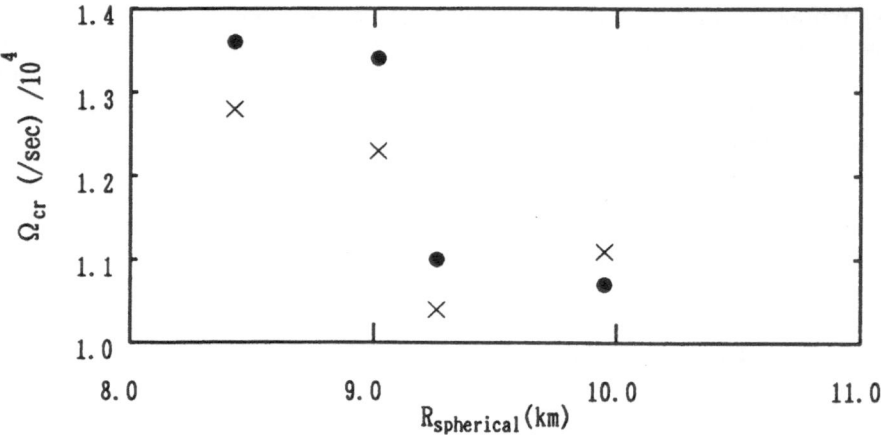

Fig. 3. The critical angular velocity is plotted against the radius of the spherical neutron stars. The radius of the spherical neutron star models is a good indicator of the
softness of the equation of state. The results of Friedman et al. [56] are shown by crosses
and those of Eriguchi et al. [61] by filled circles.

For differentially rotating models since the angular momentum can become
large, we have to consider the non-axisymmetric instability against the gravitational radiation with $m = 2$ mode, where m is the azimuthal wave number.
This instability has not been fully investigated by taking general relativity into
account yet. However using the knowledge of the Newtonian stability analysis [71,72] together with the knowledge of relativistic spherical stars [73,74] we
can roughly assume that when the ratio of the rotational energy, T, to the absolute value of the gravitational energy, W, exceeds $0.12 \sim 0.14$, the star will
become unstable against this mode of gravitational radiation.

Physical quantities obtained by Eriguchi et al. [61] are summarized in Table
1. In this table Ω_p and Ω_e are the angular velocity at the pole surface and at
the equatorial surface, respectively.

Table 1. Physical quantities for critical models for differentially rotating neutron stars
with Friedman-Pandharipande's equation of state.

$\varepsilon_c(\text{gcm}^{-3})$	$\Omega_p(\text{sec}^{-1})$	$\Omega_e(\text{sec}^{-1})$	$M/M_\odot$	$R_e(\text{km})$	$T/\mid W\mid$
2.5E15	1.33E4	1.10E4	2.29	13.2	0.148
3.2E15	1.55E4	1.25E4	2.34	12.0	0.153
4.0E15	1.75E4	1.37E4	2.34	11.2	0.158

8 Toroidal Equilibrium Configurations around Compact Objects

8.1 Compact Star – Self-Gravitating Toroid System

Thus far isolated bodies have been discussed. Rotating bodies are often surrounded by a disk or by a matter of toroidal configuration. Toroidal configurations around compact objects are relevant in modeling of quasars and active galactic nuclei. Such structures may be also formed during the evolution of close binary stars.

Although many authors [75] have studied systems with disks or toroids, the disk is usually treated as non-gravitating matter. Self-gravitating rings and disks have been considered only when the central body is a black hole [40,76,77,78,79]. Will [76,77] has studied weakly self-gravitating rings around slowly rotating black holes by treating the problem perturbationally. Chakrabarti [78] considered a static axisymmetric spacetime with a massive ring surrounding a Schwarzschild black hole. Recently, Lanza [79] has solved the Einstein equations for self-gravitating thin disks around rapidly rotating black holes by using a multigrid method. None of these studies, however, took into account the finite thickness of the disk.

Nishida et al. [80,81] have succeeded in computing the structure and the spacetime of compact star – self-gravitating toroid systems. They extended the KEH scheme so as to handle the star – toroid systems.

When the central object is a compact star, almost the same scheme explained in Sect. 6.3 can be used to obtain equilibrium states for the central star and the surrounding massive toroids. Since two disconnected matters are treated, we need 9 parameters to specify one system. Nishida et al. [80] assumed that the equations of state for the central star and the toroid were both polytropes with different polytropic indices. As for the rotation laws they chose (46a) for the central star and (46b) for the toroid. The equations of hydrostatic equilibrium were the same as (54) but for different polytropic indices. The metric components could be obtained by using (48)-(50). Therefore the same numerical scheme as the original KEH scheme can be used.

In practice they solved many equilibrium sequences with different mass ratios and different sizes of the toroid. As was shown by Will [76,77], the gravity of the toroid affects the spacetime so that stars with zero angular momentum but with non-zero angular velocity can be in equilibrium states.

If the central object is a black hole, the KEH scheme needs to be changed to include the boundary conditions (7) on the horizon. Nishida et al. [81] have revised the numerical code and succeeded in obtaining equilibrium structures of massive toroids and the spacetime of the black hole – toroid system. As mentioned before, in Will's system [76,77] the self-gravity of the toroid and the rotation of the black hole were only treated perturbationally. Although Lanza [79] included the self-gravity of the disk, the effect of the pressure in the disk was neglected. Nishida et al. [81] included both the self-gravity of the toroid and the effect of the pressure in the toroid.

8.2 Runaway Instability

In some astrophysical problems we need to know a precise shape of the star. For example let us consider a binary star system. As the evolution proceeds, one component star will expand and matter begins to overflow from the Roche lobe to the other star. In order to know when this process begins we need to know the shape of the Roche lobe as well as that of the star. In a system consisting of a compact object and the accretion disk the same problem can be considered. In particular it is important to know whether the mass overflow from the disk to the black hole continues unstably or not.

The "runaway instability" in accretion disks or toroids orbiting black holes was first investigated by Abramowicz et al. [82] Axisymmetric mass overflow from a thick toroid to a central black hole changes shapes of the toroid and its Roche lobe. If the Roche lobe shrinks faster than the change of the surface of the toroid, the mass keeps overflowing. Making use of the pseudo–Newtonian potential Abramowicz et al. [82] concluded this instability occurs if the mass of the disk was larger than a few percent of that of the central black hole. However Wilson [83] analyzed the same problem in the Kerr spacetime and found that disks were stable against the mass overflow. Although in their investigations the effect of the mass of the toroid was poorly taken into account [82] or neglected [83], we have to consider self-gravity of the toroids because self-gravity is essential in determining the shape of the Roche lobe as well as that of the toroid.

8.2.1 Equilibrium States and Critical Models

Nishida et al. [84] solved the Einstein equations for the structure of the self-gravitating toroid around a black hole as well as the spacetime outside the horizon of the black hole. In this problem the equation of state which they used was slightly different from (53). The following polytropic equation of state with $N = 3$ was chosen:

$$p = K' \tilde{\rho}^{4/3}, \tag{58}$$

$$\varepsilon = \tilde{\rho} + 3p. \tag{59}$$

As for the rotation law they chose the following law which was different from (46):

$$\ell \equiv -u_\phi/u_t = \text{constant}, \tag{60}$$

where u_t and u_ϕ are time and axial components of the four-velocity of the toroid. Under the assumptions mentioned above the distribution of

$$j \equiv u_\phi(\varepsilon + p)/\tilde{\rho}, \tag{61}$$

is also constant in space. It should be reminded that this is a conserved quantity during axisymmetric motion.

In order to determine one equilibrium state one needs to specify five parameters. Four parameters are two sets of the strength of gravity and rotation both for the black hole and the toroid. One can choose the gravitational mass M_h and the angular momentum J_h of a black hole, the total gravitational mass M_t and the total angular momentum J_t of the system. The final parameter is the distribution of the angular velocity or the angular momentum for the toroid. Since constant distribution of the specific angular momentum, (60) or (61), is assumed, only the value of j need to be specified. Therefore solutions can be represented by points in a five dimensional parameter space, i.e., (j, M_h, J_h, M_t, J_t)-space.

Among equilibrium solutions, there are critical solutions for equilibrium configurations beyond which no equilibrium states exist because a critical state is defined as an equilibrium state of a toroid whose surface coincides with its Roche lobe. Thus in five dimensional solution space, a set of critical states forms a four dimensional hypersurface which divides the five dimensional space into the allowed region for equilibrium states and the forbidden region.

When a process of an infinitesimal mass inflow from the toroid to the black hole is considered, this process can be represented by a direction from a certain point corresponding to a critical model in this parameter space. An arrow can be used to show this direction. If the arrow points toward the allowed region, the process is stable and otherwise unstable. Therefore it is important to investigate the behavior and the response of critical models on the boundary hypersurface to an infinitesimal mass overflow in this space. This can be done by computing critical equilibrium states by the code developed by Nishida et al. [81]

8.2.2 Accretion Processes and Runaway Instability

Let us consider matter overflow from the toroid which is in a critical state. During the accretion process we impose the following conditions:

$$j = \text{constant}, \tag{62}$$

$$M_t = \text{constant}, \tag{63}$$

$$J_t = \text{constant}, \tag{64}$$

$$\delta J_h = \ell \delta M_h. \tag{65}$$

Equation (62) is satisfied exactly if the accretion processes are axisymmetric because it is a conserved quantity during axisymmetric motion as mentioned before. Equations (63)-(65) are fulfilled if the gravitational radiation is neglected. In (65) δJ_h and δM_h represent the changes of the angular momentum and the gravitational mass of the black hole during the infinitesimal inflow process. These four conditions determine the direction of the arrows in the five dimensional solution space.

The results of Nishida et al. [84] can be shown in Fig.4. This figure shows the two-dimensional cross section of the five dimensional solution space which is cut by $j = \text{constant}$, $M_t = \text{constant}$, and $J_t = \text{constant}$ planes. The solid curve denotes the subspace consisting of "critical" models. The region below the critical curve is allowed for equilibrium states. Arrows display the direction determined from condition (65) alone because this plane is already restricted by conditions (62)-(64). As discussed before, equilibrium states at which arrows point towards the forbidden region are unstable against the mass overflow. Therefore models on the boundary are unstable if the ratio of the mass of the toroid to that of the black hole is larger than 0.1 or so. This result of Nishida et al. [84] is consistent with that of Abramowicz et al. [82]

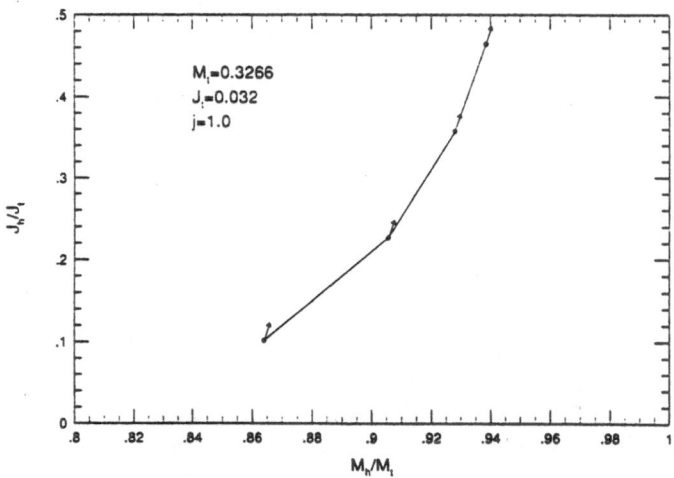

Fig. 4. The mass of the black hole, M_h, and the angular momentum of the black hole, J_h, plane. In this plane the total angular momentum, J_t, the total gravitational mass, M_t, and the specific angular momentum, j, are all constant. Critical equilibrium states are shown by the solid curve. Equilibrium states for the black hole – toroid system are forbidden in the region above the critical curve. The equilibrium configurations on the critical curve move towards the direction shown by arrows if an infinitesimal mass is transferred from the toroid to the black hole. If this direction is towards the forbidden region that equilibrium is unstable against the runaway instability.

The different conclusion from that of Wilson [83] implies the importance of the self-gravity of the toroids contrary to his discussion. Since the occurrence of the runaway instability crucially depends on the shape of the Roche lobe, the effect of the self-gravity on the shape of the toroid and the Roche lobe cannot be neglected even if the ratio of the mass of the toroid to the black hole is not so large.

Acknowledgements

I would like to express my sincere gratitude to Professor F.J. Chinea for his invitation to the Summer School on Gravitation and General Relativity (ERE 92) and his warm hospitality at El Escorial. I would also like to thank Dr. L.M. González-Romero for his hospitality. I was benefited from discussion with Professor J.M. Ipser about the rapidly rotating neutron star models.

I would also thank my coworkers, Dr. Komatsu, Mr. Nishida, Dr. Lanza and Dr. Hachisu, for their help and discussion during the development of the numerical codes and the analysis of numerical results.

References

1. R.A. James: Astrophys. J. **140** 552 (1964)
2. J.P. Ostriker, J. W-K. Mark: Astrophys.J. **151** 1075(1968)
3. J.P. Ostriker, P. Bodenheimer: Astrophys.J. **151** 1089 (1968)
4. J.P. Ostriker, F.D.A. Hartwick: Astrophys.J. **153** 797 (1968)
5. J. W-K. Mark: Astrophys.J. **154** 627 (1968)
6. S. Jackson: Astrophys.J. **161** 579 (1970)
7. P. Bodenheimer, J.P. Ostriker: Astrophys. J. **161** 1101 (1970)
8. P. Bodenheimer: Astrophys.J. **167** 153 (1971)
9. P. Bodenheimer, J.P. Ostriker: Astrophys. J. **180** 159 (1973)
10. Y. Eriguchi: Publ. Astron. Soc. Japan **30** 507 (1978)
11. Y. Eriguchi, D. Sugimoto: Prog. Theor. Phys. **65** 1870 (1981)
12. Y. Eriguchi, I. Hachisu: Prog. Theor. Phys. **67** 844 (1982)
13. Y. Eriguchi, I. Hachisu, D. Sugimoto: Prog. Theor. Phys. **67** 1068 (1982)
14. I. Hachisu, Y. Eriguchi, D. Sugimoto: Prog. Theor. Phys. **68** 191 (1982)
15. I. Hachisu, Y. Eriguchi: Prog. Theor. Phys. **68** 206(1982)
16. Y. Eriguchi, I. Hachisu: Prog. Theor. Phys. **69** 1131 (1983)
17. Y. Eriguchi, I. Hachisu: Prog. Theor. Phys. **70** 1534 (1983)
18. I. Hachisu, Y. Eriguchi: Publ. Astron. Soc. Japan **36** 239 (1984)
19. I. Hachisu, Y. Eriguchi: Publ. Astron. Soc. Japan **36** 259 (1984)
20. I. Hachisu, Y. Eriguchi: Publ. Astron. Soc. Japan **36** 497 (1984)
21. Y. Eriguchi, I. Hachisu: Astron. Astrophys. **142** 256 (1985)
22. Y. Eriguchi, E. Müller: Astron. Astrophys. **146** 260 (1985)
23. Y. Eriguchi, E. Müller: Astron. Astrophys. **147** 161 (1985)
24. Y. Eriguchi, I. Hachisu: Astron. Astrophys. **148** 289 (1985)
25. E. Müller, Y. Eriguchi: Astron. Astrophys. **152** 325 (1985)
26. I. Hachisu: Astrophys. J. Suppl. **61** 479 (1985)
27. Y. Eriguchi, E. Müller, I. Hachisu: Astron. Astrophys. **168** 130 (1986)
28. I. Hachisu: Astrophys. J. Suppl. **62** 461 (1986)
29. J.B. Hartle: Astrophys.J. **150** 1005 (1967)
30. J.M. Bardeen, R.V. Wagoner: Astrophys. J. **167** 359 (1971)
31. J.R. Wilson: Astrophys. J. **176** 195 (1972)
32. J.R. Wilson: Phys. Rev. Lett. **30** 1082 (1973)
33. S. Bonazzola, J. Schneider: Astrophys. J. **191** 273 (1974)
34. E.M. Butterworth, J.R. Ipser: Astrophys.J. **200** L103 (1975)
35. E.M. Butterworth, J.R. Ipser: Astrophys.J. **204** 200 (1976)
36. R. Stoeckly: Astrophys. J. **142** 208 (1965)
37. E.M. Butterworth: Astrophys.J. **204** 561 (1976)
38. H. Komatsu, Y. Eriguchi, I. Hachisu: Mon. Not. Roy. astr. Soc. **237** 355 (1989)
39. H. Komatsu, Y. Eriguchi, I. Hachisu: Mon. Not. Roy. astr. Soc. **239** 153 (1989)
40. J.M. Bardeen, in Black Holes (Les Houches 1972), ed. C. DeWitt and B.S. De-Witt (Gordon and Breach, New York), pp.241-289 (1973)
41. K.S. Thorne, in General Relativity and Cosmology, ed. R.K. Sachs (Academic Press, New York), pp.237-283 (1971)
42. A. Papapetrou: Ann. Inst. H. Poincaré **A-4** 83 (1966)
43. S. Chandrasekhar: Astrophys. J. **142** 1488 (1965)
44. S. Chandrasekhar, Y. Nutku: Astrophys. J. **158** 55 (1969)
45. G.G. Fahlman, S.P.S. Anand: Ap. Sp. Sci. **12** 58 (1971)

46. M.J. Miketinac, R.J. Barton: Ap. Sp. Sci. **18** 437 (1972)
47. J.B. Hartle, K.S. Thorne: Astrophys.J. **153** 807 (1968)
48. S. Chandrasekhar, J.C. Miller: Mon. Not. Roy. astr. Soc. **167** 63 (1974)
49. J.C. Miller: Mon. Not. Roy. astr. Soc. **179** 483 (1977)
50. B. Datta, A. Ray: Mon. Not. Roy. astr. Soc. **204** 75p (1983)
51. A. Ray, B. Datta: Astrophys. J. **282** 542 (1984)
52. E.E. Salpeter, R.V. Wagoner: Astrophys. J. **164** 557 (1971)
53. J.L. Friedman, J.R. Ipser, L. Parker: Nature **312** 255 (1984)
54. J.L. Friedman, J.R. Ipser, L. Parker: Astrophys. J. **304** 115 (1986)
55. J.L. Friedman, J. Imamura, J.R. Ipser, L. Parker: Nature **336** 560 (1988)
56. J.L. Friedman, J.R. Ipser, L. Parker: Phys. Rev. Lett. **62** 3015 (1989)
57. J.H. Taylor, J.M. Weisberg: Astrophys.J. **345** 434 (1989)
58. J.M. Lattimer, M. Prakash, D. Masak, A. Yahil: Astrophys.J. **355** 241 (1990)
59. X. Wu, H. Müther, M.Stoffel, H. Herold, H. Ruder: Astron. Astrophys. **246** 411 (1991)
60. G. Neugebauer, H. Herlt: Classical Quantum Gravity **1** 695 (1984)
61. Y. Eriguchi, I. Hachisu, K. Nomoto: Mon. Not. Roy. astr. Soc. submitted (1992)
62. E. Flowers, N. Itoh: Astrophys. J. **206** 218 (1976)
63. E. Flowers, N. Itoh: Astrophys. J. **230** 847 (1979)
64. C. Cutler, L. Lindblom: Astrophys. J. **314** 234 (1987)
65. R.F. Sawyer: Phys. Rev. **D39** 3804 (1989)
66. V.R. Pandharipande: Nucl. Phys. **A174** 641 (1971)
67. W.D. Arnett, R.L. Bowers: Astrophys. J. Suppl. **33** 415 (1977)
68. B. Friedman, V.R. Pandharipande: Nucl. Phys. **A361** 502 (1981)
69. H.A. Bethe, M. Johnson: Nucl. Phys. **A230** 1 (1974)
70. G. Baym, C. Pethick, P. Sutherland: Astrophys. J. **170** 299 (1972)
71. R. Managan: Astrophys. J. **294** 463 (1985)
72. J.N. Imamura, R. Durisen, J.L. Friedman: Astrophys. J. **294** 474 (1985)
73. L. Lindblom: Astrophys. J. **303** 146 (1986)
74. J.R. Ipser, L. Lindblom: Phys. Rev. Lett. **62** 2777 (1989)
75. See e.g., J. Frank, A.R. King, D.J. Raine: Accretion Power in Astrophysics (Cambridge Univ. Press) (1985)
76. C. Will: Astrophys. J. **191** 521 (1974)
77. C. Will: Astrophys. J. **194** 41 (1975)
78. S.K. Chakrabarti: J. Astrophys. Astron. **9** 49 (1988)
79. A. Lanza: Astrophys. J. **389** 141 (1992)
80. S. Nishida, Y. Eriguchi, A. Lanza Astrophys. J. in press (1992)
81. S. Nishida, Y. Eriguchi, A. Lanza: Astrophys. J. submitted (1992)
82. M.A. Abramowicz, M. Calvani, L. Nobili: Nature **302** 597 (1983)
83. D.B. Wilson: Nature **312** 620 (1984)
84. S. Nishida, A. Lanza, Y. Eriguchi, M.A. Abramowicz: in preparation (1992)

Axisymmetric Stationary Solutions of Einstein's Equations

C. Hoenselaers

Department of Mathematical Sciences
Loughborough University of Technology
Loughborough, LE11 3TU

1 Introduction

Exact solutions have a long and distinguished history within the development of General Relativity. Their study began with the exterior and interior Schwarzschild solutions which were discovered soon after Einstein published his field equations. After the derivation of the Weyl solutions a long hiatus set in as far as asymptotically flat solutions were concerned. There was, of course, considerable activity in the realm of algebraically special solutions or solutions admitting vectors with special properties. It was while looking for solutions with a particular stucture of the metric that Kerr discovered his celebrated solution. About ten years later Tomimatsu and Sato found their series of exact axisymmetric stationary solutions.

This triggered a renewed interest in stationary axisymmetric solutions, a field which had been dormant because the equations were considered wellnigh insolvable. The effort was helped by the development of solution generating techniques, Bäcklund transformations or the inverse scattering method, for other non-linear partial differential equations. Finally several groups around the world published their approaches to axisymmetric stationary vacuum solutions. For a survey the reader is referred to Ref. 1. All solution generating techniques are equivalent in the sense that almost all solutions can be generated. It should be noted, that the techniques can be applied to any situation with two commuting Killing vectors, in particular also to colliding plane gravitational waves, cf. Ref.2.

On the other hand, we who are working in the realm of exact solutions should keep in mind that writing down a metric which solves Einstein's equations is only the first part of the task. To paraphrase Kinnersley [3], we should not leave our newborn metric wobbling on its Vierbein without any visible means of interpretation.

In this series of lectures we give an overview of the HKX transformations and relate them to what, in a more general setting, are called linear problems.

As an example of a solution we desribe the double Kerr solution. This is one of the solutions which demonstrate that relativistic angular momentum interaction can balance two masses against their gravitational attraction.

In addition to the above we also outline the definition of multipole moments in General Relativity and show how the first few of them can be calculated by an expansion of the Ernst potential. Furthermore we describe some properties of metrics which can be expressed as rational functions of prolate spheroidal coordinates.

On a personal note, I should like to express my appreciation to the organisors of the meeting at El Escorial for providing such a splendid and congenial environment. The workshop was a most pleasant experience for which I wish to thank Dr. F. J. Chinea.

2 Generating Solutions by HKX Transformations

We are concerned with space-times admitting two commuting Killing vectors. One of them is assumed to be spacelike with closed orbits, the other one should be timelike. It follows that the metric can be written in the Papapetrou-Lewis form

$$ds^2 = \frac{1}{f} \left[e^{2\gamma}(dx^2 + dy^2) + \rho^2 d\phi^2 \right] - f(dt - \omega d\phi)^2 \tag{2.1}$$

φ and t are the Killing coordinates and the functions f, ω, ρ and γ depend only on the non-ignorable coordinates x and y. We shall use the derivative operators

$$\nabla = \begin{pmatrix} \partial_x \\ \partial_y \end{pmatrix}, \qquad \tilde{\nabla} = \begin{pmatrix} \partial_y \\ -\partial_x \end{pmatrix}, \qquad \partial = \partial_x + i\partial_y \tag{2.2}$$

One of the Einstein equations reads

$$\nabla^2 \rho = 0 \tag{2.3}$$

This equation implies the existence of a function z defined up to an additive constant by

$$\nabla \rho = \tilde{\nabla} z \tag{2.4}$$

Because the gradients of ρ and z are orthogonal and of equal magnitude one can use ρ and z instead of x and y as coordinates without altering the form of the metric. ρ and z, are known under the name of Weyl coordinates. It should be noted that the Weyl coordinates are uniquely determined by the geometry, ρ as being the volume element of the 2-surfaces swept out by the Killing vectors and z as the conjugate function via (2.4). Consequently any statement about metric functions formulated in Weyl coordinates is ipso facto invariant.

One of the other field equations reads

$$\nabla \frac{f^2}{\rho} \nabla \omega = 0$$

It is the integrability condition for a function ψ given by

$$\rho \tilde{\nabla} \psi = f^2 \nabla \omega \tag{2.5}$$

One combines f and ψ conveniently into the complex Ernst potential [4]

$$\mathcal{E} = f + i\psi$$

The remaining equations reduce to the Ernst equation

$$f \Delta \mathcal{E} = (\nabla \mathcal{E})^2 \tag{2.6}$$

with $\Delta = \frac{1}{\rho} \nabla \rho \nabla$. Indeed, in Weyl coordinates ρ and z this operator becomes the usual Laplacian in three dimensional Euclidian space. The solutions, of course, have to be taken independent of the azimuthal angle. The remaining function γ can be determined by a line integral from

$$\partial \gamma \partial \rho = \frac{1}{2} \partial^2 \rho + \frac{\rho}{4f^2} \partial \mathcal{E} \partial \mathcal{E}^* \tag{2.7}$$

In Weyl coordinates the first term on the right-hand side vanishes; it is, however, present when other coordinates are employed.

Another frequently used form of the Ernst equation employs the potential ξ related to $\mathcal{E}$ by

$$\mathcal{E} = \frac{1 - \xi}{1 + \xi}$$

We get as equation for ξ

$$(\xi \xi^* - 1) \Delta \xi = 2\xi^* (\nabla \xi)^2 \tag{2.8}$$

Let $\mathcal{P}$ be a point in $\mathbb{R}^3$ such that $|\xi| \neq 0$ and $\xi \in C^{2,\alpha}$ in a neighbourhood of $\mathcal{P}$. It then follows from Morrey's theorem [5] that ξ is in fact analytic in this neighbourhood. Similarly, other theorems [6] on elliptic differential equations guarantee that under the above assuptions and $\mathcal{P}$ being on the coordinate axis $\rho = 0$, ξ is uniquely determined in the neighbourhood of $\mathcal{P}$ by its values on the axis.

Let $\mathcal{E}$ be a solution of (2.6) and an analytic function of a parameter p. f and ψ will then satisfy the real and imaginary part of the Ernst equation. Now continue p into the complex plane. $f(p)$ and $\psi(p)$ will still satisfy their respective equations but they will not be real functions anymore. If, however, the continuation can be effected such that for a particular complex value p_0, say, $f(p_0)$ and $\psi(p_0)$ are real functions of the coordinates we have extended the range of the parameter p.

Once a solution of the Ernst equation either in the form (2.6) or (2.8) has been found, all other metric functions are given by quadratures. In this section we shall address the question of how to solve Ernst's equation. The interpretation of the solutions will be deferred to the subsequent chapters.

The easiest way to solve equation (2.6) is to assume that $\mathcal{E}$ is real. $\psi = 0$ implies $\omega = 0$ and we are thus dealing with static solutions, the Weyl solutions [7]. In this case it is advantageous to introduce a real function χ which satisfies Laplace's equation by

$$\mathcal{E} = e^{2\chi}, \qquad \Delta\chi = 0. \tag{2.9}$$

Solutions of Laplace's equation are reasonably well understood and we may thus, for the present purpose, consider the static axisymmetric problem solved. For recent reviews cf. refs. 8.

There is a trivial way to generate new solutions of the Ernst equation. This consists in subjecting the metric (2.1) to a linear coordinate transformation of the Killing coordinates φ and t, i.e.

$$\varphi' = a\varphi + bt, \qquad t' = c\varphi + dt, \qquad ad - bc = 1$$

which entails a transformation of the metric functions and thus of the Ernst potential. From the form of the above expressions it is obvious that the transformations of the Killing coordinates constitute a three parameter group of isometries.

In the present formulation there is another almost equally obvious transformation, namely the multiplication of ξ by a constant phase, viz.

$$\xi' = e^{i\delta}\xi$$

This transformation is known under the name of Ehlers transformation. Together with the more trivial transformations of the multiplication of $\mathcal{E}$ by a real constant and the addition of an imaginary constant there is again a three-dimensional group of transformations acting on the Ernst potential and mapping solutions of the equation into new solutions.

The natural thing to consider next is the commutator of an infinitesimal coordinate transformation with the Ehlers transformation. Far from being trivial, it yields the infinitesimal Ehlers transformation of the Ernst potential associated with t'. Hence we can symbolically write, with ET standing for the infinitesimal Ehlers transformation and CT_i for the three infinitesimal coordinate transformations

$$[ET, CT_i] = \quad \text{another} \quad ET.$$

We find three infinitesimal Ehlers transformations in this way. Commuting them yields new transformations, symbolically

$$[ET_i, ET_k] = NT_l.$$

Continuing this process, commuting Ehlers transformations with the new ones or the new ones among themselves gives yet more infinitesimal transformations and the process continues ad infinitum. This infinity of infinitesimal transformations was discovered by Geroch [9] and is referred to as the Geroch group. It can, in principle, be used to generate infinite parameter families of solutions. It has been shown [10] that the Geroch group can in fact be given the structure of a Banach Lie group.

The method used by Kinnersley et al [11] to exploit the Geroch group uses the following formulation of Einstein's equations. In terms of a matrix consisting of the $\varphi - t$ elements of the metric, viz.

$$\mathbf{g} = \begin{pmatrix} g_{\varphi t} & g_{\varphi\varphi} \\ -g_{tt} & -g_{\varphi t} \end{pmatrix} = \begin{pmatrix} f\omega & \frac{\rho^2}{f} - f\omega^2 \\ f & -f\omega \end{pmatrix}$$

the important part of the Einstein equations reads

$$\nabla(\frac{1}{\rho}\,\mathbf{g}\,\nabla\,\mathbf{g}\,) = 0.$$

Since $\mathbf{g} = \rho^2\mathbf{1}$ this equation also includes (2.4). There exists a matrix Ω defined by

$$\tilde{\nabla}\Omega = -\frac{1}{\rho}\,\mathbf{g}\,\nabla\,\mathbf{g}.$$

Note that $tr\Omega = -2z$. In analogy to the Ernst potential, $\mathbf{g}$ and Ω can be combined to a complex matrix

$$H = \mathbf{g} + i\omega$$

which satisfies

$$\nabla H = \frac{i}{\rho}\mathbf{g}\tilde{\nabla}H. \tag{2.10}$$

Indeed, $\mathcal{E}$ is just the lower left element of H. With a considerable element of hindsight one can now find a matrix F depending on a parameter u which satisfies $\epsilon = \begin{pmatrix} 0 & 1 \\ -1 & 0 \end{pmatrix}$

$$\left[1 - iu(H + \epsilon H^\mathsf{t}\epsilon)\right]\nabla F(u) = iu\nabla H F(u)$$

$$F(0) = -i\mathbf{1}, \qquad \partial_u F(u)|_{u=0} = H.$$

The integrability conditions are satisfied by virtue of (2.10) . In fact, F is the generating function for an infinite hierarchy of fields, i.e. $F(u) = \sum H_i u^i$, and satisfies the same equation as H, viz.

$$\nabla F(u) = \frac{i}{\rho}\,\mathbf{g}\,\tilde{\nabla}F(u).$$

Moreover, there exists another generating function

$$\nabla G(u,v) = \epsilon F(u)\epsilon\nabla F(v) \Rightarrow G(u,v) = -\frac{1}{u-v}\left[-u\mathbf{1} + vF^{-1}(u)F(v)\right].$$

Expanding $G(u,v) = \sum N_{nm}u^n v^m$ we get the hierarchy of potentials originally introduced by Kinnersley and Chitre. Their importance lies in the fact that the above mentioned transformations of the Geroch group act only quadratically on the N's. This description led to the original transformations and also was the starting point for Hauser and Ernst [12] to formulate their homogeneous Hilbert problem.

As $H + \epsilon H^\dagger \epsilon = g + 2iz\mathbf{1}$, we can use (2.10) to rewrite the equation for $F(u)$ as

$$\nabla F(u) = \frac{iu}{S(u)^2}[(1 - 2uz)\nabla H - 2u\rho\tilde{\nabla}H]F(u) \qquad (2.11)$$

with

$$S(u)^2 = (1 - 2uz)^2 + 4u^2\rho^2.$$

Upon multiplication with $F(u)^{-1}$, taking the trace and choosing the integration constant appropriately we arrive at

$$|F(\dot{u})| = -\frac{1}{S(u)}.$$

Let a solution H and $F(u)$ of (2.11) be given. We have argued above that there are infinitely many infinitesimal transformations which map solutions into linearized solutions. The question is now whether one can find a closed form for those linearized solutions in terms of an arbitrary given solution, the seed metric, and whether one can derive finite transformations from the infinitesimal ones.

At this stage we digress for a moment and consider the general linear problem

$$dF(u) = H(u)F(u) \qquad (2.12)$$

$H(u)$ is for the purposes of this paragraph a matrix valued 1-form. We are interested in the linerarized equations; the linearized quantities will be denoted by an overhead dot. We find

$$d\dot{F}(u) = \dot{H}(u)F(u) + H(u)\dot{F}(u). \qquad (2.13)$$

A suitable Ansatz is

$$\dot{F}(u) = \tau(u, v)P(v)F(u)$$

with to be determined quantities $P(v)$, a matrix, and $\tau(u, v)$, a function independent of the coordinates. Moreover, we assume that $H(u)$ is of the form $\frac{h(u)}{D(u)}$, $h(u)$ and $D(u)$ being polynomial in u and $\dot{D} = 0$. Upon inserting this Ansatz into (2.13) we get

$$\tau(u, v)D(u)dP(v) = \dot{h}(u) + [\tau(u, v)h(u), P(v)].$$

The aim is to arrive by an judicious choice of τ at a sufficiently simple equation for $P(v)$, i.e.

$$dP(v) = [H(v), P(v)]. \qquad (2.14)$$

To this end we expand the various functions as

$$\tau(u, v) = \sum \tau_n(v)u^n, \qquad h(u) = \sum h_n u^n, \qquad D(u) = \sum D_n u^n.$$

Taking $n > max(deg\ h(u), deg\ G(u))$ derivatives with respect to u and evaluating the result at $u = 0$ yields

$$dP(v) = \left[\frac{\sum_{i=0}^{n} \tau_{n-i}(v) h_i}{\sum_{i=0}^{n} \tau_{n-i}(v) D_i}, P(v) \right].$$

We arrive at the desired result if we set $\tau_n(v) = \sigma(v) v^{-n}$. The function $\sigma(v)$ is an arbitrary scaling of the linearized solution and can thus be disregarded without loss of generality. The solution of (2.14) is $P(v) = F(v) \alpha F(v)^{-1}$ with a constant trace-free matrix α and we get for the linearized solution

$$\dot{F}(u) = \frac{u}{u - v} F(v) \alpha F(v)^{-1} F(u).$$

As it stands the expression is not analytic at $u = v$. This defect can be rectified by adding to $\dot{F}$ a term proportional to $F(u) \alpha F(u)^{-1}$ which evidently is also a solution of (2.13). In fact, it corresponds to $F \rightarrow e^\alpha F e^{-\alpha}$. Finally we obtain

$$\dot{F}(u) = \frac{u}{u - v} \left[F(v) \alpha F(v)^{-1} F(u) - F(u) \alpha \right]. \tag{2.15}$$

Note that $|F(u)|$ remains unchanged under the transformation.

To make contact with the Einstein equations again, we note that (2.11) is of the form just discussed and that the denominator is a function of $|F(u)|$. The solution of the linearized equations is thus given by (2.15). It remains to "exponentiate" the infinitesimal transformations to finite ones. This can most easily be accomplished by chosing $\alpha = \begin{pmatrix} 0 & 1 \\ 0 & 0 \end{pmatrix}$ and the seed metric to be static.

One lets F depend on some additional parameter ϵ, say, views (2.15) as differential equation for $F(\epsilon, u)$ where the overhead dot denotes $\partial_\epsilon F$. The initial value $F(0, u)$ has to be a solution of the equations. For nilpotent α (2.15) can be integrated by first setting $u = v$, solving for $F(v) \alpha F(v)^{-1}$ and then substituting the result back into the original equation. For static seed metrics the initial $F(0, u)$ can be calculated explicitly. As we are interested primarily in the Ernst potential, it is sufficient to calculate only the lower left element of H. The process results in the so-called HKX transformations which can be given a compact form involving the original static potential χ, another potential β derived from it, the distance from n points on the coordinate axis ρ and can be expressed as the ratio of two $n \times n$ determinants. z_i and a_i are arbitrary constant parameters [13,14].

$$\mathcal{E} = e^{2\chi} \frac{D_-}{D_+} \tag{2.16}$$

$$D_\pm = |\delta_{ik} + \frac{i a_k e^{2\beta(z_k)}}{S(z_k)} (\frac{S(z_i) - S(z_k)}{z_i - z_k} \pm 1)|$$

$$S(\zeta)^2 = (\zeta - z)^2 + \rho^2$$

$$S(\zeta) \nabla \beta(\zeta) = (\zeta - z) \nabla \chi - \rho \tilde{\nabla} \chi.$$

The appearance of $S(z_i)$ which are distances from fixed points on the coordinate axis motivates the phrase "applying the transformation at a point z_i".

Even though we are resticted to static seed solutions, the above expression can be used to prove that "almost all" solutions of the Ernst equation can, in principle, be generated. To show this let us consider the expressions on a part of the ccordinate axis $\rho = 0$ such that $z_i > z$. We find

$$\beta(\zeta) = \chi \quad , \quad S(\zeta) = \zeta - z$$

$$\mathcal{E} = e^{2\chi}(1 - 2i\sum_{k=0}^{n}\frac{a_k}{z_k - z}).$$

In this expression the sum can be extended to infinity. The real and imaginary part of the Ernst potential can thus be prescribed freely by an appropriate choice of the seed solution χ and the constants a_k and z_k. Hence all solutions which are analytic in the neighbourhood of at least one point on the axis can be generated.

One can, as has been mentioned earlier, continue the parameters a_k and z_k into the complex plane and derive thereby solutions with different ranges of the parameters. Also confluence limits $z_i \rightarrow z_k$ are possible; they are the rank n HKX transformations.

3 Multipole Moments

Multipole moments give useful information about the gravitational field of an isolated source. In Newtonian theory they can be calculated by appropriately weighted integrals over the source. Such a prescription is also available in General Relativity and defines the so-called Dixon [15] moments. In Newtonian theory there is also the possibility of reading off the moments from the gravitational far field and they agree, the theory being linear, with the source moments. Again the far field moments can be defined, at least for stationary fields, in Relativity. It cannot be expected that the far field moments agree with the source moments and, indeed, no way to link them is known. This is due to the fact that the Dixon moments are extremely difficult to calculate and not many complete space-times, interior solution together with an asymptotically flat vacuum field, are known. In fact, all such space-times are spherically symmetric.

Here we shall concentrate on the far field moments for stationary gravitational fields as defined by Geroch and Hansen [16] for vacuum space-times. Let ξ be a timelike Killing vector and

$$f = -\xi_\alpha\xi^\alpha, \qquad \psi_{,\alpha} = \epsilon_{\alpha\beta\gamma\delta}\xi_\beta\xi^{\gamma;\delta}, \qquad \mathcal{E} = f + i\psi$$

its norm, twist and Ernst potential. The metric on the set of trajectories of ξ is given by

$$h_{\alpha\beta} = fg_{\alpha\beta} + \xi_\alpha\xi_\beta.$$

The Einstein equations in space-time reduce to equations in three-space involving the Ricci tensor derived from $h_{\alpha\beta}$ and the Ernst potential $\mathcal{E}$ or ξ.

A 3-dimensional manifold $\mathcal{M}$ with metric h is called asymptotically flat if there exists another manifold $\widetilde{\mathcal{M}}$ with metric $\tilde{h}$ such that

$$\widetilde{\mathcal{M}} = \mathcal{M} \cup (\text{one point} \Lambda)$$

$$\tilde{h}_{\alpha\beta} = \Omega^2 h_{\alpha\beta} \tag{3.1}$$

$$\Omega_{|\Lambda} = \Omega_{,\alpha|\Lambda} = (\Omega_{|\alpha\beta} - 2\tilde{h}_{\alpha\beta})_{|\Lambda} = 0$$

Covariant derivative with respect to $\tilde{h}$ is indicated by a solidus. The procedure amounts to a compactification with the point Λ being the point at infinity and the conditions state that the space is "almost" Euclidian there and that Ω behaves like the square of the distance from Λ. Thus $\Omega \approx r^{-2} \approx \tilde{r}^2$ for large r in the physical respectively small $\tilde{r}$ in the unphysical manifold. The conformally transformed Ernst potential $\tilde{\xi}$ is defined by

$$\tilde{\xi} = \frac{\xi}{\sqrt{\Omega}}$$

This definition models, of course, the fact in Euclidian space that if $\chi(r, \vartheta, \varphi)$ is a solution of Laplace's equations then so is $\frac{1}{r}\chi(\frac{1}{r}, \vartheta, \varphi)$.

If one writes the Einstein equations for the Ernst potential and the Ricci tensor of the three-space in the unphysical manifold $\widetilde{\mathcal{M}}$ one finds that the equations involving $\tilde{R}_{\alpha\beta}$ are formally singular due to the appearance of Ω transformation in the denominator of some terms. However, Simon and Beig [17] have shown that those formally singular terms satisfy again elliptic equations the solutions of which are analytic by virtue of Morrey's theorem.

Multipole moments have a well-defined transformation behaviour under a shift of origin in the physical space. In the unphysical manifold $\widetilde{\mathcal{M}}$ this is reflected by a change of conformal factor $\Omega \rightarrow \omega\Omega$ with $\omega_{|\Lambda} = 1$.

One calculates recursively a set of tensor fields by

$$P = \tilde{\xi}, \qquad P_\alpha = P_{,\alpha} \tag{3.2}$$

$$P_{\alpha_1 \ldots \alpha_{n+1}} = \mathcal{C}(P_{\alpha_1 \ldots \alpha_n | \alpha_{n+1}} + \frac{n}{2}(2n-1)\tilde{R}_{\alpha_1\alpha_2} P_{\alpha_3 \ldots \alpha_{n+1}})$$

where $\mathcal{C}$ denotes the operation of taking the totally symmetric trace-free part. The multipole moments are then defined by the values of those tensor fields at Λ. The Ricci tensor terms are present to ensure the correct behaviour under an additional conformal transformation with ω. The real and imaginary parts of the moments are the mass and angular momentum multipole moments.

Obviously the actual calculation of the moments for a given exact solution is quite laborious. On the other hand, all asymptotically flat solutions known to date are not only stationary but also axisymmetric. The axis, i.e. the set of fixed points of the action of the angular Killing vector, passes through Λ and the only geometric objects invariant under that action at Λ are the axis vector and the metric itself. Hence the multipole moments are proportional to the symmetrized trace-free product of the axis vector with itself and it suffices to know the numbers

$$P_n = \frac{1}{n!} P_{\alpha_1 \dots \alpha_n} n^{\alpha_1} \dots n^{\alpha_n} |_\Lambda$$

where n^α is the unit axis vector.

This reduction in the number of components to be calculated does not, unfortunately, make the actual calculation any simpler. On the other hand, we have seen that the Ernst potential is uniquely determined by its values on the axis. Hence the complex coefficients m_n to be read off in Weyl coordinates from

$$\xi(\rho = 0) = \sum_{n=0}^{\infty} \frac{m_n}{z^{n+1}}$$

can serve to characterize the solution. The metric $h_{\alpha\beta}$ of the three-dimensional space reads in Weyl coordinates

$$d\sigma^2 = e^{2\gamma}(d\rho^2 + dz^2) + \rho^2 d\varphi^2 \tag{3.3}$$

After the coordinate transformation

$$\tilde{\rho} = \frac{\rho}{\rho^2 + z^2} \quad , \quad \tilde{z} = \frac{z}{\rho^2 + z^2}$$

a conformal transformation with $\Omega = \tilde{\rho}^2 + \tilde{z}^2$ takes us to the unphysical manifold $\tilde{\mathcal{M}}$. We have now employed various programs to calculate the first P_n up to $n = 10$. The first few are [18]

$$P_n = m_n \qquad n = 0, 1, 2, 3$$
$$P_4 = m_4 + \frac{1}{7} m_0{}^*(m_1^2 - m_2 m_0) \tag{3.4}$$
$$P_5 = m_5 + \frac{1}{3} m_0{}^*(m_2 m_1 - m_3 m_0) + \frac{1}{21} m_1{}^*(m_1{}^2 - m_2 m_0).$$

The nonlinear contributions from the lower m_n become progressively worse as n increases. Unfortunately we have found no way to express those additional terms in closed form. However, the following observation might give a clue to their structure. Let us consider an infinitesimal shift of the origin, i.e. $z \to z - \epsilon$. The m_n transform under such a shift in the same manner as the P_n, viz.

$$\dot{m}_n = n m_{n-1} \quad , \quad \dot{P}_n = n P_{n-1}.$$

Hence the nonlinear terms in P_n have to reproduce those in P_{n-1} under a shift. Alas, we have not been able to exploit this fact.

It should be noted that for the Kerr metric all nonlinear terms vanish and the multipole moments are given in terms of usual parameters m, a by

$$P_n = m_n = m(ia)^n.$$

Is there a way to measure the multipole moments? To this end we consider the following Gedankenexperiment. We probe the far field of a compact object which we assume to be stationary and axisymmetric with test particles. If the field is described by the Schwarzschild solution, i.e. if it has only mass and no higher moments, we can infer the mass from the perihelion shift and the other orbital parameters of the probe. If, in addition, the field also has a quadrupole and perhaps higher moments, we should need more probes with different orbital angular momenta, i.e. different excentricities. Mass, quadrupole and higher moments enter the expression for the perihelion shift with coefficients depending differently on the excentricity of the orbit. Hence we could infer the first few moments. To infer angular momentum moments we should have to distiguish between pro- and retrograde orbits. With a sufficient number of test particles we could, of course in principle only, probe the far field moments of a compact rotating object. It is an open question whether the so measured moments agree with the moments as defined above.

In many applications we are faced with a situation in which more then one body is present. While the Geroch-Hansen moments describe the field at infinity, as far away from the sources as possible, one would like to have a local definition of moments referring to just one of the constituents of a multi-particle system. There is Penrose's definition [19] of quasi-local mass which, however, entails constructing the 2-surface twistor space for the surface in question and thus appears to be so complicated as to be unsuited for most practical applications.

The only quantities at hand are the Komar integrals [20]. They are defined in terms of a Killing vector ξ and a closed 2-surface Σ by

$$K = \int_{\Sigma} \xi_{\alpha;\beta} dS^{\alpha\beta} \tag{3.5}$$

It is well-known that the integral is independent of the surface in vacuo. We therefore define the mass and angular momentum of a single object by the Komar integrals with $\xi = \partial_t$ and $\xi = \partial_\varphi$ extended over a surface surrounding only that object. It is easy to show that the Komar mass and angular momentum for an isolated system as a whole, i.e. taken over a surface surrounding all possible sources, agree with the mass and angular momentum as defined earlier. On the other hand, the Komar integrals are clearly additive, i.e. the sum of the individual masses of a multi-particle system gives the total mass, which is not what some would expect from a local definition of mass or angular momentum in General Relativity. Yet the Komar integrals are the only practical definitions available for individual objects.

In terms of the metric (2.1) one chooses a surface with $t = const.$ which converts the Komar integrals formally into expressions in flat three-space. Mass and angular momentum are given by

$$M_i = -\frac{1}{8\pi} \int_{\Sigma_i} (\frac{1}{f} f_{,\alpha} + \frac{f^2 \omega}{\rho} \omega_{,\alpha}) k^\alpha dS \tag{3.6}$$

$$J_i = \frac{1}{16\pi} \int_{\Sigma_i} [\frac{f\omega}{\rho^2}(\frac{\rho^2}{f} - f\omega^2)_{,\alpha} - (\frac{1}{f} - \frac{f\omega^2}{\rho^2})(f\omega)_{,\alpha}] k^\alpha dS$$

where k^α denotes the unit normal vector to the chosen surface.

4 Interpretation of Exact Solutions

The fact that the Weyl coordinates are uniquely defined by the geometry does not, unfortunately, imply that they lend themselves easily to the interpretation of a given solution. It will be advantageous to introduce spherical coordinates by $\rho = r \sin \vartheta$, $z = r \cos \vartheta$.

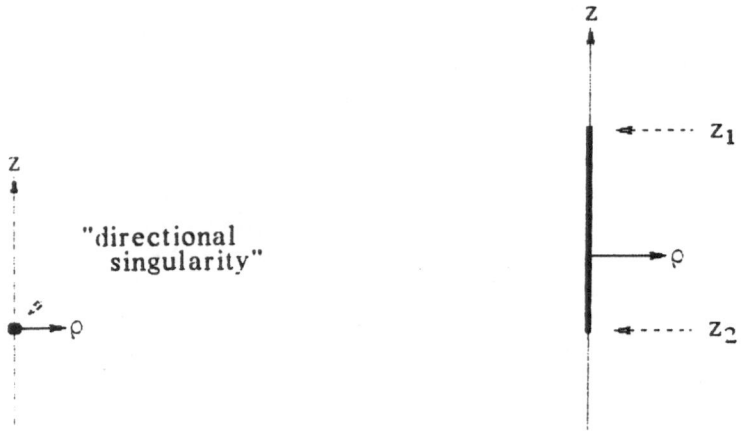

Fig. 1.   Fig. 2.

Let the potential χ be given by $\chi = -\frac{m}{r}$. The static solution corresponding to this potential is the Curzon solution. The metric derived from this simple χ is not spherically symmetric and has what appears in Weyl coordinates as a directional singularity at $r = 0$. Even though the Curzon solution has been known for a long time, it was only comparatively recently that Scott and Szekeres [21] have unravelled the rather complicated structure of this "directional singularity".

The potential χ pertaining to the Schwarzschild solution appears in Weyl coordinates as the potential of a rod of uniform density $\frac{1}{2}$ and length $2m = z_1 - z_2$. Near the "rod" for $z_1 > z > z_2$ we have $\chi \sim ln\rho$ and the coefficient of $d\varphi^2$ in (2.1) actually tends to a constant. The limit $z_2 \to -\infty$ while keeping z_1 fixed yields the C metric [22].

These two examples should serve as a caveat against taking Weyl coordinates too seriously close to singularities of the potential. Nevertheless, quite some useful information can be extracted from them.

The first question to be addressed is: Under which conditions on the Ernst potential is space-time asymptotically flat? If we want the ∂_t Killing vector to be a translation at infinity we should have $\mathcal{E} \to 1$ or equivalently $\xi \to 0$ for large r. While this ensures asymptotic flatness of the three-space as discussed above, it is not sufficient to guarantee asymptotic flatness of the four-dimensional spacetime. The axis in space-time is the set of points invariant under the action of the ∂_φ Killing vector. As ρ is the volume of the 2-surfaces spanned by ∂_φ and ∂_t , the axis will certainly be part of $\rho = 0$, the axis of the canonical Weyl coordinates. The converse, however, is not true. On the axis the ∂_φ vector should vanish which it does only if $\omega(\rho = 0) = 0$. It follows from (2.5) that ω is determined up to an additive constant and that $\omega(\rho = 0)$ is at most a step function of z. Hence we can always achieve $\omega(\rho = 0) = 0$ for sufficiently large negative z. This does not imply the $\omega(\rho = 0)$ will also vanish for large positive z. For sufficiently large r we can expand $\xi = \frac{m_0}{r} + O(r^{-2})$. This implies $\psi = -2\frac{Im\ m_0}{r} + O(r^{-2})$ and thus from (2.5) $\omega = -2Im\ m_0(1 + \cos\vartheta) + O(r^{-1})$. Hence $\omega(\vartheta = 0) \neq 0$ and consequently $\vartheta = 0$ is not a set of fixed points of the ∂_φ Killing vector. The fall-off condition on ξ together with vanishing of the angular momentum monopole or equivalently the condition $\omega(\rho = 0) = 0$ for sufficiently large $|z|$ –ensure that the four-dimensional space-time is asymptotically flat and the ∂_t Killing vector is a translation at infinity.

One could also study boost-rotation symmetric space-times, cf. Ref. 23. In this case the ∂_t Killing vector describes a boost and the form (2.1) of the metric is valid in that part of space-time in which the trajectories of the boost are timelike. The coordinate axis is composed of sets of fixed points of the action of both Killing vectors. For asymptotic flatness that part of $\rho = 0$ making up the fixed points of the boost Killing vector has to extend to infinity in Weyl coordinates. Hence the condition is that $f = \rho^2 + O(\rho^4)$ for $z < z_0$, say. This part of the coordinate axis is rather similar to a horizon, cf. the above remarks about the Schwarzschild solution. Once a solution has been found, the remaining part of space-time where ∂_t will be spacelike has to be constructed by continuation through $\rho = 0, z < z_0$.

Let us revert to stationary axisymmetric solutions and consider a system which we should like to interpret as composed of two (or more) individual constituents. We are not interested in the precise structure of the objects; they may be thought of as singularities or being made of matter of unknown composition. Our interest lies in the vacuum field they produce. For two objects we have two conditions on ω. Firstly, it should vanish on $\rho = 0$ above the objects for reasons of asymptotic flatness. Secondly, ω should vanish on the coordinate axis between the constituents. For were it not to vanish there, that part of $\rho = 0$ would not be part of the axis and we should have no justification of speaking about two seperated objects.

The other function in the metric to which we have to pay attention is γ. In Weyl coordinates γ is also defined up to an additive constant and at most a step function of z at $\rho = 0$. It follows easily from (2.7) that the constant can be chosen such that $\gamma(\rho = 0) = 0$ for sufficiently large $|z|$, i.e. on the axis outside the objects. Between them γ will in general not vanish. To examine the effect of a non-vanishing γ let us consider a small circle in a plane $z = const.$ around a point $\mathcal{P}$ on the axis. The radius of the circle is $\rho(f^{-\frac{1}{2}}e^\gamma)_{|\mathcal{P}}$ and the circumference is $2\pi\rho f^{\frac{1}{2}}_{\mathcal{P}}$, each to first order in ρ. Their ratio is thus $2\pi e^\gamma_{\mathcal{P}}$ and the geometry of the $z = const.$ plane thus resembles the geometry near the tip

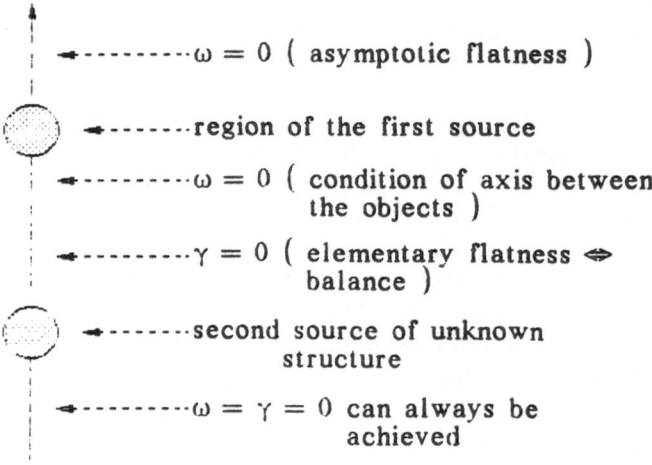

Fig. 3.

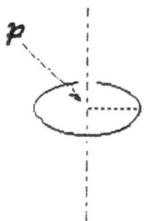

Fig. 4.

of a cone or the central point of a ruff as $\gamma_{|\mathcal{P}}$ is positive or negative. This lack of elementary flatness, i.e. lack of differentiability as the metric is only C_0 at $\mathcal{P}$, gives via the Einstein equations rise to a non-vanishing energy-momentum tensor [24]. This energy-momentum tensor has the form of a δ function and describes consequently a line source. It has the peculiar property that the energy density equals the absolute value of the pressure - hardly the material you can buy in your neighbourhood hardware shop. On the other hand, ropes of that material are the only ones which one can hang right down to the horizon of a black hole without breaking. By chosing the Ernst potential we have chosen the sources of the gravitational field. γ will, in general, not vanish between the objects and Einstein's equations force the line source, frequently called the Weyl strut, upon us. This is an instance for the fact that Einstein's equations imply the equations of motion for the sources, for without the Weyl strut space-time could not be stationary. The pressure in the strut is given by [25]

$$p = \frac{1}{4}(e^{\gamma}|_{\mathcal{P}} - 1) \tag{4.1}$$

and we shall use its negative as definition of the force acting between the objects. The question now arises whether one can choose the parameters of a two-body solution, i.e. masses, angular momenta and distance, such that the Weyl strut is absent. Of course, the masses involved should be positive. If one allows negative masses to be present, it is easy to construct even a static solution involving three masses without struts. Permitting only positive masses would imply that gravitational attraction can be balanced by the relativistic interaction of angular momenta. The effect of gravitational angular momentum interaction is closely akin to the Lense-Thirring effect and the discussion dates back even to pre-relativistic days [26].

The first solution constructed explicitly with the aim of investigating the possibility of balance was derived by acting with two HKX transformations on a static seed metric describing two Curzon particles, i.e. (2.16) with $n = 2$. It has been shown that the parameters can be chosen such that the solution is symmetric with respect to $z = 0$. It has no singularities except at the places of the Curzon particles and thus describes two objects balanced by their angular momentum against gravitational attraction [27]. The internal structure of those Curzon objects is, of course, obscure.

Earlier the so-called double Kerr solution had been derived by Kramer and Neugebauer [28]. The solution describing n Kerr objects can be cast into the neat form [29]

$$\mathcal{E} = \frac{D_-}{D_+} \tag{4.2}$$

with the $n \times n$ determinants and $2n$ constants w_k and z_k

$$D_\pm = |\frac{e^{i\omega_k} S(z_k) + e^{i\omega_l} S(z_l)}{z_k - z_l} \pm 1|$$

where $k = 1, 3, \ldots, 2n - 1$ denotes the rows and $k = 2, 4, \ldots, 2n$ the columns. One introduces the parameters

$$\omega_k = \alpha_k + \lambda_k, \qquad \omega_{k+1} = \alpha_k - \lambda_k$$
$$e^{i\lambda_k} = p_k + iq_k.$$

The α's and q's are called NUT and Kerr parameters, respectively, and mass parameters m_k are defined in terms of those and the z_k as indicated in the diagram. It should be kept in mind that those names are nothing but words chosen in analogy to the Kerr metric which arises from (4.2) for $n = 1$.

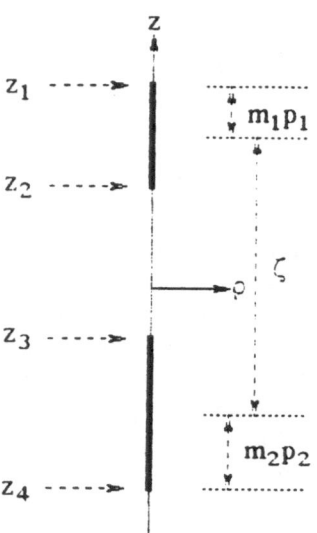

Fig. 5.

Here we shall give only a brief description [30] of the double Kerr solution, a longer treatment can be found in Ref. 31. A prerequisite for interpreting the solution as describing two seperated objects is obviously $\zeta > m_1 p_1 + m_2 p_2$. It can be shown that in this case the regions $\rho = 0, z_1 > z > z_2$ and $z_3 > z > z_4$ are null hypersurfaces on which the ∂_φ Killing vector does not vanish and which have signature $(+ + 0)$. The metric can be extendend through them. Those hypersurfaces are thus horizons. The precise nature of the points where the Weyl strut meets the horizons has not yet been resolved.

The Komar masses M_i and angular momenta J_i can be calculated easily by extending the integrals (3.6) over the horizons and using (2.5) to convert ρ- into z-derivatives.

The condition of asymptotic flatness can be written as a quadratic polynomial for ζ with coefficients depending on the other parameters. The axis condition, i.e. the condition that there be an axis between the objects, yields a fourth order polynomial for ζ. One could solve the quadratic equation and insert the solutions into the fourth order one to end up with a rather complicated expression for the other parameters which one would have to solve. Analytically this is clearly quite hopeless. On the other hand, the balance condition, i.e. $\gamma(\rho = 0, z_2 > z > z_3) = 0$, is the trivial equation $\cos(\alpha_1 - \alpha_2) = 0$. Assuming balance the equations for

ζ become somewhat more manageable and one can express $\frac{m_i}{\zeta}$ in terms of the remaining three parameters q_i and α_i. Checking numerically over the range of the q_i and α_i we found that no solution which would make both Komar masses positive appears to exist. Thus we conclude tentatively that black holes cannot be balanced.

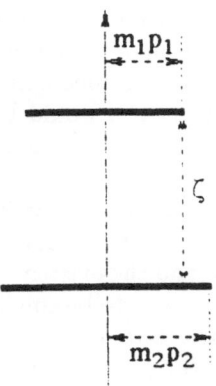

Fig. 6.

We can extend the parameters p_i individually or together into the complex plane and arrive after $p_i \to -ip_i$ at the superposition of two hyperextreme Kerr objects. z_i and z_{i+1} now become complex conjugate and the solution may be visualized as the superposition of two rotating discs with curvature singularities at the rim. The symmetric case, i.e. $p_1 = p_2$ and $m_1 = m_2$, can be treated exactly and balance can be achieved [14] . In terms of the Komar quantities and the dimensionless parameter

$$Q = \frac{J}{M^2}$$

the distance at which the two objects balance is given by

$$\zeta_b = M[Q(1 + \sqrt{1 - \frac{2}{Q}}) - 2] \qquad (4.3)$$

Clearly $Q > 2$. For large Q we find $\zeta_b \sim 2MQ$ with corrections of the order M.

Let us look at some typical values for Q. For an object of size R, mass M and angular velocity ω –the typical velocity would then be $v = R\omega$– we may estimate $J \sim MRv$ in non-geometrized units or $Q = \frac{R}{M}v$ in geometrized units. For the sun we have $Q \approx 0.2$ and stars in the early stages of their evolution should have Q values greater by a factor of about 10. Neutron stars also cannot attain Q values larger than 1, in fact they break up much earlier [32]. On the other hand, non-gravitationally bound objects of everyday size and mass have amazingly high values for Q. An album, for instance, rotating with 33 rpm on its turntable —not something which one would describe as rotating rapidly— has $Q \approx 10^{18}$. Any rotating object we come across in daily life has such huge values for Q that, were we to require $Q < 1$, we should not recognize it as rotating at all. Of course, everyday objects are of enormous size compared to their Schwarzschild radius. For a neutron $(J = \frac{\hbar}{2})$ we find, only formally of course, $Q \approx 8 \times 10^{37}$. Bodies of laboratory dimensions would thus balance at $\zeta_b \sim 2R\, v \approx R \times 10^{-6}$. Two neutrons, just to quote the figure, would balance at $\zeta_b \approx 2 \times 10^{-14}$ cm. These numbers indicate that in practical situations balance is quite impossible to achieve.

If we abandon the balance condition $\gamma(\rho = 0) = 0$ between the objects we are not only stuck with the Weyl strut but also with the inability to solve the conditions of asymptotic flatness and the existence of an axis between the objects exactly. However, assuming the masses to be small compared to the distance, i.e. $\frac{m_i}{\zeta} \ll 1$, we can solve the conditions to any desired order. The attractive force, it was defined as the negative of the pressure in the strut, becomes a function of the two masses, angular momenta and the distance. We can then compare two systems, a static and a stationary one, such that the masses and the distance in one system have the same values as in the other one. The relative change of force is

$$\delta F = \frac{F_{\text{stat}} - F_{\text{rot}}}{F_{\text{stat}}} = \frac{3}{\zeta^2}(Q_1 M_1 + Q_2 M_2)^2$$

It appears surprising that δF should always be positive, no matter whether the objects are co- or counterrotating. Calculating the total quadrupole moment of the solution we find [33]

$$\Re P_2 = (\frac{\zeta^2}{4} - M_1 M_2)(M_1 + M_2) - Q_1{}^2 M_1{}^3 - Q_2{}^2 M_2{}^3$$

The last two terms describe the difference in the quadrupole moment between a static and a stationary configuration. In the approximation used here the changes in the individual quadrupole momenta should be additive and we can therefore ascribe to each object a moment of $-Q_i^2 M_i^3$. The fact that the relative change of force is a perfect square can thus be seen as arising from a conspiracy of the mass-quadrupole interaction with the interaction of the angular momenta. Had we considered a solution other than the double Kerr solution, no perfect square would have appeared. Finally the force due only to angular momentum interaction between two rotating bodies is given by

$$\delta F_{qq} = -\frac{6}{\zeta^4} J_1 J_2 \tag{4.4}$$

For objects which could be handled in an experiment we can estimate $\delta F \approx 10^{-7}$. On the other hand, the force itself is $F_{stat} \approx 10^{-8} N$. In an experiment of the Cavendish type, angular momentum interaction would show up as a different gravitational constant for the rotating and non-rotating configuration. If only we knew G to seven decimal places!

5 Factor Structure of Rational Vacuum Solutions

It has been more or less common knowledge that the polynomials appearing in certain rational metrics like the Kerr or Tomimatsu-Sato metrics can be factorized. In this chapter we shall give an outline of our recent research into this topic [34]. Factorizability is, of course, a property which holds only in special coordinate systems. Here we shall consider prolate or oblate speroidal and spherical coordinates related to the Weyl coordinates by

$$\rho = \begin{pmatrix} \sinh x \\ e^x \\ \cosh x \end{pmatrix} \sin y, \qquad z = \begin{pmatrix} \cosh x \\ e^x \\ \sinh x \end{pmatrix} \cos y$$

The metric (2.1) will be used in the form

$$ds^2 = \frac{B}{A} e^{2\gamma} (dx^2 + dy^2) + \frac{1}{B}(G d\varphi^2 + 2C d\varphi dt - A dt^2) \qquad (5.1)$$

The functions A, B, etc., are supposed to be polynomials in either $\sinh(x)$, e^x, or $\cosh(x)$ and $\cos(y)$, depending on whether oblate spheroidal, spherical or prolate spheroidal coordinates are used. All common factors are assumed to have been cancelled. The usual forms are recovered by the coordinate transformation $\xi = [\sinh(x), e^x, \cosh(x)]$, $\eta = \cos(y)$; A, B, etc. then become polynomials in ξ and η. The assumption of the polynomials depending only on $\cosh(x)$, which we shall from now on use as representative for the other cases –it can easily be seen that the arguments given below are also valid in the other cases–, and $\cos(y)$ is justified by the field equations being elliptic and thus their solutions being analytic at $\rho = 0$.

We have the condition on the determinant of the $\varphi - t$ part of the metric

$$AG + C^2 = \rho^2 B^2 \qquad \Rightarrow \qquad AG = (\rho B + C)(\rho B - C).$$

Even though polynomials in trigonometric or hyperbolic functions are not uniquely factorizable $-\sin^2(y) = \sin(y)\sin(y) = [1 + \cos(y)][1 - \cos(y)]-$ one can show from the fact that A and C contain no linear sin or sinh terms that they have to be factorizable as, e.g., $A = (\lambda_1 \tau + \lambda_2 \nu)(\lambda_1 \tau - \lambda_2 \nu)$. Finally the argument [35] shows that the metric can be written as

$$ds^2 = \frac{B}{A} e^{2\gamma}(dx^2 + dy^2) + \frac{1}{B}[\lambda_2^2(\mu d\varphi + \nu dt)^2 - \lambda_1^2(\sigma d\varphi + \tau dt)^2] \qquad (5.2)$$

with

$$\lambda_1 \lambda_2 = \rho, \qquad i.\ e. \qquad \lambda_1 = \begin{pmatrix} \sinh x \\ 1 \end{pmatrix}, \lambda_2 = \begin{pmatrix} \sin y \\ \sinh x \, \sin y \end{pmatrix}$$

and μ, ν etc. being polynomials in $\cosh x$ and $\cos y$.

Some "experimental facts" —we are using this term to indicate that these properties have been found to hold for a sufficiently large number of solutions checked, often by algebraic computer programmes, to make one wonder whether there is a general principle behind it— are surprising; for instance: Let the Ernst potential be $\mathcal{E} = (\alpha - \beta)(\alpha + \beta)^{-1}$ with α and β being polynomials in $\cosh(x)$ and $\cos(y)$ as indicated above, then $\alpha \partial \beta - \beta \partial \alpha$ is divisible by one of the factors of A. Or: The nonvanishing spin components of the Weyl tensor, C_0 and $C_{\pm 2}$, are essentially second derivatives of the Ernst potential; they are thus proportional to $(\alpha + \beta)^{-3}$. Why is it then that $I_2 = C_2 C_{-2} - 9 C_0^2$ is proportional to $(\alpha + \beta)^{-5}$ and not to $(\alpha + \beta)^{-6}$ as one would have assumed?

In what follows we shall adhere essentially to the formalism developed by Ernst and Hauser [35]. We shall use the notation for the derivative operator

$$\partial_\pm = \partial_x \pm i \partial_y$$

for reasons which will become apparent below. Note that $\partial_+{}^* = \partial_-$; as the appearing polynomials are polynomials in hyperbolic respectively trigonometric functions, $\partial_\pm$ does not lower the degree of the polynomial to which it is applied.

The invariants of the Weyl tensor are given by

$$C_0 = \frac{1}{8} e^{-2\gamma} [2 \partial_+ \partial_- \mathcal{E} - \frac{1}{f} \partial_+ \mathcal{E} \partial_- \mathcal{E}] \tag{5.3}$$

$$C_{\pm 2} = -\frac{1}{8} e^{-2\gamma} [2 \partial_\pm{}^2 \mathcal{E} - 4 \partial_\pm \gamma \partial_\pm \mathcal{E} + \frac{1}{f} (\partial_\pm \mathcal{E})^2]$$

from which one can form two spin invariant quantities

$$I_1 = C_0 \quad , \qquad I_2 = C_2 C_{-2} - 9 C_0^2.$$

The solution is of Petrov type D if either $C_{\pm 2} = 0$ or $I_2 = 0$. By use of the Ernst equation (2.6) C_0 can be rewritten as

$$C_0 = \frac{1}{8} e^{-2\gamma} [\frac{1}{f} \partial_+ \mathcal{E} \partial_- \mathcal{E} - \frac{1}{\rho} (\partial_+ \rho \partial_- \mathcal{E} + \partial_- \rho \partial_+ \mathcal{E})].$$

We assume that $\mathcal{E}$ can be written as

$$\mathcal{E} = \frac{\alpha - \beta}{\alpha + \beta} = \frac{A + iI}{B} \tag{5.4}$$

with complex polynomials α and β. The various polynomials in the metric (5.1) are given by

$$A = \lambda_1^2 \tau^2 - \lambda_2^2 \nu^2 = \alpha \alpha^* - \beta \beta^* = A_+ A_-$$

$$B = \mu\tau - \nu\sigma = (\alpha + \beta)(\alpha^* + \beta^*)$$
$$C = \lambda_2{}^2\mu\nu - \lambda_1{}^2\sigma\tau, \qquad \rho = \lambda_1\lambda_2$$

Thus

$$f = \frac{A}{B}, \qquad \omega = \frac{C}{A}, \qquad \rho B \pm C = A_\pm B_\mp, \qquad \psi = \frac{I}{B}$$

First we shall turn out attention to equation (2.5) relating ω and ψ. It becomes

$$-i\rho(B\partial_+ I - I\partial_+ B) = A\partial_+ C - C\partial_+ A$$

From the two expressions for $\partial_\pm \mathcal{E}$ to be derived from (5.4) we get

$$-2\rho(\alpha^* + \beta^*)^2(\alpha\partial_\pm\beta - \beta\partial_\pm\alpha) = A_\pm[A_\pm(N_\mp\partial_\pm A_\mp - A_\mp\partial_\pm N_\mp) - A_\mp B\partial_\pm\rho]$$

As $A_\pm$ have no common factor with $(\alpha^* + \beta^*)$, we conclude that they have to divide $\alpha\partial_\pm\beta - \beta\partial_\pm\alpha$. Thus we define $K_\pm$ by

$$\alpha\partial_\pm\beta - \beta\partial_\pm\alpha = K_\pm A_\pm \qquad (5.5)$$

This is one of the divisibility properties which had been noticed earlier. There is, of course, a sign ambiguity due to the sign in λ_1 and λ_2; we thus define A_+ as that factor of A which divides $\alpha\partial_+\beta - \beta\partial_+\alpha$. Having cancelled $A_\pm$ we can use the definition of B and repeat the previous argument to deduce the existence of polynomials $P_\pm$ by

$$(\alpha^* + \beta^*)P_\pm = A_\mp\partial_\pm N_\mp - N_\mp\partial_\pm A_\mp \qquad (5.6)$$

The Ernst equation (2.6) written in terms of α and β and using (5.5) reads

$$A_+ A_-(\alpha\Delta\beta - \beta\Delta\alpha) = (\alpha^*\partial_+\alpha - \beta^*\partial_+\alpha)K_- A_- + (\alpha^*\partial_-\alpha - \beta^*\partial_-\beta)K_+ A_+$$

Again we find new polynomials $L_\pm$ defined by

$$\alpha^*\partial_\pm\alpha - \beta^*\partial_\pm\beta = L_\pm A_\pm \qquad (5.7)$$

From the definition of A in terms of α and β we find

$$A_+(\partial_+ A_- - L_+) - A_-(\partial_+ A_+ - L_-{}^*) = 0$$

whence we conclude

$$\partial_+ A_- = L_+ - cA_-, \qquad \partial_+ A_+ = L_-{}^* + cA_+$$

c has to be constant because A_+ and A_- are of the same degree. We shall now show that c vanishes. From the above equations we derive

$$A_+\partial_+ A_- - A_-\partial_+ A_+ - A_+ L_+ + A_- L_-{}^* = 2cA_+ A_-.$$

The terms with L contain expressions of the form $\alpha^*\partial_+\alpha - \alpha\partial_+\alpha^*$. Consider the leading terms in $\cosh x$ and among these the one leading term in $\cos y$; this term will drop out. A similar argument shows that the leading term in the $A_\pm$ terms drop out as well. The left-hand side of the equation is thus of lower degree than the right-hand side. Therefore c has to vanish and we find

$$\partial_\pm A_\mp = L_\pm \tag{5.8}$$

Eliminating α and β between (5.5) and (5.7) we get for the derivatives

$$A_\mp \partial_\mp \alpha = \beta^* K_\pm + \alpha L_\pm$$
$$A_\mp \partial_\mp \beta = \alpha^* K_\pm + \beta L_\pm.$$

With the polynomials K and L the Ernst equation can be rewritten as

$$A_\pm \partial_\mp K_\pm + \frac{1}{2\rho}(A_+ K_+ \partial_- \rho + A_- K_- \partial_+ \rho) - L_\mp K_\pm = 0 \tag{5.9}$$

Using "leading term" arguments similar to the one which was used to prove (5.8) one can show from (2.7) that $e^{2\gamma}$ has to be of the form

$$e^{2\gamma} = c\frac{A}{D}, \qquad D = (\partial_+ \rho \, \partial_- \rho)^{\frac{n}{2}-1} \tag{5.10}$$

where n is the degree of A in $\sinh x$ and c is at most a step function of z.

With this expression for e^γ the invariants of the Weyl tensor I_1 and I_2 appear to have a factor of A in the denominator. Of course, this factor should not be there; $A = 0$ indicates the limit of stationarity and we do not expect this surface to be singular. Calculating C_0 from (5.3) we get

$$C_0 = \frac{1}{4}\frac{D}{(\alpha+\beta)^3}\frac{1}{A\rho}T_0$$

$$T_0 = (\alpha+\beta)[K_+ A_+ \partial_- \rho + K_- A_- \partial_+ \rho] + 2\rho K_+ K_-(\alpha^* + \beta^*).$$

The square bracket contains a factor of ρ; cf. the Ernst equation (5.9). T_0 can be rewritten as

$$T_0 = 2\rho A_-[K_- \partial_+(\alpha+\beta) - (\alpha+\beta)\partial_+ K_-]$$
$$= 2\rho A_+[K_+ \partial_-(\alpha+\beta) - (\alpha+\beta)\partial_- K_+].$$

Hence it contains both factors $A_\pm$ and is thus divisible by A. With $T_0 = 2A\rho M$ we get for C_0

$$C_0 = \frac{1}{2}\frac{DM}{(\alpha+\beta)^3}.$$

The calculation of I_2 is a bit more involved. First we observe that (5.3) yields

$$C_{\pm 2} = \frac{1}{2}\frac{D}{(\alpha+\beta)^3}\frac{1}{\partial_\pm \rho A_\mp}T_{\pm 2}$$

$$T_{\pm 2} = (\alpha+\beta)[\partial_\pm \rho \partial_\pm K_\pm - K_\pm \partial_\pm{}^2 \rho(\frac{n}{2}-1)] - 3K_\pm \partial_\pm \rho \partial_\pm(\alpha+\beta).$$

We need not worry about the $\partial_+\rho$ in the denominator; such a factor is found either in D given by (5.10) or in $T_{\pm 2}$ if $n = 2$. On the other hand, there is $A_\mp$.

To avoid an A in the denominator of I_2, $T_\pm$ has to contain a factor of either $A_\mp$ or $A_\pm$. Using the Bianchi identities we get an alternative formula for $C_{\pm 2}$

$$C_{\pm 2} = \frac{(\alpha + \beta)A_\pm}{P_\mp A_\mp}(3\partial_\pm \rho C_0 + 2\rho\partial_\pm C_0).$$

Comparing this with the above expression we see the factor $A_\pm$ explicitly. To show that I_2 is proportional to $(\alpha + \beta)$ we need to consider only those terms in T_0 and $T_{\pm 2}$ which are not explicitly multiplied by $(\alpha + \beta)$. Evidently these terms drop out and I_2 is proportional to $(\alpha + \beta)^{-5}$ as has been observed in all cases where the invariants have been calculated.

As an instance we quote the polynomials for the Kerr metric:

$$\alpha = p\cosh(x) - iq\cos(y), \qquad \beta = 1$$
$$A_\pm = p\sinh(x) \mp q\sin(y), \qquad K_\pm = -1, \qquad L_\pm = \alpha^*$$
$$P_\pm = -\cosh(x)\sin(y) + i\sinh(x)\sin(y) +$$
$$\qquad A_\mp[\sinh(x)\sin(y) \pm i\cosh(x)\cos(y)] \pm q\sinh(x) - p\sin(y)$$
$$M = 1.$$

Note that one of the rules used by Tomimatsu and Sato [37] to derive their series of solutions is $K_\pm = real$. Moreover, by studying the behaviour of the various polynomials under an Ehlers transformation $(\alpha, \beta) \to (\alpha, e^{i\delta}\beta)$ we find that $P_\pm$ has to split up into terms which are individually divisible.

There are reasons to suspect that the factors discussed here exist for an even wider class of solutions. Inspection of the double Kerr solution with its Ernst potential being rational in S_i leads to the conjecture that metrics rational in distances from given points on the coordinate axis are factorizable. This is currently under investigation [38].

Moreover, there is at least one metric not rational in prolate spheroidal coordinates which is factorizable. One of the solutions derived by Yamazaki which is a generalization of the Kinnersley-Chitre solution to a noninteger parameter can be shown by direct calculations to admit all the factors mentioned above. One may conjecture that solutions depending continuously on a parameter such that they are rational for certain some of its values are also factorizable in the above sense.

References

1. Solutions of Einstein's Equations: Techniques and Results, Eds. C. Hoenselaers and W. Dietz, Springer Verlag, Heidelberg (1984)
2. J. B. Griffiths: Colliding Plane Waves in General Relativity, Oxford Science Publications, Oxford (1991)
3. W. Kinnersley: in General Relativity and Gravitation (Proceedings of GR7) Eds. G. Shaviv and J. Rosen, Wiley, New York (1975)
4. F.J. Ernst: Phys. Rev. **167**, 1175 (1968)
5. C. B. Morrey: Am. J. Math. 80, 198 (1958)
6. H. O. Cordes: Nachr. Akad. Wiss. Göttingen 11, (1956); N. Aronszajn: Journ. de Math. **36**, 235 (1957)
7. H. Weyl: Annalen Physik **54**, 117 (1917)
8. S. M. Scott: loc. cit. 35; W. B. Bonnor: Gen. Rel. Grav. **24**, 551 (1992)
9. R. Geroch: J. Math. Phys. **13**, 394 (1972)
10. B. G. Schmidt: loc. cit. 1
11. W. Kinnersley: J. Math. Phys. **18**, 1529 (1977); **21**, 2231 (1980); W. Kinnersley, D. Chitre: J. Math. Phys. **18**, 1538 (1977); **19**, 1926 & 2037 (1978); C.Hoenselaers,W. Kinnersley,B.Xanthopoulos: J.Math.Phys. **20**, 2530(1979); C. Hoenselaers: loc. cit. 1
12. I. Hauser, F. J. Ernst: Phys. Rev. **20**, 362 & 1783 (1979); J. Math. Phys. **21**, 1126 & 1418 (1980); 22, 1051 (1981); I. Hauser: loc. cit. 1; F. J. Ernst: loc. cit. 1
13. W. Dietz: loc. cit. 1
14. W. Dietz: Habilitationsschrift, Universität Würzburg (1984)
15. W. G. Dixon: Phil. Trans. Roy. Soc. Lond. **227**, 59 (1974)
16. R. Geroch: J. Math. Phys. **11**, 2580 (1970); R. O. Hansen: J. Math.Phys. **15**, 46 (1974)
17. W. Simon, R. Beig: J. Math. Phys. **24**, 1163 (1983)
18. I. Hauser: private communication; G. Fodor, C. Hoenselaers, Z. Perjés: J. Math. Phys. **30**, 2252 (1989)
19. R. Penrose: in Gravitational Collapse and Relativity, Eds. H. Sato & T. Nakamura, World Scientific, Singapore (1986)
20. A. Komar: Phys. Rev. **113**, 934 (1958)
21. S. M. Scott, P. Szekeres: Gen. Rel. Grav. **18**, 557 & 571 (1986); S. M. Scott: loc. cit. 19
22. W. B. Bonnor: Gen. Rel. Grav. **15**, 535 (1983); C. Hoenselaers: in Essays in Relativity, Ed. M. A. H. MacCallum, Cambridge University Press, 1985
23. J. Bičák, B. G. Schmidt: Phys. Rev. D **40**, 1827 (1989)
24. R. Bach, H. Weyl: Math. Z. **13**, 134 (1922)
25. W. Israel: Phys. Rev. D **15**, 935 (1977)
26. A. Einstein: Vierteljahresschrift für gerichtliche Medizin und öffentliches Sanitärwesen **44**, Juliheft (1912)

27. W. Dietz, C. Hoenselaers: Phys. Rev. Lett. **48**, 778 (1982);
     C. Hoenselaers: in Proc. 3rd Marcel Grossmann Meeting, Ed. Hu Ning, North
     Holland, Amsterdam (1983)
28. D. Kramer, G. Neugebauer: Phys. Lett. **75A**, 259 (1980)
29. M. Yamazaki: loc. cit. 1
30. C. Hoenselaers: in Proc. 4th Marcel Grossmann Meeting, Ed. R. Ruffini, Elsevier
     Science Publishers (1986)
31. W. Dietz, C. Hoenselaers: Ann. Physics **165**, 319 (1985)
32. J. Ipser: This volume
33. C. Hoenselaers: loc. cit. 19
34. C. Hoenselaers, Z. Perjés: Class. Quant. Grav. **7**, 2215 (1990)
35. R. P. Kerr: in Proceedings of the Centre for Mathematical Analysis, Australian
     National University, 19 (1989)
36. F. J. Ernst: J. Math. Phys. **15**, 1409 (1974)
37. A. Tomimatsu, H. Sato: Prog. Theor. Phys. **50**, 93 (1973)
38. A. Lun: in Proc. 4th Hungarian Relativity Meeting, Ed. Z. Perjés & R. P. Kerr,
     World Scientific, Singapore, to appear

The Dyadic Approach to Solutions for Rotating Rigid Bodies

H. D. Wahlquist

Jet Propulsion Laboratory
California Institute of Technology
4800 Oak Grove Drive
Pasadena, CA 91109

I. Introduction

That a renaissance is occurring in the field of rotating interior solutions is attested by the increasing number of recent publications on the subject, many of them authored by participants in this conference [1]. One reason for this resurgence is undoubtedly the achievement over the last two decades of a general solution to the problem of stationary, axisymmetric vacuum spacetimes, accomplished using the techniques developed for integrable nonlinear equations. Given the plethora of rotating exterior solutions now available, many including infinite numbers of arbitrary mass and electromagnetic multipole moments, it is no longer just a matter of "finding a source for the Kerr solution". It seems that we should be on the threshold of obtaining a complete, asymptotically flat, spacetime to represent a finite relativistic rotating body.

Constructing a complete interior and exterior solution is an exceedingly difficult problem, however–an observation which is underscored by the paucity of comparable solutions in Newtonian gravitation. On the other hand, having such a solution is far more crucial in relativistic, than in Newtonian, gravitation because of the importance of exact solutions in uncovering the full implications of the nonlinear field equations. While a great variety of interior solutions already exists, almost all have additional symmetries which forbid using them to describe finite bodies. Success in finding a complete solution will probably require the discovery of new, more general, interior solutions which include some arbitrary functions to accommodate the matching procedure.

Soon after the 3 + 1 dyadic formalism for spacetime congruences [2] was developed, I believed it might provide a new method to search for rotating interior solutions. The dyadic equations are particularly simple for material undergoing a stationary rigid motion where the problem can be posed as differential equations in the quotient 3-space, the space of trajectories. A distinguishing feature of the dyadic equations is that they involve the Weyl tensor of the gravitational field, as well as the Ricci tensor of the matter. By so doing, they invite a solution approach that is not readily apparent if one looks only at the Einstein field equations where all explicit reference to the Weyl tensor is eliminated. In fact, the original solution for a finite rotating body obtained using the dyadic equations [3] resulted primarily from making an assumption about the form of the tidal gravitational field (the Weyl tensor) inside the body, rather than from assumptions about the kind of matter constituting the body. Section II of this paper briefly re-capitulates the 3 + 1 dyadic formalism and introduces the 3-space covariant vector and dyadic quantities. Section III presents the dyadic equations as they appear for stationary axisymmetric rigid rotation. Section IV describes how the Weyl tensor approach has been used to obtain solutions of these equations. Section V examines some of the ordinary, and some of the peculiar, configurations of finite rotating bodies that appear to be encompassed by the original solution.

II. The 3 + 1 Dyadic Formalism

We use the coordinate basis vectors and 1-forms

$$\mathbf{e}_\mu = \partial/\partial x^\mu, \quad \mathbf{e}^\mu = \mathrm{d}x^\mu \qquad (\mu, \nu = 0, 1, 2, 3) \tag{1}$$

with metric products

$$\mathbf{e}^\mu \cdot \mathbf{e}_\nu = \delta^\mu_\nu, \quad \mathbf{e}_\mu \cdot \mathbf{e}_\nu = g_{\mu\nu}, \quad \mathbf{e}^\mu \cdot \mathbf{e}^\nu = g^{\mu\nu}, \tag{2}$$

and define orthonormal tetrads by

$$\mathbf{u}_r = {}_r\lambda^\mu \mathbf{e}_\mu, \quad \mathbf{u}^r = {}^r\lambda_\mu \mathbf{e}^\mu \qquad (r, s = 0, 1, 2, 3) \tag{3}$$

with metric products

$$\mathbf{u}_r \cdot \mathbf{u}^s = \delta^s_r, \quad \mathbf{u}_r \cdot \mathbf{u}_s = \eta_{rs}, \quad \mathbf{u}^r \cdot \mathbf{u}^s = \eta^{rs} \tag{4}$$

$$\mathbf{u}_r \cdot \mathbf{e}^\mu = {}_r\lambda^\mu, \quad \mathbf{u}^r \cdot \mathbf{e}_\mu = {}^r\lambda_\mu$$

where η_{rs} is the Minkowski metric.

The dyadic formalism [2] results from aligning the timelike basis vector $\mathbf{u}_0$ with a preferred timelike congruence and projecting all tensor quantities into components along $\mathbf{u}_0$ or into the 3-space orthogonal to $\mathbf{u}_0$. Thus, for example, a second-rank 4-tensor

$$T = T_{rs}\mathbf{u}^r\mathbf{u}^s = T_{00}\mathbf{u}^0\mathbf{u}^0 + \mathbf{L}\mathbf{u}^0 + \mathbf{u}^0\mathbf{R} + \mathbf{T} \tag{5}$$

leads to one scalar, T_{00}, two 3-vectors,

$$\mathbf{L} = T_{a0}\mathbf{u}^a, \quad \mathbf{R} = T_{0b}\mathbf{u}^b \qquad (a, b, \ldots, m = 1, 2, 3) \tag{6}$$

and a dyadic

$$\mathbf{T} = T_{ab}\,\mathbf{u}^a\,\mathbf{u}^b,\tag{7}$$

where the tensor product of adjacent vectors is implied.

The covariant derivatives of the orthonormal bases are given by

$$\mathbf{D}\,\mathbf{u}_s = \Gamma_{rst}\,\mathbf{u}^r\mathbf{u}^t,\tag{8}$$

where the Ricci rotation coefficients are skew-symmetric on the last two indices

$$\Gamma_{rst} + \Gamma_{rts} = 0.\tag{9}$$

Exterior derivatives are obtained by antisymmetrizing; i.e.,

$$\mathbf{d}\,\mathbf{u}_s = \mathbf{D}\,\mathbf{u}_s - \mathbf{u}_s\mathbf{D} = \Gamma_{rst}\,(\mathbf{u}^r\mathbf{u}^t - \mathbf{u}^t\mathbf{u}^r).\tag{10}$$

Alternatively, by defining the skew-symmetric matrix of connection 1-forms,

$$\boldsymbol{\omega}_{st} = -\,\Gamma_{rst}\,\mathbf{u}^r,\tag{11}$$

we can write the Cartan moving-frame equation

$$\mathbf{d}\,\mathbf{u}_s + \boldsymbol{\omega}_{st} \wedge \mathbf{u}^t = 0.\tag{12}$$

For the timelike vector we have

$$\mathbf{D}\,\mathbf{u}_0 = \mathbf{u}^0\mathbf{a} + \mathbf{S} - \boldsymbol{\Omega} \times \mathbf{I}.\tag{13}$$

The 3-space covariant dynamical variables of the timelike congruence

$$\begin{aligned}
\mathbf{a} &\equiv \Gamma_{00b}\,\mathbf{u}^b,\\
\mathbf{S} &\equiv \frac{1}{2}\Gamma_{a0b}\,(\mathbf{u}^a\mathbf{u}^b + \mathbf{u}^b\mathbf{u}^a),\\
-\,\boldsymbol{\Omega} \times \mathbf{I} &\equiv \frac{1}{2}\Gamma_{a0b}\,(\mathbf{u}^a\mathbf{u}^b - \mathbf{u}^b\mathbf{u}^a),
\end{aligned}\tag{14}$$

are, respectively, the acceleration vector $\mathbf{a}$, the symmetric rate-of-strain dyadic $\mathbf{S}$, and the skew-symmetric vorticity dyadic which has been expressed by using the dual angular velocity vector $\boldsymbol{\Omega}$, and the unit dyadic, $\mathbf{I} = \mathbf{u}^c\,\mathbf{u}_c$.

A similar expansion for the spatial triad vectors can be written

$$\mathbf{D}\,\mathbf{u}_b = -(\mathbf{a} \cdot \mathbf{u}_b)\,\mathbf{u}^0\mathbf{u}^0 - (S - \Omega \times I) \cdot \mathbf{u}_b\mathbf{u}^0 + \mathbf{u}^0\boldsymbol{\omega} \times \mathbf{u}_b + \nabla\mathbf{u}_b,\tag{15}$$

where, in analogy to Eq. (8), the 3-space covariant derivative $\nabla\mathbf{u}_b$ is defined by

$$\nabla\mathbf{u}_b \equiv \Gamma_{abc}\mathbf{u}^a\mathbf{u}^c = -\,\mathbf{N} \times \mathbf{u}_b.\tag{16}$$

These equations introduce two new quantities; the angular velocity 3-vector of the spatial triad itself,

$$\boldsymbol{\omega} \equiv \frac{1}{2}\,\varepsilon_{abc}\,\Gamma_{0bc}\mathbf{u}^a,\tag{17}$$

and the 3-space dyadic affinity,

$$\mathbf{N} \equiv -\frac{1}{2}\, \varepsilon_{bcd}\, \Gamma_{acd} \mathbf{u}^a \mathbf{u}^b\,. \tag{18}$$

It can be verified quickly that the orthonormality relations

$$\mathbf{D}\,\mathbf{u}_r \cdot \mathbf{u}_s + \mathbf{D}\,\mathbf{u}_s \cdot \mathbf{u}_r = 0 \tag{19}$$

are satisfied by these expressions. The five vector-dyadic quantities $(\mathbf{a},\, \mathbf{S},\, \boldsymbol{\Omega},\, \boldsymbol{\omega},\, \mathbf{N})$ incorporate all 24 independent components of Γ_{rst}.

Expressing the Ricci identity

$$^r\lambda_{\mu\,;[\nu\,\sigma]} = \frac{1}{2}\, R^\tau_{.\,\mu\nu\sigma}\,{}^r\lambda_\tau \tag{20}$$

in terms of rotation coefficients gives

$$\Gamma_{[s\,|\,r\,t\,|,p]} = \Gamma_{[p|tq|}\,\Gamma_{s]r}{}^q + \Gamma_{qrt}\,\Gamma_{[p\,s]}{}^q + \frac{1}{2}\,R_{rts\,,p}\,. \tag{21}$$

Taking all possible independent combinations of timelike and spacelike indices, and using the dyadic representation of Γ_{rst}, gives four, 3-space covariant, first-order, dyadic differential equations. The general equations were first presented in Ref. [2] and were summarized again in Ref. [4]. In the next section the pertinent equations are written out explicitly for the special situation being considered here.

These equations also require a projected 3-space representation of the curvature components; viz., two symmetric dyadics

$$\begin{aligned}\mathbf{Q} &\equiv R_{0a0b}\,\mathbf{u}^a\mathbf{u}^b\,, \\ \mathbf{P} &\equiv -\frac{1}{4}\varepsilon_{acd}\,\varepsilon_{bfg}\,R^{cdfg}\,\mathbf{u}^a\mathbf{u}^b\,,\end{aligned} \tag{22}$$

and a trace-free dyadic

$$\mathbf{B} - \mathbf{t} \times \mathbf{I} \equiv -\frac{1}{2}\varepsilon_{acd}\,R_{0bcd}\,\mathbf{u}^a\mathbf{u}^b\,. \tag{23}$$

The symmetric dyadic $\mathbf{B}$ incorporates the "magnetic" components of the Weyl tensor, whereas the 3-vector $\mathbf{t}$ in the skew-symmetric part is the momentum density associated with the Ricci tensor, i.e.,

$$\mathbf{t} \equiv \frac{1}{2}\,R_{0a}\,\mathbf{u}^a\,. \tag{24}$$

We use units such that $4\pi G = c = 1$. The remainder of the Ricci tensor comprises the energy density

$$\rho \equiv -\frac{1}{2}\,(R_{00} + \frac{1}{2}\,R^s_s) \tag{25}$$

and the 3-space stress dyadic

$$\mathbf{T} = \frac{1}{2}\,(R_{ab} - \frac{1}{2}\,R^s_s\,\delta_{ab})\,\mathbf{u}^a\mathbf{u}^b\,. \tag{26}$$

It is convenient to extract the trace of each of these curvature dyadics. Defining the pressure by

$$3p \equiv -Tr\,\mathbf{T} \tag{27}$$

and the trace-free, or anisotropic, stress dyadic,

$$\mathbf{M} \equiv \mathbf{T} - \frac{1}{3}(Tr\,\mathbf{T})\mathbf{I} = \mathbf{T} + p\mathbf{I}, \tag{28}$$

we find

$$
\begin{aligned}
Tr\,\mathbf{Q} &\equiv \mu = \rho + 3p, \\
Tr\,\mathbf{P} &= -2\rho,
\end{aligned}
\tag{29}
$$

so we can write

$$
\begin{aligned}
\mathbf{Q} &= \mathbf{A} + \mathbf{M} + \frac{1}{3}\mu\mathbf{I}, \\
\mathbf{P} &= \mathbf{A} - \mathbf{M} - \frac{2}{3}\rho\mathbf{I},
\end{aligned}
\tag{30}
$$

where the symmetric, trace-free, dyadic $\mathbf{A}$ incorporates the "electric" components of the Weyl tensor. The entire Weyl tensor is comprehended in the complex, symmetric, trace-free dyadic

$$\mathbf{C} = \mathbf{A} + i\mathbf{B}. \tag{31}$$

III. The Dyadic Equations for Stationary, Axisymmetric, Rigid Bodies

To apply the formalism in this situation, we take the preferred congruence as the timelike Killing congruence of matter worldlines with unit tangent $\mathbf{u}_0$ and magnitude $1/\phi$, so

$$\mathbf{e}_0 = \partial/\partial t = \frac{1}{\phi}\mathbf{u}_0, \tag{32}$$

and $\mathbf{e}_0$ satisfies Killing's equation $\mathbf{D}\,\mathbf{e}_0 + \mathbf{e}_0\mathbf{D} = 0$. Axial symmetry is imposed by postulating a spacelike Killing vector field with closed orbits, say $\mathbf{e}_3 = \partial/\partial\theta$, with $\mathbf{D}\,\mathbf{e}_3 + \mathbf{e}_3\mathbf{D} = 0$. We can write the expansion

$$\mathbf{e}_3 = -\frac{A}{\phi}\mathbf{u}_0 + r\hat{\mathbf{K}}, \tag{33}$$

where $\mathbf{u}_3 = \hat{\mathbf{K}}$ is a spatial triad vector in the 2-spaces spanned by $\mathbf{e}_0$ and $\mathbf{e}_3$, and the functions ϕ, A, r are independent of the symmetry coordinates t and θ. Assuming "orthogonal transitivity" [5], we can solve for the corresponding coordinate 1-forms as

$$\mathbf{e}^0 = dt = \phi \mathbf{u}^0 + \frac{A}{r} \hat{\mathbf{K}},$$

$$\mathbf{e}^3 = d\theta = \frac{1}{r} \hat{\mathbf{K}},$$

(34)

and the line element of the orbit 2-spaces reads

$$-\frac{1}{\phi^2} (dt - A\, d\theta)^2 + r^2\, d\theta^2 .$$

(35)

Combining the Killing's equations for $(\mathbf{e}^0, \mathbf{e}^3)$ with the exterior derivatives of the exact 1-forms $(\mathbf{e}_0, \mathbf{e}_3)$, and using (13) for $\mathbf{D}\,\mathbf{u}_0$ and (15) for $\mathbf{D}\,\hat{\mathbf{K}}$, we can extract several consequences for the dyadic quantities: firstly, vanishing of the rate-of-strain dyadic, $\mathbf{S} = 0$, which is the condition for rigid motions. Other relations exhibit ϕ and A as potentials for the acceleration and angular velocity vectors

$$\nabla\phi + \phi\mathbf{a} = 0, \qquad \nabla A + 2r\phi\ \boldsymbol{\Omega} \times \hat{\mathbf{K}} = 0,$$

(36)

where time-independence of the scalars allows replacing the spacetime exterior derivative $\mathbf{d}$ with the 3-space gradient operator ∇. The orthogonality conditions

$$\mathbf{a} \cdot \hat{\mathbf{K}}, \qquad \boldsymbol{\Omega} \cdot \hat{\mathbf{K}} = 0,$$

(37)

also result, in addition to,

$$(\boldsymbol{\Omega} - \boldsymbol{\omega}) \times \hat{\mathbf{K}} = 0,$$

(38)

which allows the choice of body-fixed spatial triads co-rotating with the matter; i.e., $\boldsymbol{\omega} = \boldsymbol{\Omega}$. In consequence, the time derivatives of all vector-dyadic quantities vanish, as well as of the scalar variables. Finally, we find an equation for the gradient of $\hat{\mathbf{K}}$

$$\nabla\hat{\mathbf{K}} + \hat{\mathbf{K}}\frac{\nabla r}{r} = 0,$$

(39)

which implies $\hat{\mathbf{K}} \cdot \nabla r = 0$, and then also

$$\nabla \times \hat{\mathbf{K}} + \hat{\mathbf{K}} \times \left(\frac{\nabla r}{r}\right) = 0,$$

$$\nabla \cdot \hat{\mathbf{K}} = 0 .$$

(40)

When these reductions are applied to the general dyadic equations, [2, 4] we find one equation for the gradient of the acceleration vector

$$\nabla\mathbf{a} = -\mathbf{a}\mathbf{a} + \boldsymbol{\Omega}\,\boldsymbol{\Omega} - (\Omega^2)\mathbf{I} + \mathbf{Q},$$

(41)

and another for the gradient of the angular velocity vector

$$\nabla\boldsymbol{\Omega} = -2\mathbf{a}\boldsymbol{\Omega} + (\mathbf{a}\,.\,\boldsymbol{\Omega})\,\mathbf{I} + \mathbf{B},$$

(42)

where $\mathbf{Q}$ and $\mathbf{B}$ are the curvature dyadics defined in (22) and (23). The curls of these gradient equations give a pair of Bianchi identities

$$\nabla \times \mathbf{Q} = -\mathbf{a} \times \mathbf{Q} + \mathbf{P} \times \mathbf{a} - \mathbf{B} \times \boldsymbol{\Omega} + (\boldsymbol{\Omega} \cdot \mathbf{B}) \times \mathbf{I},$$

$$\nabla \times \mathbf{B} = -\mathbf{a} \times \mathbf{B} + \mathbf{B} \times \mathbf{a} + \mathbf{P} \times \boldsymbol{\Omega} - (\boldsymbol{\Omega} \cdot \mathbf{Q}) \times \mathbf{I} .$$

(43)

The dyadic $\mathbf{P}$ enters these equations by way of the 3-space Ricci identity for any vector $\mathbf{V}$ [the 3-dimensional analog of Eq. (20)]

$$\nabla \times \nabla \mathbf{V} = -\mathbf{E} \times \mathbf{V}, \tag{44}$$

where $\mathbf{E}$ is the symmetric, 3-space, curvature dyadic defined by

$$\mathbf{E} \equiv \frac{1}{4} \varepsilon_{acd}\, \varepsilon_{bfg}\, {}^{(3)}R^{cdfg}\, u^a u^b, \tag{45}$$

and satisfies the Bianchi identity

$$\nabla \cdot \mathbf{E} = 0. \tag{46}$$

In terms of the dyadic affinity $\mathbf{N}$, $\mathbf{E}$ is given by

$$\mathbf{E} = \nabla \times \mathbf{N} + \frac{1}{2} \mathbf{N}^T \times \mathbf{N}, \tag{47}$$

which represents the 3-dimensional analog of Eq. (21). The Gauss equation, relating the 3- and 4-dimensional curvatures, in this case ($\mathbf{S} = 0$, $\boldsymbol{\omega} = \boldsymbol{\Omega}$) is

$$\mathbf{E} = -\mathbf{P} - 3\,\boldsymbol{\Omega}\,\boldsymbol{\Omega}. \tag{48}$$

Combining the skew-symmetric part of the $\nabla \times \mathbf{Q}$ equation with Eqns. (46) and (48) gives the contracted Bianchi identity for the material stress dyadic

$$\nabla \cdot \mathbf{M} = \nabla p + (\rho + p)\mathbf{a} - \mathbf{a} \cdot \mathbf{M}. \tag{49}$$

If the stress is isotropic, $\mathbf{M} = 0$, this becomes the usual relativistic equation of hydrostatic equilibrium.

For rotating rigid bodies, when $\mathbf{a}$ and $\boldsymbol{\Omega}$ are both non-zero and linearly independent, it is convenient to define the complex dynamical vector

$$\mathbf{Z} \equiv \mathbf{a} + i\boldsymbol{\Omega}, \quad \bar{\mathbf{Z}} = \mathbf{a} - i\boldsymbol{\Omega}. \tag{50}$$

Combining the gradient equations then gives

$$\nabla \mathbf{Z} = -\mathbf{Z}\,\mathbf{Z} + \frac{1}{2}\mathbf{Z}\bar{\mathbf{Z}} - \frac{1}{2}\bar{\mathbf{Z}}\mathbf{Z} + \left(\frac{1}{2}\mathbf{Z}^2 - \frac{1}{2}\mathbf{Z}\cdot\bar{\mathbf{Z}} + \frac{1}{3}\mu\right)\mathbf{I} + \mathbf{R}, \tag{51}$$

where $\mathbf{R}$ is the symmetric, tracefree, complex curvature dyadic defined by

$$\mathbf{R} \equiv \mathbf{Q} - \frac{1}{3}\mu\mathbf{I} + i\mathbf{B} = \mathbf{A} + \mathbf{M} + i\mathbf{B} = \mathbf{C} + \mathbf{M}. \tag{52}$$

The trace and skew-symmetric parts of (51) give

$$\nabla \cdot \mathbf{Z} = \frac{1}{2}\mathbf{Z}^2 - \frac{3}{2}\mathbf{Z}\cdot\bar{\mathbf{Z}} + \mu, \tag{53}$$

$$\nabla \times \mathbf{Z} = \mathbf{Z} \times \bar{\mathbf{Z}}. \tag{54}$$

Correspondingly, the complex Bianchi identity resulting from the curl of Eq. (51) can be written

$$\nabla \times \mathbf{R} = -\frac{1}{4}(3\mathbf{Z} + \bar{\mathbf{Z}}) \times \mathbf{R} + \frac{1}{4}\mathbf{R} \times (3\mathbf{Z} + \bar{\mathbf{Z}}) + \mathbf{Z} \times \mathbf{M} - \mathbf{M} \times \mathbf{Z}$$
$$+ [\mathbf{Z} \cdot \mathbf{M} - \frac{3}{4}(\mathbf{Z} - \bar{\mathbf{Z}}) \cdot \mathbf{R} - \frac{1}{3}\nabla\mu - (\mu - 2p)\mathbf{Z}] \times \mathbf{I} \ . \tag{55}$$

The first line on the right side of this equation is symmetric; the entire skew-symmetric part is contained in the bracket on the second line.

Equations (51) and (55) are the starting point for this approach to finding solutions for rotating rigid bodies.

IV. Generating Solutions of the Dyadic Equations

As an immediate consequence of the equations for $\mathbf{Z}$, Eqns. (36) and (54) give

$$\nabla \times (\frac{\mathbf{Z}}{\phi^2}) = 0, \tag{56}$$

implying the existence of a complex potential

$$\mathbf{Z}/\phi^2 = \nabla F, \tag{57}$$

with $F + \bar{F} = 1/\phi^2$, analogous to the Ernst potential for stationary vacuum solutions. Indeed, with an appropriate choice of coordinates in the vacuum case, Eq. (53) for $\nabla \cdot \mathbf{Z}$ becomes the Ernst equation [6] for F. Treating F and $\bar{F}$ as independent coordinates; i.e.,

$$\phi^4 \nabla F \times \nabla \bar{F} = \mathbf{Z} \times \bar{\mathbf{Z}} = -2i\mathbf{a} \times \mathbf{\Omega} \neq 0, \tag{58}$$

the gradient of an arbitrary scalar function ψ can be expanded as

$$\nabla\psi = \frac{1}{\phi^2}[\frac{\partial\psi}{\partial F} \mathbf{Z} + \frac{\partial\psi}{\partial \bar{F}} \bar{\mathbf{Z}}] \ . \tag{59}$$

Then for any ψ the equation

$$\nabla\psi \times \mathbf{Z} = 0 \tag{60}$$

implies that ψ is an analytic function of F, $\partial\psi/\partial\bar{F} = 0$, and further if ψ is real, it must be constant.

One approach to finding solutions is to postulate simple forms for the curvature dyadics, including some unspecified scalar coefficients, and to substitute the forms into the dyadic equations to derive differential equations for the scalars. The first solution of these equations that led to a finite rotating body [3] was generated from dyadic forms suggested by the vacuum Schwarzschild and Kerr solutions. In the static ($\mathbf{\Omega} = 0$) Schwarzschild solution, the curvature dyadic $\mathbf{B}$ vanishes and the dyadic $\mathbf{A}$ has the value

$$\mathbf{A} = \frac{m}{r^3} (\mathbf{I} - 3\hat{r}\hat{r}). \tag{61}$$

Up to the scalar coefficient, this is the unique, symmetric, trace-free, form for spherical symmetry, but with $\mathbf{a} = a\hat{r}$, it can also be written

$$\mathbf{A} = \frac{m}{r^3}\left(\mathbf{I} - 3\frac{\mathbf{a}\,\mathbf{a}}{a^2}\right). \tag{62}$$

It is suggestive to try the replacement, $\mathbf{a} \rightarrow \mathbf{Z} = \mathbf{a} + i\boldsymbol{\Omega}$, while allowing the coefficient to be arbitrary. This leads to an Ansatz for the Weyl dyadic of rotating solutions

$$\mathbf{C} = \mathbf{A} + i\mathbf{B} = \alpha\left(\mathbf{I} - 3\frac{\mathbf{Z}\,\mathbf{Z}}{Z^2}\right), \tag{63}$$

with α an unspecified complex variable. This form has several attractive properties; 1) it is quite simple and introduces only one new variable into the equations, 2) in vacuum, it is the correct form of $\mathbf{C}$ for the Kerr solution, and 3) the dynamical vector $\mathbf{Z}$ is an eigenvector

$$\mathbf{Z} \cdot \mathbf{C} = -2\,\alpha\mathbf{Z}. \tag{64}$$

In the original solution this Ansatz was supplemented with the assumption of isotropic stress, $\mathbf{M} = 0$, so $\mathbf{R} = \mathbf{C}$.

When these values for $\mathbf{C}$ and $\mathbf{M}$ are inserted in the $\nabla\mathbf{Z}$ equation, Eq. (51), and $\mathbf{Z}$ is dotted from the right, one finds

$$\nabla\left(\frac{Z^2}{\phi^2}\right) = -2\left(2\alpha - \frac{1}{3}\mu\right)\frac{\mathbf{Z}}{\phi^2}\,, \tag{65}$$

where the potential equation for a has been used again. Thus,

$$\nabla\left(\frac{Z^2}{\phi^2}\right) \times \mathbf{Z} = 0\,, \tag{66}$$

and taking account of Eq. (56), the curl integrability condition gives

$$\nabla\left(2\alpha - \frac{1}{3}\mu\right) \times \mathbf{Z} = 0\,, \tag{67}$$

so that both of these quantities must be analytic functions of F.

The symmetric part of the Bianchi identity, Eq. (55), shows that these postulates also demand

$$\nabla\alpha \times \mathbf{Z} = 0\,, \tag{68}$$

so Eq. (67) now requires

$$\nabla\mu \times \mathbf{Z} = 0\,,$$

and since μ is real, $\mu = \rho + 3p = \mu_0 =$ const.

Thus, the integrability conditions may force spatial relations among the energy density and stress components of the matter. It should be noted, however, that since the solutions obtained here describe only *stationary* configurations of some kind of material undergoing a rigid motion (i.e., not involving any expansion, shear, or other time-dependent process), these spatial relations need not imply universal properties of the matter in the usual sense of a thermodynamic equation-of-state which would hold throughout non-stationary processes as well.

When the skew-symmetric part of Eq. (55) is examined, one finds a gradient equation for α

$$\nabla \alpha = - [3\alpha(\alpha + \frac{1}{3}\mu_0) + Z^2(\mu_0 - 2p)]\frac{\mathbf{Z}}{Z^2} , \tag{69}$$

consistent with Eq. (68), and the integrability condition here can be written

$$\nabla [\phi^2(\mu_0 - 2p)] \times \mathbf{Z} = 0 , \tag{70}$$

or

$$\phi^2(\mu_0 - 2p) = \mu_0 \kappa , \tag{71}$$

where κ is a new integration constant. Now we have

$$p = \frac{1}{2}\mu_0(1 - \frac{\kappa}{\phi^2}) , \qquad \rho = \frac{1}{2}\mu_0(3\frac{\kappa}{\phi^2} - 1) , \tag{72}$$

and it can be verified that these results satisfy the stationary equilibrium equation, Eq. (49) with $\mathbf{M} = 0$.

Combining Eq. (65) and Eq. (69) gives a first-order, analytic, ordinary differential equation

$$[3\alpha(\alpha + \frac{1}{3}\mu_0) + \mu_0 \kappa \xi^2] d\xi - \xi (2\alpha - \frac{1}{3}\mu_0)d\alpha = 0 \tag{73}$$

for the analytic variables α and $\xi^2 \equiv Z^2/\phi^2$. As it stands this Abel's equation is not integrable in these variables, but the solution can be written in terms of a parametric variable λ introduced by setting

$$\alpha + \frac{1}{3}\mu_0 = - (\mu_0\kappa)^{\frac{1}{2}}\xi \tan \lambda . \tag{74}$$

The solution for F is [4]

$$F = \frac{1}{2\kappa} [\varepsilon_0 + (\lambda + \lambda_0) \tan \lambda] , \tag{75}$$

where ε_0 and λ_0 are integration constants, and the other complex variables are given by

$$\xi = (\mu_0\kappa)^{\frac{1}{2}}\frac{dF}{d\lambda} = \frac{1}{2} (\frac{\mu_0}{\kappa})^{\frac{1}{2}}[\tan \lambda + (\lambda + \lambda_0) \sec^2 \lambda] , \tag{76}$$

$$2\alpha - \frac{1}{3}\mu_0 = -(\mu_0\kappa)^{\frac{1}{2}}\frac{d\xi}{d\lambda} = -\frac{1}{2}\mu_0 \sec^2 \lambda [1 + (\lambda + \lambda_0)\tan \lambda]. \tag{77}$$

It can be shown from the square of Eq. (57) that orthogonal real coordinates $\{u, v\}$ are obtained by setting $\lambda = u + iv$. With these results and a few additional integrations, we arrive at the line element

$$d s^2 = -\frac{1}{\phi^2}(dt - A d\theta)^2 + r^2 d\theta^2 + \frac{g}{\mu_0}[\frac{du^2}{h_1} + \frac{dv^2}{h_2}] , \qquad (78)$$

where

$$
\begin{aligned}
&g = \cos 2u + \cosh 2v , \\
&h_1(u) = h_0 + \varepsilon_0 \cos 2u + (u + u_0) \sin 2u , \\
&h_2(v) = h_0 - \varepsilon_0 \cosh 2v + (v + v_0) \sinh 2v , \\
&\frac{1}{\phi^2} = \frac{1}{\kappa} \frac{(h_1 - h_2)}{g} , \qquad\qquad (79) \\
&r^2 = 4 r_0^2 \phi^2 h_1 h_2 , \\
&A = \frac{2\kappa r_0}{(h_1 - h_2)}[h_2(\cos 2u + \cosh 2v_A) + h_1(\cosh 2v - \cosh 2v_A)] .
\end{aligned}
$$

Symbols with a zero subscript are arbitrary real constants. The axis of rotation ($r = A = 0$) is given by a constant value of the v coordinate, $v = v_A$, determined by $h_2(v_A) = 0$.

In Ref. [4], a less restrictive Ansatz was treated that included anisotropic stresses of the form

$$\mathbf{M} = \tau(3\hat{\mathbf{K}}\hat{\mathbf{K}} - \mathbf{I}) , \qquad (80)$$

together with the most general form of $\mathbf{C}$ having $\hat{\mathbf{K}}$ and $\mathbf{Z}$ as eigenvectors.

V. Configurations of Relativistic Rotating Rigid Bodies

The variety of configurations of rotating bodies that may be encompassed in this solution has recently been explored. Embedding diagrams to illustrate the intrinsic body shapes have been computed and plotted. The calculations assume the outside surface of the bodies to be given by $p = 0$, which of course begs the question as to the existence of matching vacuum exterior solutions, a problem that remains unsolved. Nevertheless, it is interesting to explore the possibilities, some of which are quite unexpected and intriguing.

Among the configurations are ordinary simple bodies with approximately spheroidal shapes of both oblate and prolate types. These calculations, however, do confirm the analytical conclusion reached in Ref. [3] that, in the slowly-rotating, Newtonian, limit, the intrinsic shapes are always slightly prolate. More unusual configurations are found by searching farther regimes of the solution parameters. Some bodies become stretched along the axis of rotation into dumbbell-like shapes, and these can develop further into configurations with an arbitrary number of disjoint bodies symmetrically distributed on the rotation axis. These latter configurations are certainly reminiscent of the stationary vacuum solutions with multiple Kerr black holes aligned along the axis of rotation [7]. This suggests that the ability of rotation to balance gravitational attraction may operate also in these non-singular stationary interior solutions. The analysis leading to these results will be described in detail in a forthcoming paper [8]; only a brief discussion is presented here.

Note that the pressure p is a linear function of the real part of $F(\lambda)$, so that the pressure is a solution of Laplace's equation

$$p_{uu} + p_{vv} = 0 .$$ (81)

In general, for arbitrary values of $\{u_0, v_0\}$, the pressure will be singular whenever the function $g = 0$; i.e., for

$$u = \pi(n + \frac{1}{2}), \quad v = 0 .$$ (82)

One of the singularities can be removed by a choice of $\{u_0, v_0\}$; for instance, by setting $u_0 = -\pi/2$, $v_0 = 0$. Defining a new coordinate, $\tilde{u} = u - \pi/2$, the ring singularity of p at the new origin, $\tilde{u} = 0$, $v = 0$, is eliminated.

The entire solution is now symmetric under reflections about $\tilde{u} = 0$ and $v = 0$, and near the origin the $\{\tilde{u}, v\}$–coordinates are analogous to oblate spheroidal coordinates confocal at the origin, which corresponds to a circle of proper circumference $2\pi r(0, 0)$. See Fig. 1. As usual for confocal coordinates, the double symmetry gives a double covering, and the entire manifold is contained in the upper half-plane, $v \geq 0$, where the left and right halves of $v = 0$, are identified; i.e., $\{\tilde{u}, 0\} \approx \{-\tilde{u}, 0\}$.

We can choose units such that the scaling parameter, $\mu_0 = \rho + 3p = 1$. Then by defining a function $\tilde{x}(\tilde{u}, v)$ by

$$\tilde{x}(\tilde{u}, v) \equiv p(\tilde{u}, v) - p(0, 0),$$ (83)

so that $\tilde{x}$ is the pressure difference from the ring at the origin (not the body center!), we find

$$\tilde{x}(\tilde{u}, v) = \frac{1}{2}[\frac{(\tilde{u} \sin 2\tilde{u} + v \sinh 2v)}{(\cosh 2v - \cos 2\tilde{u})} - 1]$$ (84)

which is the solution of Laplace's equation involved. Taking the limit confirms that $\tilde{x}$ vanishes at the origin, which is now a saddle point of the function. Notably, $\tilde{x}$ is independent of *all* solution parameters; a single contour plot of $\tilde{x}$, as in Fig. 2, therefore includes (with interpolation) all possible $p = $ const. surfaces. Further, any one of them can be taken as the $p = 0$ surface by adjusting the value of p at the origin, $p(0, 0)$, which is equivalent to the arbitrary parameter ε_0 in the solution.

For example, if we choose $\tilde{x} = 0$ in Fig. 2 to be the $p = 0$ surface, the equator of the body will be situated at the origin. The axis of rotation is a horizontal line at $v = v_A > 0$, and the center of the body is located at $\tilde{u} = 0$, $v = v_A$. The poles will be given by the intersections of $v = v_A$ with the two branches of $\tilde{x} = 0$. Raising the axis to larger values of v_A clearly gives a larger body with greater central pressure, $p_c = p(0, v_A)$, and also higher rotational speed at the equator, V_e.

While it appears from Fig. 2 that the pressure has cusps at the poles and equator, this is simply an artifact of the $\{\tilde{u}, v\}$-coordinate plane, which does not incorporate the metric geometry and the required topological identification, $\{\tilde{u}, 0\} \approx \{-\tilde{u}, 0\}$. In fact, the interior and surface pressures have the proper behavior and the space is locally flat at the axis when the solution parameter r_0 is chosen appropriately. This can be demonstrated by integrating the proper distance along $\tilde{x} = 0$ as a function of radius r, from equator to pole, to obtain a representation of the intrinsic geometry, and then plotting a surface in Euclidean space with the same geometry.

Figure 3 shows a set of such embedding diagrams for $\tilde{x} = 0$ and $0.514 \le v_A \le 1.288$. The smaller bodies are slightly prolate, but at $p_c = 1/6$ the shape becomes spherical, and larger ones are oblate. The rotational speeds of the larger bodies, and their central pressures, are highly relativistic, however.

Qualitatively similar results are obtained for most simple bodies obtained by using contours $\tilde{x} < 0.45$. But, as can be seen in Fig. 2, the contour $\tilde{x} = 0.45$ reaches new saddle points, and the axis of rotation can extend beyond the singularities at $v = 0$, $\tilde{u} = \pi, -\pi$. A contour such as $\tilde{x} = 0.5$ appears to allow two or three disjoint bodies, or one elongated body, as the value of v_A is increased. For $v_A = 2$, Fig. 2 shows three disjoint bodies along the axis of rotation.

Another set of embedding diagrams illustrating this situation, based on the $\tilde{x} = 0.52$ contour, is included in Fig. 4. The first diagram, when completed by reflection symmetry, shows a two-body configuration; the others show three-body configurations, except for the last where they merge into a single elongated body.

A new pair of saddle points occurs in each region, $n\pi < |\tilde{u}| < (n+1)\pi$, corresponding to successively larger values of $\tilde{x}$. Choosing $p = 0$ surfaces above these saddle point contours and varying the axis of rotation, $v = v_A$, as in Fig. 4, appears to allow configurations with an arbitrary number of disjoint bodies distributed along the axis of symmetry.

Acknowledgment

The research described in this paper was carried out by the Jet Propulsion Laboratory, California Institute of Technology, under contract to the National Aeronautics and Space Administration.

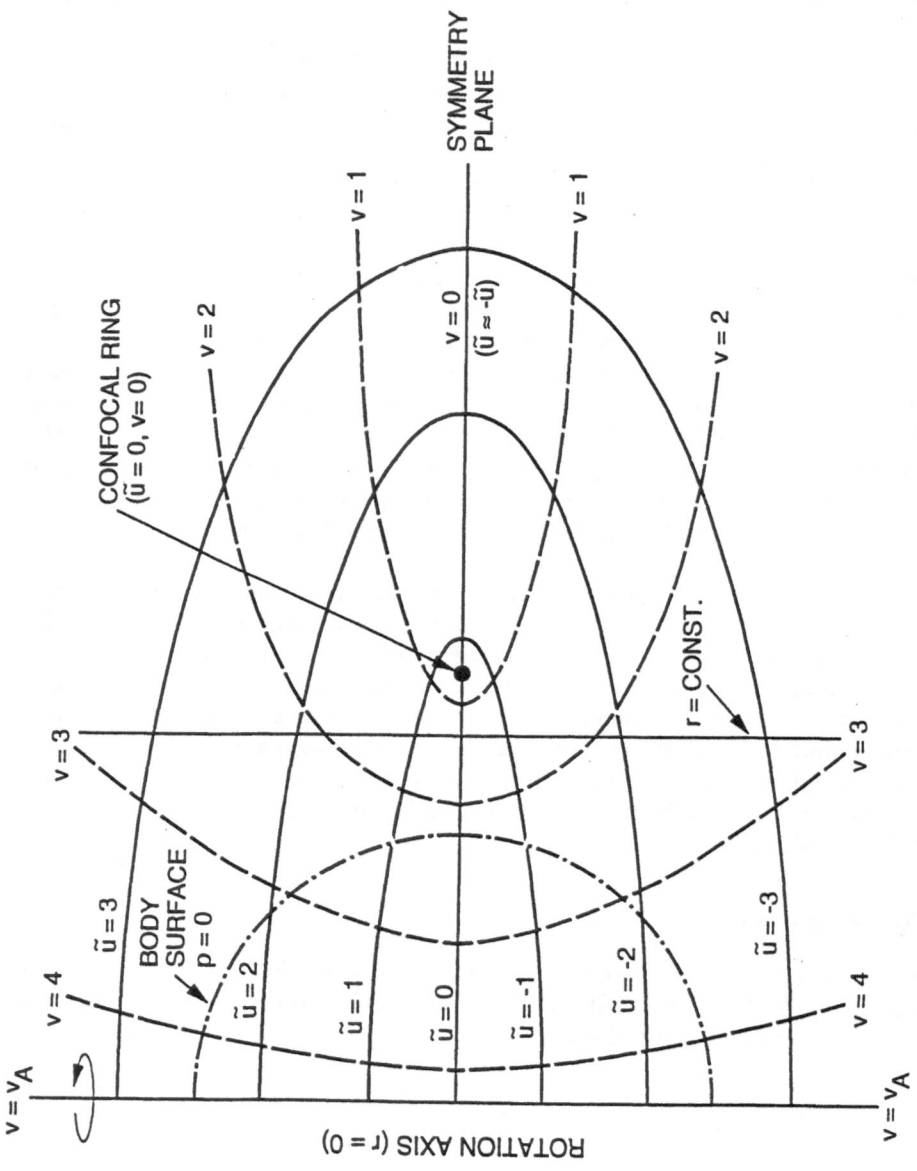

Fig. 1. Topology of the confocal coordinates $\{\tilde{u},\ v\}$.

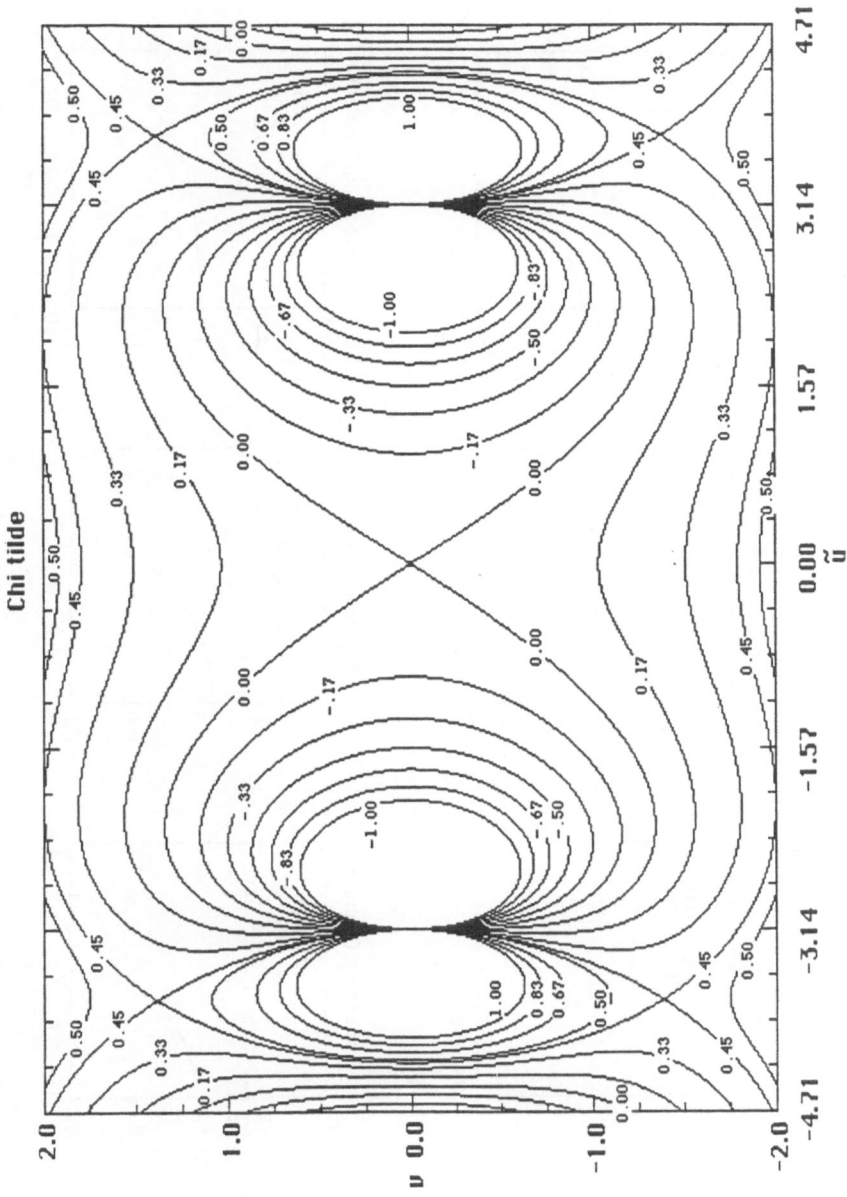

Fig. 2. Contours of constant $\tilde{x}$ in the $\{\tilde{u},\ v\}$ plane. The upper half plane, $v \geq 0$, covers the physical space with symmetric points on the left and right halves of the $\tilde{u}$-axis identified: $\{\tilde{u},\ 0\} \approx \{-\tilde{u},\ 0\}$. The rotation axis is given by a horizontal line, $v = v_A > 0$.

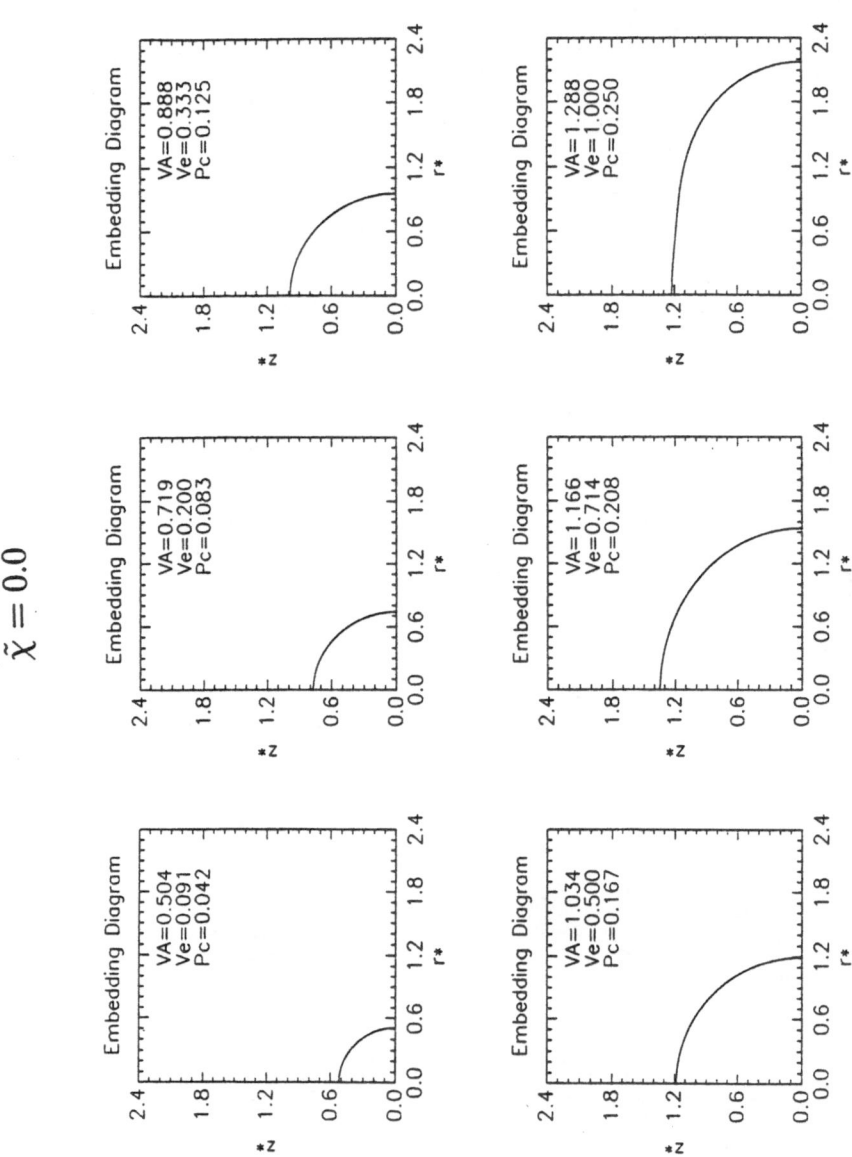

Fig. 3. A set of embedding diagrams showing one quadrant of the $p = 0$ surface for six simple bodies based on the $\tilde{x} = 0$ contour. Increasing levels of the rotation axis, $v = v_A$, on Fig. 2 were chosen to give equal increments of the central pressure in the range $1/24 \leq p_c \leq 1/4$.

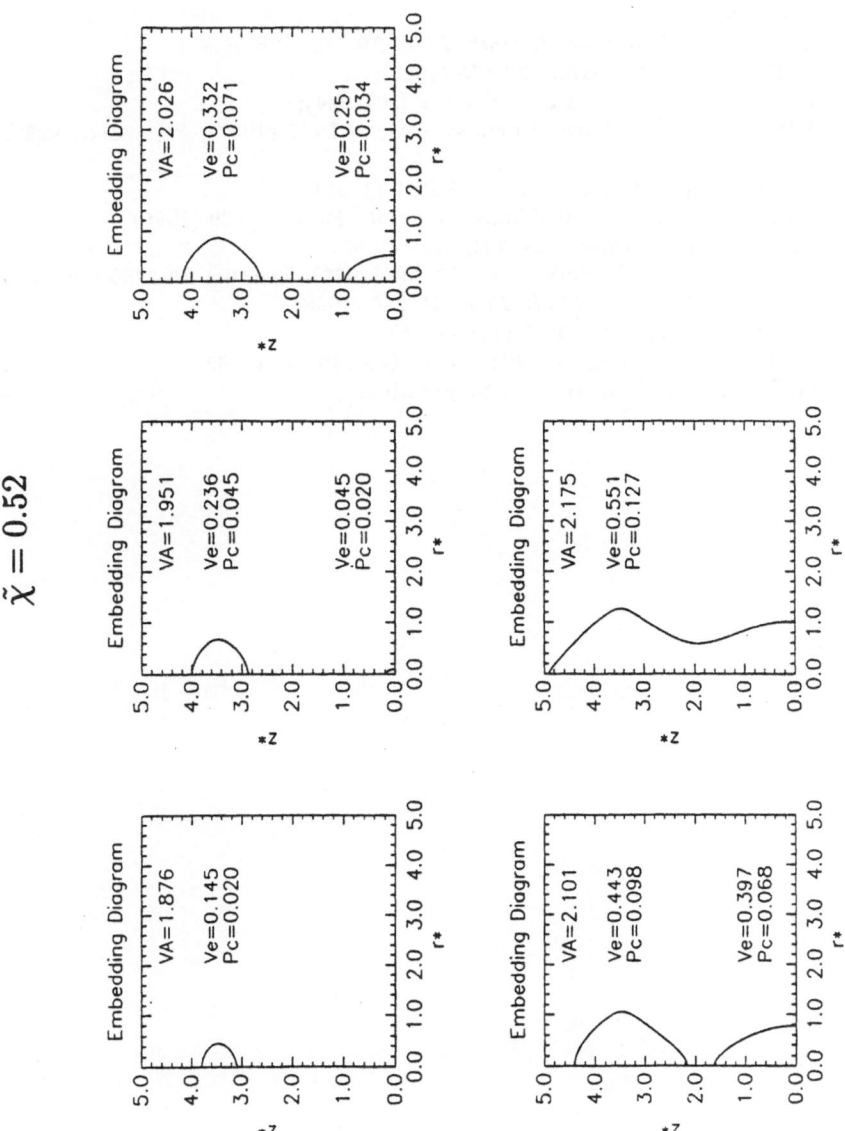

Fig. 4. Embedding diagrams for some multiple bodies based on the $\tilde{x} = 0.52$ contour. The complete configurations are obtained by reflections about the vertical rotation axis and the horizontal symmetry plane.

References

1. F.J. Chinea, L.M. Gonzalez-Romero: *Class. Quantum Grav.* **7** L99 (1990); **9** 1271 (1992);
    D.A. García, I. Hauser: *J. Math. Phys.* **29** 175 (1987);
    E. Herlt: *Gen. Rel. Grav.* **20** 635 (1988);
    D. Kramer: *Class. Quantum Grav.* **1** L3 (1984);
    J.M.M. Senovilla: *Class. Quantum Grav.* **4** L115 (1987); *Phys. Lett.* **123A** 211 (1987);
    D. Sklavenites: *J. Math. Phys.* **26** 2279 (1985)
2. F.B. Estabrook, H.D. Wahlquist: *J. Math. Phys.* **5** 1629 (1965)
3. H.D. Wahlquist: *Phys. Rev.* **172** 1291 (1968)
4. H.D. Wahlquist: *J. Math. Phys.* **33** 304 (1992); Erratum **33** 3255 (1992)
5. B. Carter: *Commun. Math. Phys.* **10** 280 (1968)
6. F.J. Ernst: *Phys. Rev.* **167** 1175 (1968)
7. W. Dietz, C. Hoenselaers: *Phys. Rev. Lett.* **48** 778 (1982)
8. M. Tinto, H.D. Wahlquist: in preparation.

Stationary and Axisymmetric Perfect-Fluid Solutions to Einstein's Equations

José M. M. Senovilla

Departament de Física Fonamental, Universitat de Barcelona,
Diagonal 647, 08028 Barcelona
and
Laboratori de Física Matemàtica, Societat Catalana de Física, I.E.C.,
Barcelona, Spain.

1 Introduction

This contribution deals with stationary and axisymmetric perfect-fluid solutions of Einstein's equations, which are the best candidates to describe the interior of isolated rotating bodies. There are many such bodies in nature and the description of their gravitational field is one of the main topics in General Relativity. The gravitational field *outside* the body is usually represented by the so-called exterior solutions (that is, solutions to the vacuum field equations), and a lot of them are explicitly known at present. There exist also several generation techniques to obtain new solutions from those already known as well as a more or less established definition of the multipole moments associated with these exterior solutions (see, for instance, Hoenselaers' contribution in this book). However, the complete interpretation of the exterior solutions and their multipole moments will not be achieved until we relate them with the solution *inside* the body, usually called the interior solution. Furthermore, the whole gravitational field of an isolated body is not well described until both the interior and exterior solutions are given and correctly matched across the body's timelike limit surface. As a matter of fact, no exterior solution is known to be matchable with any interior solution. This is due, in our opinion, to two main reasons. First, because there are very few explicitly known interior solutions. And second, due to the difficulty in looking for the interior solution given the exterior one, for we do not know the limit surface of the body so that we cannot impose the matching conditions as initial conditions, and also because many interior solutions will give rise to the same exterior (as happens in spherically symmetric bodies).

The only way out to these problems is, in our view, to find as many as possible interior solutions given that in this case we can define the limit surface uniquely (for example, by the condition of vanishing of the pressure in perfect fluids) and also the exterior solution will be determined unambigiously by the initial conditions at the matching surface together with the vacuum field equations (in relation with this, see [1] in this book). But if we wish to search interior solutions it is necessary to specify the type of the energy-momentum tensor inside the body. It is widely accepted in this respect that the matter inside most bodies we are interested in is, to very high precision, a perfect fluid (see, for the case of stars, [2]).

In this paper we consider, therefore, stationary and axisymmetric perfect-fluid solutions. Our purpose is twofold. First, we shall classify the solutions by combining different properties such as rigid or differential rotation, Petrov type, etc. As we proceed with this classification we will find where the very few known solutions are placed in the scheme and thereby it will become clear, in most cases, why they have been actually found. Together with this, we shall see which are the relevant parts of the scheme that have not been studied already and are likely to contain *easily obtainable* solutions. Second, we wish to present the ideas and clues that, in our view, will serve to reduce the problem under consideration in such a way that new solutions can be actually found.

Let us start by fixing our notations and assumptions in a clear manner. The spacetime is stationary so that there exists a timelike Killing vector field, say ξ. That is to say

$$\mathcal{L}_\xi \, g = 0 \, , \qquad \xi^\alpha \xi_\alpha < 0 \, ,$$

where $\mathcal{L}_\xi$ stands for the Lie derivative with respect to ξ and g is the metric tensor field in the spacetime. Similarly, spacetime is axially symmetric, so that there is a spacelike Killing vector field η with closed orbits and vanishing at a timelike 2-surface called the axis of symmetry. As before, we have

$$\mathcal{L}_\eta \, g = 0 \, , \qquad \eta^\alpha \eta_\alpha > 0 \, .$$

It follows then from a theorem due to Carter [3] that both Killing vector fields commute

$$[\xi, \eta] = 0 \, ,$$

and, therefore, we can choose coordinates t, ϕ such that $0 \leq \phi \leq 2\pi$ with 0 and 2π identified and

$$\xi = \frac{\partial}{\partial t} \, , \qquad \eta = \frac{\partial}{\partial \phi} \implies \partial_t \, g_{\alpha\beta} = \partial_\phi \, g_{\alpha\beta} = 0 \, .$$

The energy-momentum tensor is that of a perfect fluid:

$$T_{\alpha\beta} = (\rho + p) u_\alpha u_\beta + p g_{\alpha\beta} \, , \quad u^\alpha u_\alpha = -1 \tag{1}$$

where ρ is the energy density, p the pressure and u is the velocity vector field of the fluid. Throughout this paper we shall assume that there is *no convection* in the motion of the fluid or, in other words, that the velocity vector lies in the two-planes spanned by the two Killing vectors

$$u = F \left(\xi + \Omega \eta \right) \tag{2}$$

where F and Ω are functions not depending on t and ϕ. Thus, only axial motions are allowed and the fluid elements cannot move in the directions orthogonal and parallel to the axis of symmetry. This is obviously a true restrictive assumption, but it simplifies the problem substantially and also, due to a theorem of Papapetrou and others [4] [5], implies that if the axis is regular then both Killing vectors are orthogonal to a family of two-surfaces. Thus, we can choose coordinates $\{t, x^1, x^2, \phi\}$ in the spacetime such that the metric splits up into two 2×2 blocks as follows

$$ds^2 = -e^{2U}(dt + A \ d\phi)^2 + W^2 d\phi^2 + G_{ab}dx^a dx^b \tag{3}$$

where U, A, W and G_{ab} depend only on the coordinates x^a and $a, b, \ldots = 1, 2$.

The first classification we make depends on whether or not the function Ω in (2) is constant. If $\partial_\alpha \Omega \neq 0$, the fluid is said to have *differential rotation* (DR from now on), while if $\partial_\alpha \Omega = 0$, the fluid has *rigid rotation* (RR). In both cases, from formula (2) it follows that the velocity vector of the fluid is expansion-free. However, in the RR case, this vector is proportional to a (necessarily timelike) Killing vector and therefore u is also shear-free. It can be shown then that in the RR case there always exists a barotropic *equation of state*, that is, a relation between the pressure and the energy density of the form $p = p(\rho)$. These last properties are not necessarily true in the general DR case.

Trying to solve in general Einstein's field equations with the energy-momentum tensor (1) and for a line-element of type (3) is a very difficult (and often useless) task. Therefore, we must refine the classification and restrict our search by adding some suitable characterization or assumption. Our own proposition is to use the *Petrov type* of the Weyl tensor and look for solutions from lower to greater complexity in a sense that will be precised in the next section. To justify our proposition, let us remind that a desired property of the solutions we are seeking is that they should contain *static* limits (and, if possible, spherically symmetric too). This is because we can think of switching off the rotation in the fluid and then we are left with a static and axially symmetric perfect fluid. But a very well-known result is that static spacetimes can only be of very particular Petrov types (see, for instance, [6]), namely

$$\text{STATIC} \begin{cases} \text{type 0} \\ \text{type } D \\ \text{type I} \end{cases}$$

and therefore we are interested in solutions with Petrov types that can degenerate into one of the above. Thus, the list of Petrov types we shall consider is

$$\text{STATIONARY AXISYMMETRIC} \begin{cases} \text{type 0} \\ \text{type } D \\ \text{type II} \\ \text{type I} \end{cases} \downarrow$$

where we can see that neither type III nor type N appears, as should be expected. We have also shown an arrow pointing downwards which indicates the searching direction if we want to go from the easiest case to those more complicated. In the next section, we shall divide each of these cases into appropriate subcases and will analyze exhaustively every one of them. As we proceed on, we shall also present the explicitly known solutions we are aware of (discarding those with higher symmetry) and will state clearly to which case they belong. This will allow us to ascertain which special properties they have in order to possibly use them in looking for new solutions.

2 General Classification and Explicit Solutions

After the definition of searching direction (represented by the arrow) given in the previous paragraph, we start with the simplest case of conformally flat metrics.

2.1 Type 0 Solutions

This case is uninteresting because of a theorem due to Collinson [7] stating that the only conformally flat stationary and axisymmetric perfect-fluid solution is the Schwarzschild interior metric. Therefore, all the solutions in this class are static and spherically symmetric with constant energy density. Let us go then to the following case.

2.2 Type D Solutions

Petrov type D solutions have two multiple principal null directions of the Weyl tensor, and we shall represent them here by the two null vector fields ℓ and k ($\ell^\alpha \ell_\alpha = k^\alpha k_\alpha = 0$). This fact was used by Wainwright to set up a classification of Petrov type D perfect-fluid solutions [8]. For our purposes, we only consider the main part of this classification, which divides the metrics into two classes depending on whether or not the velocity vector of the fluid lies in the two-space generated by ℓ and k. Thus we have

$$
\begin{array}{c|c}
Class\ 1 & Class\ 2 \\
u = a\ell + bk & u \neq a\ell + bk
\end{array}
$$

for some scalar functions a and b. These two classes can be split up into subclasses by adding the property of rigid or differential rotation. We finally obtain the following table for Petrov type D solutions:

Table 1. The four cases of Petrov type D stationary axisymmetric perfect-fluid metrics. The arrows indicate the searching direction of increasing difficulty.

	Class 1		Class 2
RR	D1RR	$\rightarrow$	D2RR
		$\swarrow$	
DR	D1DR	$\rightarrow$	D2DR

Here, D2DR means, for example, type D solutions belonging to class 2 of Wainwright's classification and with differential rotation. From the position of the arrows we see that the increasing complexity of the solutions goes from left to right and from top to bottom. As always, we start our analysis with the simplest case.

D1RR Solutions.

All solutions in this class are explicitly known [9]. They are given by the following line-element

$$ds^2 = M^{-2}\left[-x^2\left(dt + \frac{S}{x^2}d\phi\right)^2 + Fd\phi^2 + F^{-1}dx^2 + dz^2\right] \qquad (4)$$

where the function F depends only on x and reads

$$F = C\log\frac{x}{b} + \frac{S^2}{x^2} - c\,x^2 \ ,$$

the function M depends only on z and is given by

$$M(z) = \begin{cases} A\cosh\left(\sqrt{c}\,z\right) + B\sinh\left(\sqrt{c}\,z\right) & \text{if } c > 0 \\ Az + B & \text{if } c = 0 \\ A\cos\left(\sqrt{-c}\,z\right) + B\sin\left(\sqrt{-c}\,z\right) & \text{if } c < 0 \end{cases} \qquad (5)$$

and A, B, C, S, b and c are constants. The velocity vector of the fluid is the unit timelike vector proportional to ξ, and the energy density and pressure of the fluid are given by

$$\rho + p = C\frac{M^2}{x^2} \ , \qquad (6)$$

$$\rho = p + \begin{cases} 6c\left(A^2 - B^2\right) & \text{if } c > 0 \\ -6A^2 & \text{if } c = 0 \\ 6c\left(A^2 + B^2\right) & \text{if } c < 0 \end{cases} \qquad (7)$$

this last expression being the equation of state of the fluid. From (6) we see that C must be positive and from (7) it follows that the energy condition $\rho > p$ is satisfied only when $c > 0$ and also $A^2 > B^2$.

With respect to the properties of the solution, it can be shown that there is a regular axis of symmetry, but the above coordinate ϕ is not appropriate for that. The solutions have a static limit when the rotation parameter S vanishes, and this limit leads to the solutions found by Barnes in [10]. None of them is spherically symmetric though. From (6) and (7) we see that there is a timelike surface of vanishing pressure interpretable as a limit surface. However, it is not clear if this surface is compact at each instant of time. The magnetic part of the Weyl tensor (with respect to the velocity of the fluid) vanishes, and the two multiple null directions defined by ℓ adn k are *not* geodesic. As was shown by Kramer [11], solution (4) possesses also a proper conformal Killing vector field ζ, such that

$$\zeta = \frac{\partial}{\partial z}, \quad \mathcal{L}_\zeta g = 2\psi g, \quad \text{where } \psi = -\frac{M_{,z}}{M}. \tag{8}$$

The conformal Lie group defined by $\{\xi, \eta, \zeta\}$ is Abelian. All in all, it seems hardly possible to interpret this solution as the interior of a well-defined exterior one.

D2RR Solutions.

Contrarily to the previous cases, in this rather simple case the general solution is not known yet. In fact, we only know explicitly what will be called here the Wahlquist family of solutions. We present it in the form that appeared in [12] because then its limit cases appear in a simple way. The line-element can be written as

$$ds^2 = -e^{2U}(dt + A d\phi)^2 + (V - W)\left[G^{-1}dx^2 + H^{-1}dy^2 + c^2\frac{GH}{G - H}d\phi^2\right] \tag{9}$$

where we have

$$e^{2U} = \frac{G - H}{V - W},$$

$$A = c\frac{HV - GW}{H - G},$$

functions V, G depend only on x and H, W depend only on y. The explicit form for V and W is

$$V(x) = \frac{1}{2b^2}\left[m + \sqrt{m^2 - 4b^2n} \, \sin(4bx + C_2)\right],$$

$$W(y) = \frac{1}{4b^2}\left[2m + (m^2 - 4b^2n)e^{-4by} + C_1 e^{4by}\right]$$

and b, c, m, n, C_1 and C_2 are constants. It only remains to give the explicit form of functions G and H, which is obtained very easily. However, the solutions split up into two different classes:

Case 1. This case is defined by

$$m^2 - 4b^2n \neq 0$$

and gives rise to the original Wahlquist solution [13], the functions G and H being now

$$G(x) = \frac{3}{2}\left[\frac{am^2b^{-2} - 2an + sm + 2b^2h}{m - 4b^2n}V(x) - \frac{amnb^{-2} + 2ns + hm}{m^2 - 4b^2n}\right.$$
$$\left. + \frac{a}{4b^4}\sqrt{m^2 - 4b^2n}\,(4bx + C_4)\cos(4bx + C_4)\right],$$

$$H(y) = \frac{3}{2}\left[\frac{am^2b^{-2} - 2an + sm + 2b^2h}{m - 4b^2n}W(y) - \frac{amnb^{-2} + 2ns + hm}{m^2 - 4b^2n}\right.$$
$$\left. - \frac{a}{4b^4}\sqrt{m^2 - 4b^2n}\,(4by + C_3)\left(m - 2b^2W(y) + C_1e^{4by}\right)\right]$$

where a, s, h, C_3 and C_4 are new constants. How to get the Wahlquist solution in its original form from the above expressions can be seen in [12].

Case 2. The remaining metrics are characterized by

$$m^2 - 4b^2n = 0$$

which produces some solutions found by Kramer in [14] [15]. In this case, the solution is completed with

$$G(x) = \frac{h}{16b^2} - C_3\cos(4bx) - C_4\sin(4bx),$$

$$H(y) = \frac{h}{16b^2} - e^{4by}\left(\frac{3aC_1}{4b^3}y + C_2\right)$$

using the same notation as above.

However, the solutions in case 1 can be considered as the general solution in the sense that every solution in case 2 can be obtained from those in case 1 by a limiting procedure. The explicit limiting procedure can be seen in [12] [15].

With regard to the properties of these solutions, the velocity vector of the fluid is the unit timelike vector proportional to $\boldsymbol{\xi}$ and the density and pressure are defined by

$$\rho + p = b^2e^{2U},$$
$$\rho + 3p = 12a, \tag{10}$$

this last expression being the equation of state. From (10) we see that both ρ and p cannot decrease simultaneously from the center outside. The solutions have no singularities in their domain of definition if we set $C_3 = C_4 = 0$, and the axis of symmetry is regular once the constant c has been chosen suitably. The solutions in case 2 have no static limit, while the solutions in case 1 do have a static *and spherically symmetric* limit given by $m^2 - 4b^2n \to \infty$ which leads to a metric due to Whittaker [16]. In general, there is a limit timelike surface of vanishing pressure. In case 2 this surface is oblate. However, in case 1 it seems that it can adopt several different shapes, some of them rather astonishing, as can be seen in Wahlquist's contribution in this volume. Both principal null directions of the Weyl tensor are geodesic and shear-free. In case 1 they are also expanding, while in case 2 they are expansion-free. There is no exterior solution known to be matchable with any of the solutions in the Wahlquist family.

Although this family is not the most general solution in class D2RR, it appears to be rather unique and, in fact, it seems difficult to find other D2RR solutions. This is due to the following set of results, which prove how most of the restrictions one can imagine to solve the equations in this case lead, in general, to the Wahlquist family.

Theorem 1 *For a stationary perfect-fluid solution with the velocity vector proportional to a timelike Killing vector (RR), the following statements are equivalent:*
1) The metric is given by the Wahlquist family.
2) The Simon tensor vanishes [15].
3) The metric has a Killing tensor of type [(11)(11)] (see [15], where it is cited).
4) The solution is type D, axisymmetric and with equation of state (10) [12].

In fact, this theorem can be completed with the following result that has not been proven yet:

Conjecture 2. *5) The metric is type D, axisymmetric and both principal null directions are geodesic and shear-free.*

Therefore, there remains very little room to find new D2RR solutions, specially if the above conjecture is true. Let us move on to the next class.

D1DR Solutions.

Again in this class the general solution is as yet unknown. Recently, however, we have been able to construct the family of D1DR solutions with *vanishing magnetic Weyl tensor* [17]. This family is a straightforward generalization of the D1RR solutions, and its line-element reads

$$ds^2 = M^{-2}\left[-m\left(dt + \frac{s}{m}d\phi\right)^2 + \frac{hm + s^2}{m}d\phi^2 + (hm + s^2)^{-1}dx^2 + dz^2\right] \quad (11)$$

where h, m, s are functions of only x and M is a function of only z given by

$$M_{,z}^2 = v + \epsilon a^2 M^2, \qquad \epsilon = \pm 1, 0, \qquad v, a = \text{const.}$$

or equivalently

$$\epsilon = 1: \quad M(z) = A\cosh(az) + B\sinh(az), \qquad v = a^2(B^2 - A^2),$$
$$\epsilon = -1: \quad M(z) = A\cos(az) + B\sin(az), \qquad v = a^2(B^2 + A^2),$$
$$\epsilon = 0: \quad M(z) = Az + B, \qquad v = A^2,$$

where A and B are constants of integration (with v depending on them). The line-element (11) is a solution of the field equations for a perfect-fluid energy-momentum tensor if and only if the functions h, m and s satisfy

$$h''m'' + s''^2 = 0 \ , \tag{12}$$

$$(hm + s^2)'' + 4\epsilon a^2 = h'm' + s'^2 \ , \tag{13}$$

where a prime stands for derivative with respect to x. As we can check, there are two ordinary differential equations for three unknown functions, and thus the general solution depends on an arbitrary function.

The velocity vector of the fluid is

$$\mathbf{u} = M^{-1}\left[-\sqrt{1+f^2}\sqrt{m}\left(dt + \frac{s}{m}d\phi\right) + f\frac{\sqrt{hm + s^2}}{\sqrt{m}}d\phi\right] \ ,$$

$$f^2 = \frac{m''}{m}\frac{hm + s^2}{h'm' + s'^2 + 4\epsilon a^2} > 0 \ ,$$

and the energy density and pressure can be obtained from

$$\rho + p = \frac{1}{2}M^2(h'm' + s'^2 + 4\epsilon a^2) \ ,$$

$$\rho = p - 6v \ .$$

This last expression is the equation of state of the fluid. We see that the energy condition $p \le \rho$ is satisfied only if $v \le 0$, which is possible in the case $\epsilon = 1$ with $A^2 \ge B^2$. In this case, there can be a timelike hypersurface of vanishing pressure that could be interpreted as the fluid limit surface.

In general the solutions have differential rotation, except for the case $\left(\frac{s''}{m''}\right)' = 0$ which always leads to the D1RR solutions previously studied and contained here as the general RR limit. Therefore, the static limit of these D1DR solutions is again the Barnes metrics [10]. The solutions (11) have a proper conformal Killing vector field ζ given exactly by (8), and the conformal group is again Bianchi I. The general solution for (12,13) has not been found explicitly yet. However, some particular solutions have been presented in [17].

D2DR Solutions.

As far as we know, no explicit solution has been found in Class 2 with differential rotation. In fact, this class has not been even explored. We have exhausted the type D solutions and now pass to the study of the Petrov type II solutions.

2.3 Type II Solutions

Type II solutions can be classified in a similar manner to that for type D solutions. Now we have three principal null directions of the Weyl tensor $\boldsymbol{\ell}, \boldsymbol{k}$ and $\boldsymbol{n}$ ($\ell^\alpha \ell_\alpha = k^\alpha k_\alpha = n^\alpha n_\alpha = 0$), one of them multiple, say $\boldsymbol{\ell}$. Thus we have four different classes depending on whether or not the velocity vector of the fluid is a linear combination of the principal null vector fields. Namely, we have:

Class 1 if $\boldsymbol{u} = a\boldsymbol{\ell} + b\boldsymbol{k}$ or $\boldsymbol{u} = a\boldsymbol{\ell} + b\boldsymbol{n}$
Class 2 if $\boldsymbol{u} = a\boldsymbol{k} + b\boldsymbol{n}$
Class 3 if $\boldsymbol{u} = a\boldsymbol{\ell} + b\boldsymbol{k} + c\boldsymbol{n}$
Class 4 if $\boldsymbol{u} = a\boldsymbol{\ell} + b\boldsymbol{s}$ with $\boldsymbol{s}$ a null vector independent of $\boldsymbol{k}$ and $\boldsymbol{n}$

where a, b and c are non-zero functions. Each of these classes can be divided into two subclasses depending on RR or DR. We finally get the following table

Table 2. The eight cases of Petrov type II stationary axisymmetric perfect-fluid metrics. The arrows indicate the searching direction of increasing difficulty.

	Class 1		Class 2		Class 3		Class 4
RR	II1RR		II2RR		II3RR		II4RR
	↓	↗	↓	↗	↓	↗	↓
DR	II1DR		II2DR		II3DR		II4DR

We use again self-evident notations so that, for instance, case II3RR means Petrov type II solutions of Class 3 with rigid rotation. On the contrary to what happenned in Table 1, arrows of increasing difficulty point first from top to bottom and then from left to right, that is to say, solutions in Class N (N=1,2,3,4) are simpler in general (irrespective of RR or DR) than solutions in Class M with M greater than N. Thus, the easiest case is II1RR followed by II1DR. Then cases II2RR and II2DR come, but these two cases are not physically good for they cannot have Petrov type D limits. This follows because type D limits are those in which both $\boldsymbol{k}$ and $\boldsymbol{n}$ collapse to a unique repeated principal null vector, which implies for Class 2 solutions that the velocity vector of the fluid becomes a null vector. Therefore, solutions in Class 2 can be studied, but there is very little hope that they can have interesting static and spherically symmetric limits. Next cases are II3RR and II3DR, which can have good properties and static limits, they are much more complicated solutions than those in Class 1 though. Finally, II4RR and II4DR are the most difficult cases.

The above classification can be better understood and even improved by combining it with the study of the following invariantly defined vector $\boldsymbol{v}$:

$$v^\alpha \equiv \eta^{\alpha\beta\mu\nu} \ell_\beta k_\mu n_\nu \ , \tag{14}$$

where $\eta^{\alpha\beta\mu\nu}$ is the contravariant volume element in spacetime. This vector has very interesting properties. First of all, as follows immediately from its definition

(14), v is orthogonal to the three principal null directions, that is to say,

$$v^\mu \ell_\mu = v^\mu k_\mu = v^\mu n_\mu = 0 \ ,$$

and therefore it is spacelike everywhere

$$v^\mu v_\mu > 0 \ .$$

Class 4 above is unambigously defined by the condition

$$\text{Class 4} \iff v^\mu u_\mu \neq 0 \ ,$$

while in classes 1, 2 and 3 we have that v is orthogonal to the velocity vector of the fluid u. Furthermore, from (14) we deduce that v is invariant under both Killing vectors

$$\mathcal{L}_\xi v = \mathcal{L}_\eta v = 0 \ .$$

Finally, from (14) follows also that if there is a type D limit, then v becomes the zero vector in this limit, because as explained before type D limits are defined by the proportionality of k and n. As we shall see in the next section, these properties of v may be of some importance in trying to find explicit solutions.

To our knowledge, no explicit Petrov type II stationary and axisymmetric perfect-fluid solution has been found yet.

2.4 Type I Solutions

This is the generic case, and therefore no simplification is achieved by assuming Petrov type I, which is like not assuming anything at all regarding Petrov types. One could think that no solution would be found under these circumstances, due to its difficulty. Surprisingly enough, however, there is an explicitly known solution which was presented in [18]. The line-element for this solution reads

$$ds^2 = \left(-U^2\frac{p_o}{p} + \delta^2\Omega^2 r^2 \frac{p}{p_o}\right) dt^2 - 2\delta^2\Omega r^2 \frac{p}{p_o} dt d\phi + \delta^2 r^2 \frac{p}{p_o} d\phi^2$$

$$+ \frac{TT'}{2pr}\left[\frac{k}{2} - \frac{r^4}{4}\left(\frac{p}{p_o}\right)^2\right]^{-1}$$

$$\times \left\{\left[\frac{1}{r^2} + \frac{k}{2T^2} - \frac{r^4}{4T^2}\left(\frac{p}{p_o}\right)^2\right] dr^2 + \frac{1}{pr}dpdr + \frac{1}{4p^2}dp^2\right\}$$

where the pressure p has been used as a coordinate, the function Ω depends only on the coordinate r and is given by

$$\Omega = \Omega_o + \epsilon\frac{U}{\delta}\int_{r_o}^r \frac{R}{T(R)}dR \ ,$$

and the function T depends only on r and satisfies the following ordinary differential equation

$$T'' + \left(\frac{1}{r} + \frac{kr}{T^2}\right) T' = 0 \ . \tag{15}$$

A prime denotes derivative with respect to r, and $k, U, \delta, \epsilon, r_o, p_o$ and Ω_o are arbitrary constants.

The velocity vector of the fluid takes the very simple form

$$\mathbf{u} = -U\sqrt{\frac{p_o}{p}}\, dt$$

and therefore the rotation tensor vanishes. The fluid has differential rotation and there is no limit with rigid rotation apart from the uninteresting case with $\delta = 0$. The equation of state is that of a stiff fluid

$$p = \rho \ .$$

There is a timelike surface of vanishing pressure (and density). This surface is compact at each instant of time, but it seems that it is toroidal in shape.

Explicit metrics can be constructed by solving (15), which is the only remaining equation. However, no explicit solution for $T(r)$ in terms of elementary functions is known. No exterior matchable solution has been found either.

3 Clues for Seeking New Solutions

In this last section, our aim is to present our ideas concerning how to look for new solutions, where we should search for them and why there are possibilities of actually finding them under some appropriate restrictions. To that end, we follow the same division as that used in the previous section.

3.1 Seeking Type D Solutions

This is the most important part: Petrov type D solutions should be exhausted. As is well known, *all* spherically symmetric solutions are type D (apart from the very particular case of the Schwarzschild interior solution), and therefore, as has been explained several times in this paper, stationary and axisymmetric type D perfect-fluid solutions are likely to have good static limits. Moreover, they constitute the simplest case and the possibilities of finding general new solutions are big enough to hope for some successes. Finally, experience shows that Petrov type D solutions *have* been found in several occasions (see Sect.2), although in most cases it has been necessary to assume some other property. In fact, this is the procedure we propose in looking for new solutions. But, which kind of assumptions are suitable for that matter? Experience tells us which assumptions have worked previously, and a good try would be to assume the properties that have been proven themselves fruitful. Let us analyze this in detail.

First of all, it is obvious that the searching direction shown in Table 1 by means of arrows must be taken into account. In other words, if some assumption does not work in some case, then there is very little hope that it should work in more complicated cases. Therefore, the first case to be studied is D2RR, despite the fact that Theorem 1 does not permit much to be done. However, one important try in this case is the study of Conjecture 2. If this conjecture is true, there will not appear new solutions but we shall be even more convinced of the rather strange uniqueness of the Wahlquist family of solutions. On the other hand, if the conjecture is false and this can be proven, then a whole family of new solutions will come out quite naturally along with the proof.

The next steps in seeking D2RR solutions are the following (without order of priority): First, a linear equation of state $p = c\rho + b$ with c, b =consts. may be assumed. It can be checked in Sect.2 that all found solutions have this type of equation of state, and thus it seems that this is somehow important in allowing the integration of the field equations. Second, the magnetic part of the Weyl tensor can be set equal to zero, because this is a property that has worked in two other cases (namely D1RR and D1DR). As we are talking about rigid rotation solutions (D2RR), they are shear-free and then it follows that if they are also rotation-free, the magnetic part of the Weyl tensor vanishes. The converse is not true, though, for type D solutions, but in any case this is a good indication that the solutions with vanishing magnetic Weyl tensor may include static limits. Third, one may also assume the existence of conformal Killing vectors as happens with the solutions in cases D1RR and D1DR. One should bear in mind, in this case, that the Bianchi types of the conformal group are severely restricted if one wishes to have a well defined axial symmetry. In fact, the axial Killing vector must commute with both the timelike Killing vector and the assumed conformal Killing vector. This has been recently proven as can be seen in another contribution in this book [19]. One important consequence of the existence of a proper conformal Killing vector is that the components of the metric usually take a separated form as product of functions of one single coordinate, and thereby the integration of the field equations may be substantially simplified leading to the resolution of simple ordinary differential equations. Needless to say, the three properties mentioned above can be combined and assumed to hold simultaneously, the simplications achieved by these means being even greater.

The next case to be considered is D1DR, which, in fact, is not much more complicated than the previous one. Here, the three properties mentioned above have already been used to find the only family of known solutions presented in Sect.2 (vanishing of magnetic Weyl tensor, linear equation of state and conformal Killing vectors). However, these last two properties have not been exhausted and may still be used for other cases. Apart from that, another possible assumption is the existence of a Killing tensor, which has worked for the D2RR case and also leads to separation of variables in most cases.

Finally, the most complicated type D case is D2DR. Given that no solution in this case has been found yet, we can use all the previously mentioned assumptions including every combination of them. Another possibility is trying to generalize the D2RR solutions to having differential rotation (something that was used in finding the D1DR solutions, which are an obvious generalization of the D1RR solutions). This might be achieved by allowing the rotation parameter (defined as the one vanishing when going to the static limits) to be a function rather than a constant. In the case under consideration this means that the parameter $m^2 - 4b^2n$ appearing in solution (9) should be redefined to be a function (possibly of one coordinate). Whether or not this will work is uncertain, anyway.

3.2 Seeking Type II Solutions

The same comments made for type D solutions are also in order for type II solutions. Thus, the existence of non-isometric symmetries (conformal Killings or Killing tensors), special properties of the Weyl tensor (such as vanishing of its magnetic part or geodesic and shear-free principal null directions) and a linear equation of state are, all of them, interesting possible assumptions in this case. We should keep in mind, however, that if these assumptions fail in helping us to find new type D solutions, then they will probably fail for type II solutions too. Table 2 shows the arrows of increasing difficulty, and thus the easiest cases (and perhaps the only relevant) are II1RR and II1DR. As was explained in Sect.2, the next cases in difficulty are II2RR and II2DR but they will not produce solutions with good static limits. Unfortunately, the next case in the list II3RR is, in our opinion, too complicated and if the previous program does not produce results for previous cases then there are very little hopes for cases as difficult as II3RR (and those even more complicated). In our view, the efforts in seeking type II solutions should be focused, at least in the beginning, to just cases II1RR and II1DR.

Nevertheless, type II solutions have their own properties, not shared by type D solutions, which could be of great help in finding solutions. We refer, principally, to the existence of the vector field v defined in (14). In Sect.2 we saw that this vector is spacelike, is left invariant by the Killing vectors, is orthogonal to the velocity vector of the fluid (unless in Class 4 metrics) and, also, it vanishes in static limits (if there is such a limit). All these properties are nothing but properties that the rotation vector of the fluid does have. One may assume then, from the very beginning, that the vector v is proportional to the rotation vector, and this will lead to some simplifications. As always, whether or not this interesting property is helpful in any way will not be clear until it has been explicitly used. The combination of this property with, for example, the existence of conformal Killing vectors and a linear equation of state for case II1RR should, in principle, simplify the equations drastically and enable us to find new solutions. If this is not the case, then we would be in real trouble and finding type II solutions would become a task far from easy.

3.3 Seeking Type I Solutions

Type I solutions are generic, and therefore the algebraic classification used in this paper is not relevant at all. As this is the general case, studying it is equivalent to studying everything from the Petrov type point of view. The question reduces then to imagine some other desirable properties, as those presented in the previous cases, without making any assumption regarding the Petrov type of the solution.

Final Comments. The classification presented in this paper is just one among many possibilities. It can also be improved or refined by means of some properties other than the Petrov type and RR or DR. Thus, for example, the classification in [8] for type D perfect-fluid metrics is much more elaborated than has been used here. Nevertheless, the solutions explicitly found hitherto are so few that this classification seems appropriate at the present stage.

On the other hand, a new solution has been found by Chinea very *recently*; so recently, in fact, that we have not been able to include it here, despite the fact that our intention was to present *every* solution known at present. Fortunately, this new solution is presented in this same volume.

References

1.   Martín-Prats, M., Senovilla, J.M.M.: Matching of stationary axisymmetric space-times. Contribution to this volume (1992)
2.   Misner, C.W., Thorne, K.S., Wheeler, J.A.: Gravitation. W. H. Freeman and Company (1973) San Francisco
3.   Carter, B.: The commutation property of a stationary axisymmetric system. Commun. Math. Phys. **17** (1970) 233-238
4.   Papapetrou, A.: Champs gravitationnels stationnaires à simmétrie axiale. Ann. Inst. H. Poincaré **4** (1966) 83-90
5.   Kundt, W., Trümper, M.: Orthogonal decomposition of axi-symmetric stationary spacetimes. Z. Phys. **192** (1966) 419-425
6.   Kramer, D., Stephani, H., MacCallum, M., Herlt, E.: Exact Solutions of Einstein's Field Equations. Cambridge University Press (1980) Cambridge
7.   Collinson, C.D.: The uniqueness of the Schwarzschild interior metric. Gen. Rel. Gravit. **7** (1976) 419-425
8.   Wainwright, J.: Classification of the type D perfect fluid solutions of the Einstein equations. Gen. Rel. Grav. **8** (1977) 797-802
9.   Senovilla, J.M.M.: On Petrov type-D stationary axisymmetric rigidly rotating perfect-fluid metrics. Class. Quantum Grav. **4** (1987) L115-L119
10.   Barnes, A.: Static perfect fluids in general relativity. J. Phys. A **5** (1972) 374-383
11.   Kramer, D.: Perfect fluids with conformal motion. Gen. Rel. Gravit. **22** (1990) 1157-1162
12.   Senovilla, J.M.M.: Stationary axisymmetric perfect-fluid metrics with $\rho + 3p$ =const. Phys. Lett. A **123** (1987) 211-214
13.   Wahlquist, H.D.: Interior solution for a finite rotating body of perfect fluid. Phys. Rev. **172** (1968) 1291-1296

14.   Kramer, D.: A new solution for a rotating perfect fluid in general relativity. Class. Quantum Grav. **1** (1984) L3-L7
15.   Kramer, D.: Rigidly rotating perfect fluids. Astron. Nachr. **307** (1986) 309-312
16.   Whittaker, J.M.: An interior solution in general relativity. Proc. Roy. Soc. Lond. **A 306** (1968) 1-5
17.   Senovilla, J.M.M.: New family of stationary and axisymmetric perfect-fluid solutions. Class. Quantum Grav. (1992) (to appear)
18.   Chinea, F.J., González-Romero, L.M.: Interior gravitational field of a stationary, axially symmetric perfect fluid in irrotational motion. Class. Quantum Grav. **7** (1990) L99-L102
19.   Mars M., Senovilla, J.M.M.: Axial symmetry and conformal Killings. Contribution to this volume (1992)

Black-Holes in X-Ray Binaries

Jorge Casares

Instituto de Astrofísica de Canarias, 38200-La Laguna, Tenerife, SPAIN

1 Introduction

The study of interacting binaries has attracted the interest of many astrophysicists due to the exhibition of spectacular variability over a wide range of timescales (from milliseconds to several years). The new window opened in the 1970's by X-ray satellites (UHURU, Ariel V, Einstein and several others)[1] made possible the discovery of a new class of objects that produced vast amounts of high-energy radiation: the *X-ray binary systems*. The advent of a second generation of X-ray detectors has increased substantially the number of constituents and the diversity of this group, enabling statistical studies to be performed and a subclassification of observed properties. The X-ray emission is naturally generated by the acceleration of accreted matter in a deep gravitational well. Therefore it is believed that one of the binary's components must be a very compact object, i.e. the endpoint of stellar evolution.

Despite its youth, X-ray astronomy has made major contributions in almost all astronomical fields and, in particular, it can now offer the most compelling evidence for the existence of the elusive astrophysical objects termed *black holes*. In this paper I present the current status of the search for stellar-size blackholes in the galaxy. However, in order to associate the reader with this topic, a brief review on accretion physics and X-ray binaries is given first. To conclude the paper I summarize the most recent experimental results on the V404 Cyg system, which is considered the strongest black-hole candidate yet found.

---

[1] without forgetting the historic rocket flights in 1962 [1] and 1966 [2] which represent the discovery and optical identification of the first extra-solar X-ray source Sco X-1.

2 Accretion geometry

X-ray binaries are bright Galactic X-ray sources containing a compact object and a non-degenerate optical companion. The compact object can either be a neutron star, a black hole or a white dwarf in which case the system is called a *Cataclysmic Variable* (an excellent updated review is available in [3]). I will not consider cataclysmic variables hereafter and will only deal with neutron stars and black hole systems. They are characterized by $L_x \sim 10^{36} - 10^{38} ergs^{-1}$.

Both components spin around their common centre of mass in very close orbits, so the gravitational potential is described by the canonical *Roche model*. If we take a reference frame centered on the system's center of mass and corotating with the binary, the potential can be expressed by:

$$\Phi_R(r) = -\frac{GM_1}{|r - r_1|} - \frac{GM_2}{|r - r_2|} - \frac{1}{2} (\omega \wedge r)^2 \qquad (1)$$

where ω is the angular velocity of the binary ($= \frac{2\pi}{P}$), M_1, M_2 are the component masses and r_1, r_2 the stellar position vectors. This expression results from summing the potentials of each component plus the centrifugal term. Strictly speaking the Roche Potential is only an approximate solution of Poisson's equation since it associates the stars with mass points and thus neglects the distortion effects produced by rotation and the proximity of a companion. However, it is considered a good approximation, correct to a few percent, because stars are centrally condensed [4].

For a critical value of Φ_R the equipotential surfaces define two oblate spheroids called the *Roche lobes* which intersect at L_1, the *inner Lagrangian point* (Fig.1). This saddle-point is particularly important since for certain phases of stellar evolution the star's envelope expands until it fills its Roche lobe. The binary system is then said to be *semi-detached*. The use of the Roche geometry imposes very useful constraints on the system parameters. In the *Roche filling* hypothesis the radius of a spherical star occupying the same volume as its corresponding Roche lobe is given by [5]

$$\frac{R_2}{a} = 0.38 + 0.20 \, log \, q \qquad\qquad 0.3 < q < 2.0 \qquad (2)$$

$$\frac{R_2}{a} = 0.462 \left(\frac{q}{1 + q}\right)^{1/3} \qquad\qquad 0 < q < 0.8 \qquad (3)$$

with a being the separation between the two stars and q=M_2/M_1 the mass ratio. These formulae are found to be accurate to within 2 percent (a more accurate algorithm was derived by Eggleton [6]).

L_1 is unstable to perturbations so matter that has reached it can easily flow into the compact object's lobe. Given its high angular momentum, acquired from the system's rotation, the transferred material will orbit the central object at a certain distance, the *circularization radius* r_c. The action of viscous forces, associated with differential rotation, will produce the conversion of gravitational

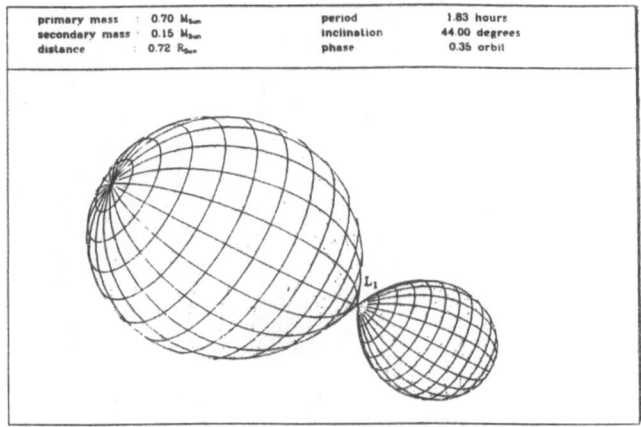

primary mass	:	0.70 M_sun	period	1.83 hours
secondary mass	:	0.15 M_sun	inclination	44.00 degrees
distance	:	0.72 R_sun	phase	0.35 orbit

Fig. 1. The *Roche lobes* of a close binary system and the *inner Lagrangian point* L_1. In the case of a semi-detached system matter will be transferred through this saddle point towards the compact objet's lobe.

potential energy into heat, which is radiated away. As a result of this energy loss matter drops closer towards the compact object and hence angular momentum is passed to the outer material. The incipient ring formed at r_c will be spread into a disc-like configuration, the *accretion disc* (Fig.2). The net effect of the disc is the accretion of material from the donor star to the compact object through redistribution of angular momentum [7]. The collision between the gas stream and the outer rim produces a bright shock front, the *hot-spot*.

There is a single unique analytical solution for the structural equations of an accretion disc (i.e. conservation of mass, angular momentum and energy) which is obtained through the prescriptions of Shakura & Sunyaev [9], the *α-discs*. Their model is based on the assumption that the mass transfer rate through the disc $\dot{M}$ is constant (a *stationary* disc) in addition to three great simplifications. The first one supposes that the disc is *axisymmetric*. We are therefore neglecting the action of tidal forces from the non-degenerate star as well as centrifugal and coriolis forces, which is a good approximation for the inner part of the disc. The second approach is that the disc is geometrically thin, i.e. the radius R is much greater than the thickness H. This implies that the energy transport needs only be considered in the direction perpendicular to the disc resulting in a slightly concave shape, expressed as $H \propto R^{9/8}$. The final assumption is the proportionality between the local viscosity and the pressure for the whole disc. The constant of proportionality α wraps our ignorance of the viscosity mechanism in a dimensionless parameter. At present it is only clear that molecular viscosity is insufficient to account for the estimated diffusion timescales and therefore some other mechanism has to be invoked, such as turbulent motions or small-scale magnetic fields.

92   Jorge Casares

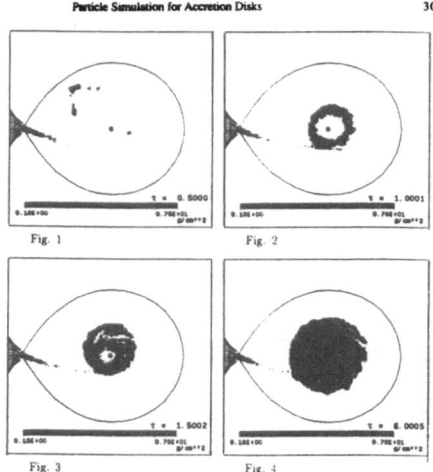

Fig. 2. Particle simulation model representing the formation of an accretion disc by Roche lobe overflow. From Ref.8.

The disc models constructed under these assumptions are characterized by a radial distribution of the effective temperature T_{eff} according to

$$T_{\rm eff} \propto M_1^{1/4} \, \dot{M}^{1/4} \, R^{-3/4} \, \left(1 - \left(\frac{R_1}{R} \right)^{1/2} \right)^{1/4} \tag{4}$$

where R_1 is the radius of the compact object[2]. The outer parts of the disc are rather cool ($T_{eff} \leq 10^4$ K) and radiate in the optical whereas ultraviolet and X-ray emission, on the other hand, are expected to be produced in the hot inner parts of the disc ($T_{eff} \geq 10^6$ K) and the *boundary layer* between the accretion disc and the compact object. The virial theorem predicts that half of the disc gravitational potential energy is radiated away at

$$L_{\rm acc} = \frac{G \, M_{\rm X} \, \dot{M}}{2 R_{\rm X}} \tag{5}$$

while the other half is preserved in the form of kinetic energy which will finally be released at the boundary layer. Only for black hole systems the absence of a solid surface prevents the recovery of this contribution as radiated energy.

A peculiar situation, characterized by the equality of the radiation pressure and gravity, can be reached for sufficiently high accretion rates. Such a critical luminosity, defined as *the Eddington luminosity*, is given by

---

[2] For the matter of simplicity we will hereafter adopt the common nomenclature which uses the subscripts x and c for the parameters refering to the *X-ray source* and the *non-degenerate companion*, respectively.

$$L_{\text{Edd}} = \frac{2\,\pi\,R_g\,m_p\,c^3}{\kappa_d} = 1.4\,\frac{M_X}{M_\odot}\,10^{38}\ \text{erg s}^{-1} \tag{6}$$

where m_p is the proton mass, κ_d the Thomson and/or Compton scattering opacity and $R_g = 2\frac{GM_X}{c^2}$ the Schwarzschild radius. This expression only provides an approximation to the real situation since, while assuming that the accretion flow is spherical, it neglects anisotropic effects such as radial accretion through the boundary layer as well as possible accretion columns.

Observational support for the existence of accretion discs comes from the detection of broad double-peaked emission lines [10] [11]. The profile corresponding to a rotating flat layer of gas results from summing the contributions of equal-velocity surfaces and thus depends merely on the disc size and inclination (defined as the angle between the normal to the orbital plane and the observer). In particular, the half-separation of the double peaks gives an estimate of the projected outer velocity of the disc and hence of the outer radius (assuming Keplerian rotation). Moreover the broad wings are produced in the higher velocity inner parts of the disc and presumably reflect the orbital motion of the compact object.

The core of the line usually suffers great periodic distortions as a result of the orbital motion of a narrow component arising in the hot-spot. Since it is located in the outer disc this component moves back-and- forth between the two peaks, giving the visual effect of an *S-wave* on *trailed spectra*. Other "S-waves" may also be produced by other localized sources of line emission. This is the case, for instance, when intense X-ray emission heats the hemisphere of the companion star facing the compact object. The resulting S-wave will be opposite in phase to the accretion disc. As a matter of convention, orbital phase zero is defined as the inferior conjunction of the optical star.

Accretion discs experience quasi-periodic outbursts in which the optical luminosity typically rises by 2-10 magnitudes. These outbursts are attributed to some kind of mass transfer instability triggered by two possible mechanisms:

- **The Disc Instability Model (DI):** Material is transferred at a constant rate from the companion star and is somehow stored in the accretion disc. Thus the surface density grows continuously until the viscosity suddenly increases and matter is dumped onto the compact object [12]. A possible mechanism, based on a change from radiative to convective energy transport, has been invoked by Meyer & Meyer-Hoffmeister [13] to explain such an instability.

- **The Mass Transfer Instability Model (MTI):** An instability, associated with the companion star's surface, modulates the mass-injection rate on the outburst timescale. This model predicts locally steady properties for the disc (i.e. surface density or T_{eff}) until the burst takes place [14] [15]. The response of the accretion disc to the episodic deposition of matter is an initial shrinking followed by the subsequent expansion of the outer radius [16] [17]. Bath [18] proposed the liberation of energy upon recombination of hydrogen in the vicinity of the inner Lagrangian point as the likely process responsible for the instability.

Despite the very different processes involved in each model no conclusive test has yet been devised to decide which mechanism is operating in the outbursts. This is partially due to the variety of properties exhibited during different outbursts of the same system[3], as well as from one system to another.

3 Classification of X-Ray Binaries

Two main categories of X-ray binaries can be distinguished according to their optical properties [19] [20]. Only later on was it found that they are associated with two different physical structures (see Fig.3). They are as follows:

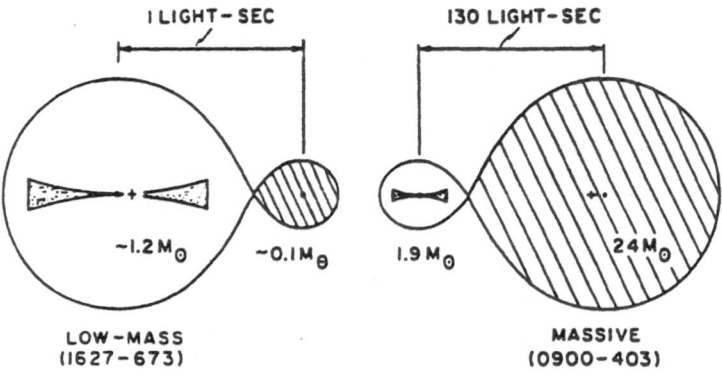

Fig. 3. Cartoon comparison of two X-ray binary systems: the LMXB 4U1627-673 and the HMXB 4U0900-403. From Ref.19.

3.1 Class I or High Mass X-ray Binaries (HMXB)

The companion star is a massive ($\geq 10 M_\odot$) early-type (O-B) star which transfers mass to the compact object either via stellar wind or incipient Roche lobe overflow. Given the large size of the optical star, HMXB are also characterized by the low ratio of X-ray to optical luminosity ($L_x/L_{opt} \leq 1$) and the presence of X-ray eclipses. The optical luminosity is thus dominated by the bright companion star. HMXB systems are concentrated towards the galactic plane and the Magellanic Clouds indicating that they belong to relatively young ($\leq 10^7$ years) Population I stars, as expected for their massive early type companions.

---

[3] Indeed, there is evidence for both mechanisms operating, sometimes in the same Cataclismic Variable (e.g. the SU UMa systems during normal and super outbursts).

Most HMXB exhibit X-ray pulsations. This peculiarity is interpreted under the assumption of a strong magnetic field ($\sim 10^{12}$ G), anchored in the neutron star, which channels the accreted material towards the magnetic poles of the star. The frequent misalignment of the rotation axis and the magnetic axis of the neutron star will produce a rotating beam of X-ray emission wich periodically sweeps the Earth. As a result of the orbital motion, the pulse arrival times are Doppler shifted, which can be used to derive the radial velocity curve of the compact object, the projected size of the orbit and the eccentricity. On longer timescales a slow decrease in the pulse-periods has been detected, which implies an overall spin-up for most of the sources [21]. This is considered as confirmation for the accretion hypothesis, through transfer of angular momentum. The very hard X-ray spectra (kT$\geq$15 keV) observed are also attributed to the extremely hot areas near the base of the *accretion columns*. HMXB are divided into two further groups according to the nature of the optical star [22] and their evolutionary history (see [23] and references therein):

– The companion star is a close main sequence Be star which underfills its Roche lobe and rotates very rapidly. Episodes of variable accretion are produced as a result of long-period (>10 days), highly eccentric orbits. Based on such a property these systems are considered as *Hard X-ray transients* (for reviews see [24] and [25]).

– The companion star has a spectral type earlier than B2 I-III and fills (or nearly fills) its Roche lobe. Orbital periods are generally <10 days. They are sometimes called the *standard HMXB*.

3.2 Class II or Low Mass X-ray Binaries (LMXB)

The companion star is a low-mass ($\leq 1 M_\odot$) late-type (F-M) star[4] which transfers material via Roche lobe overflow. The ratio of X-ray to optical luminosity is in the range 10^2-10^3 and the dominant source of optical light is believed to be the reprocessing of X-rays into the accretion disc. LMXB systems are concentrated towards the galactic center but many have also been found in globular clusters. This evidence, in addition to the presence of late-type optical companions, strongly suggests that LMXB systems belong to the old ($\geq 10^9$ years) galactic Population II. Cowley et al. [26] have combined the galactic distribution and the kinematics of a sample of 39 LMXB to derive a likely age of $\sim$15 x 10^9 years.

The X-ray emission is rather soft (kT$\leq$10 keV) whilst the absence of X-ray pulsations probably indicates that the magnetic field of the compact object is rather weak ($\leq 10^{10}$ G). Combining this with the old population argument has led to the idea that neutron star magnetic fields in X-ray binaries decay.

Whereas the study of HMXBs proceeded rapidly, the investigation of LMXBs was much more difficult due to the lack of orbital features in the X-ray emission,

---

[4] This follows a common nomenclature whereby the non-degenerate star is called the *secondary*. We will use it frequently through the sections devoted to V404 Cyg.

such as eclipses or orbital modulation. A few notable exceptions, such as Her X-1, were not sufficient to explain such behaviour on the basis of random values of the inclination. Milgrom [27] finally solved the discrepancy by invoking a simple selection effect. The accretion disc was proposed to have enough vertical size so as to shield efficiently the companion star from the X-ray emission coming from the vicinity of the compact object.

The observations compiled since Milgrom's work has enabled the detection of partial eclipses and periodic X-ray features, such as irregular dips, for several systems. Depending on the disc azimuthal structure (Fig.4) and the nature of the obscured X-ray source the following classification is provided for high inclination sources[5]:

- **Dipping Sources:** Azimuthal structure placed at the edge of the disc will obscure a point-like X-ray source to produce irregular dips in the X-ray light curve. They usually take place at about orbital phase 0.8 (X1755-338 [30]; XB1254-690 [31]) suggesting a possible link with the bulge created by the gas-stream. Sometimes a secondary dip is detected almost exactly out of phase with the main dip, indicating the existence of more structure or bulge stable at about phase 0.2 (XBT0748-676 [32]). To date there are nine sources which exhibit periodic dips with periods between 0.83 hours (X1916-053) and 235 hours (Cyg X-2).
- **Accretion Disc Corona sources (ADC):** These are similar to the dipping sources but are at higher inclinations. The point-like X-ray source is, therefore, completely obscured by the vertical structure of the disc. Faint X-ray emission is however detected through scattering into a corona or wind surrounding the accretion disc, which appears as an extended X-ray source. The vertical structure of the disc and the companion star will produce smooth orbital modulations and partial eclipses on the light curve of this extended source. These systems are characterized by $L_x/L_{opt} \sim 1-20$ which is peculiar among LMXB. Succesful modelling of the complex X-ray and optical light curves has provided the relative contribution and sizes of the different emitting regions (e.g. X1822-371 [33]). There are currently at least eight sources which exhibit ADC properties.

According to their X-ray behaviour two further categories of LMXB can be listed:

- **X-Ray Bursters:** Many of the LMXBs exhibit intense X-ray bursts with typical rise times of 1 second and decay times of several seconds to minutes. When bursts are separated by intervals of hours or days (*type I*) they experience a considerable spectral softening throughout the decay, consistent with a cooling blackbody. In addition the integrated flux is preserved throughout the burst. They are then thought to be powered by thermonuclear flashes on the surface of the neutron star [35] which is considered as additional

---
[5] an updated review with separate treatement to every single system is given in [28]; also see [29].

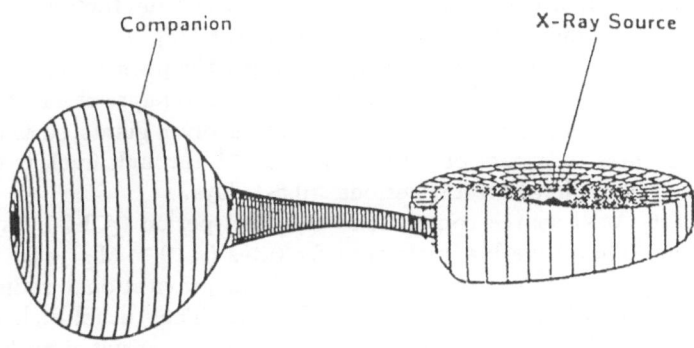

Companion   X-Ray Source

Fig. 4. Simplified model of a LMXB with representation of the vertical structure of the accretion disc. Note the bulge produced where the gas-stream collides with the outer rim of the disc. From Ref.34.

evidence for the weakness of the magnetic field on neutron star LMXB. Alternatively, when the burst occur at intervals of seconds or minutes (*type II* or *rapid bursters*), they are attributed to an instability in the accretion flow onto the neutron star. In this case no significant spectral change is observed throughout the burst. X-ray bursters are generally lower luminosity sources ($L_x \sim 10^{37}$). More information can be found in the review by Lewin et al. [36].

- **Soft X-Ray Transients:** A subset of LMXB undergo dramatic changes in brightness with recurrence times in the range 10-100 years. They are believed to be driven by any of the two accretion instability models proposed in the previous section. The morphology of the light-curve is characterized by a fast rise ($\sim$ days) followed by a short maximum ($\sim$ weeks) and a slow decline ($\sim$ months) until the system returns to quiescence. At this phase the system is said to be in an *off state* because the X-ray intensity normally settles down below any X-ray satellite detection threshold. The typical optical amplitude for the outburst is $\sim$ 8 magnitudes. In quiescence the optical counterpart is dominated by the companion star spectrum showing classical late-type absorption features. As we will see in Sect.5, V404 Cyg belongs to this category.

LMXB systems are also divided into two groups according to the shapes of their X-ray "colour-colour" diagrams: the **Z sources** and the **Atoll sources** [37]. This division is believed to enclose a different evolutionary status which provides dissimilarities concerning the companion star type, the accretion rate and the magnetic field of the neutron star. Z sources are characterized by X-ray luminosities close to the Eddington limit and the exhibition of quasi periodic oscillations (QPOs), which could be the result of the interaction of inflowing matter and the magnetosphere of the neutron star [38]. The relatively strong magnetic fields required to account for QPOs imply the presence of young ($\sim 10^7$ years) neutron stars, formed by the *accretion-induced collapse* of a massive white dwarf [39]. The companion stars are thought to be evolved giants, so as to provide mass-transfer rates in excess of $\sim 10^{-8}$ $M_\odot$ year^{-1} through nuclear evolution. Hence these systems have orbital periods $\geq$0.5-1 days.

In contrast, Atoll source companions are short-period ($\leq$0.5-1 days) main-sequence stars. They have low mass-transfer rates ($<10^{-8}$ $M_\odot$ year^{-1}), driven by angular momentum loss mechanisms, such as gravitational radiation and magnetic braking, which last for up to 109 years. They usually show low X-ray luminosities (0.005-0.3 L_{Edd}) and regular bursts, associated with a weakly magnetized and very old neutron star. Further information can be found in the works by van Paradijs & Lewin [40] and van den Heuvel [23] and references therein.

A different evolutionary history has also been invoked to explain some exotic properties exhibited by LMXB from globular clusters [41] [42] and the Magellanic Clouds [43] appealing to the effects of tidal capture and lower metallicity, respectively. Some authors consider them to be unique categories in their own right.

Table 1. Classification

Class	Companion Mass ($M_\odot$)	Companion Type	L_x/L_{opt}	X-ray Features	X-ray Eclipses	kT keV	Prototype
HMXB	≥ 10	O-B	≤ 1	pulses	often	≥ 15	Cyg X-1
LMXB	≤ 1	F-M	10^2-10^3	bursts	rare	≤ 10	Sco X-1

4 Black-Hole Signatures

Conversely to the case of neutron stars, the existence of black holes as members of X-ray binaries is much more difficult to assert. While the former are rapidly recognized by the exhibition of X-ray pulses or bursts, no specific black hole feature has been yet proposed and/or detected. In fact it is almost imposible to distinguish between a 10 $M_\odot$ black hole and a 1.4 $M_\odot$ non-magnetized neutron star since the properties derived from their gravitational potential wells are nearly identical (e.g. the innermost radius, 15 km, and the frequency of the last stable orbit, 2 x 10^3 s^{-1}). However, two groups of properties are frequently drawn as indirect evidence in the search for new black hole candidates. These are summarized below.

4.1 Spectral X-ray features

Since 1976 the HMXB Cyg X-1 has been considered as one of the strongest black-hole candidates and the only one until 1983. This is the reason why its peculiar X-ray properties have long been considered as "fingerprints" in the search for new black-holes; e.g. the energy distribution in the 1-100 keV range can be modelled by a single flat power-law, resembling the Crab Nebula spectrum[6]. These fingerprints are summarized below:

The bimodal spectral behaviour: The X-ray spectrum of Cyg X-1 usually remains in a *low state* ($L_X \sim 0.3 \times 10^{37}$ erg s^{-1}) which, as we mentioned before, is well described by a single power-law with spectral photon index $\alpha \sim 0.5$ for the whole 1-100 keV range. However, on timescales of weeks to years, it switches into a high state ($L_X \sim 3.0 \ 10^{37}$ erg s^{-1}), which is accompanied by a substantial softening ($\alpha \sim 2$-3) of the low energy (≤ 10 keV) spectrum (Fig.5). Hard X-rays (≥ 10 keV) are not seriously affected by these transitions and still preserve the flat power-law tail [44].

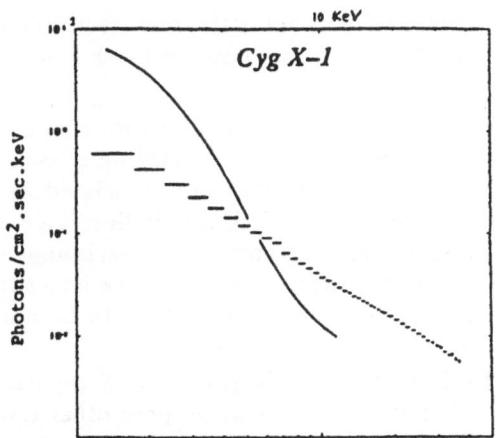

Fig. 5. Bimodal behaviour of the energy distribution of Cyg X-1. Note the anti-correlation between *intensity* and *hardness* exhibited by the soft X-rays (≤ 10 keV). From Ref.45.

Although the physics involved is very complicated it is generally accepted that the bimodal behaviour reflects optically-thin/optically-thick transitions of the innermost accretion disc which are powered by changes in the accretion rate [46]. The ultrasoft component detected in the high state is associated with the optically thick regions of the accretion disc. On the other hand, from equation 1.3

---

[6] This led to an obsolete X-ray classification which distinguished between the *supernova* or *Crab Nebula sources* and the *Sco X-1 like sources*, characterized by an exponential or thermal spectrum.

we see that the maximum temperature achieved by an optically thick accretion disc is insufficient to account for the observed hard X-ray flux. Therefore the hard spectrum is attributed to inverse Compton scattering (Comptonization) of soft photons in optically thin regions of hot electrons. Several models propose a variety of geometries for the source of hot electrons, such as a torus [47], ADC [44] and clouds in the vicinity of the compact object [48]. The disapperance of the ultrasoft component during the low state is interpreted as the receding of the inner boundary of the optically thick accretion disc [49].

It is assumed that the energy spectrum of a neutron star LMXB system is made up of two components: the ultrasoft component, attributed to the optically thick accretion disc[7], and a 1-2 keV blackbody component which originates in the boundary layer of the neutron star. Furthermore we must note that an additional third component, the single power-law hard spectrum, is usually present when the luminosity fades below $\sim 10^{37}$ erg s^{-1}. The important thing here is the absence of the 1-2 keV blackbody component, especially for high luminosities (or high accretion rates) which is taken as the hallmark for the non-existence of a material surface. Therefore, either the exhibition of the ultrasoft distribution in high states, the single power-law in low states (in particular for $L_X > 10^{37}$ erg s^{-1} and extending up to 200 keV without a thermal frequency cutoff) or both (the bimodal behaviour) is considered as a possible black hole indication.

Rapid flickering: The X-ray intensity of Cyg X-1 shows an erratic variability or flickering on timescales of 1 ms to 1 s. The flickering is seen to fade almost completely at the high state, so it is thought to be associated with the optically thin component. The typical frequency of flickering indicates a ~ 100 km size for the emitting region. At this location the accreted material is unable to distinguish whether the compact object is a neutron star or a black hole so it is surprising that fast flickering was actually considered for a time to be a solid black hole signature.

The huge volume of information provided by new X-ray satellites (such as Ginga, Granat or Rosat) led some authors to propose other temporal features as additional evidence. These include the exhibition of violent variability, the relation between the amplitude and the timescale of the flickering and several features detected in the power spectrum such as hard X-ray lags, low frequency noise or the absence of QPOs. These are presumably related to the hard component, in a similar manner to the flickering. Nevertheless, since the timescale of QPOs is longer than flickering this makes it a very weak indication. In particular QPOs could be detected for some black hole candidates (e.g. LMC X-1 or GX339-4 [50]) but not for others (Cyg X-1 and GS2023+338; same Ref.). Extensive reviews about the observations and models of QPO in LMXB have been presented in [51] and [52].

---

[7] Since the temperature of the accretion disc decreases with distance to the central object this component will consist of a sum of blackbodies produced from each concentric annulus of the disc.

Recently, though, a collection of discoveries seems to refute the established paradigm, introducing ambiguity into the uniqueness of the X-ray black hole diagnostics; e.g. Stella et al. [53] and Tennant et al. [54] revealed that the fast flickering sources X0331+53 and Cir X-1 are indeed an X-ray pulsar and burster, respectively. Furthermore, the source GX339-4, although exhibiting characteristic soft/hard transitions, has been recognized as a neutron star system after the upper limit of 2.1 $M_\odot$ was established for the mass of the compact object [55]. Also Cir X-1 was found to mimic the Cyg X-1 bimodal behaviour [56]. A list of the remaining black hole candidates, according to their X-ray properties, is given in Table.2 (adapted from [45] and [58]).

Table 2. X-Ray Properties of Black Hole Candidates

Name	Type	Transient/ Persistent	Variability	X-Spectrum[†]	References
Cyg X-1	HMXB	Persistent	Flickering	Bimodal	[69]
LMC X-3	HMXB	Persistent	No Flickering	High State	[45]
LMC X-1	HMXB	Persistent	QPO+No Flickering	High State	[45]
A0620-00	LMXB	Transient	No Flickering	High State	[45]
GS2023+338	LMXB	Transient	Flickering	Low State	[45]
GS2000+25	LMXB	Transient	No Flickering	High State	[45]
GS1124-683	LMXB	Transient	No Flickering	High State	[57]
GS1826-24	?	Transient	Flickering	High State	[45]

† For transient sources this refers to the outburst phase, when X-rays are in the *on state*.

Following Ilovaisky [22] it is clear that the above collection of temporal and spectral X-ray properties indicates something about the compactness of the emitting region and the accretion process but they are not conclusive black-hole diagnostics. Instead, the situation is entirely different for neutron star systems where the detection of X-ray pulses or bursts is given as a sufficient and reliable test. This seems to leave us with the most pure Popper's epistemology which one would summarize in the sentence *Science is everything that is liable to be refuted but will never be confirmed.* For an updated review of those and other new black hole candidates, on the basis of their X-ray properties, I address the reader to the works published by Ilovaisky [22] and Tanaka [45].

4.2 Lower limit to M_x

This can be obtained by measuring the *radial velocity curve* of the optical star using spectroscopic techniques. The *mass function*

$$f(M) = \frac{P\,K_C^3}{2\pi G} = \frac{(M_X\,sini)^3}{(M_X + M_C)^2} \qquad (7)$$

is a cubic equation which relates the masses of the compact object (M_X) and the companion star (M_C) with the inclination (i) and two measurable quantities: the orbital period (P) and semi-amplitude of the radial velocity curve (K_C). The value of the mass function represents a firm lower limit to M_X which can be obtained for the case of a zero-mass companion ($M_C=0$) in an edge-on system (i=90°).

The strength of this argument relies on the theoretical assessment, based upon General Relativity, that there exists a maximum mass for neutron stars to be stable. Actually such a limit will depend on the equation of state employed. The use of the stiffest known equation of state predicts an upper bound of 2.7 $M_\odot$, which can be increased further by 20 percent if the star is considered to rotate rapidly enough [59]. On the other hand, Rhoades & Ruffini [60] establish a solid upper limit of 3 $M_\odot$, almost independent of the equation of state employed, by assuming that: i) General Relativity is the correct theory of gravity and ii) causality is not violated (i.e. speed of sound < speed of light). Upper limits have been predicted through a wide collection of models which combine several types of exotic assumptions and equations of state (e.g. [61] [62]). However, the 3 $M_\odot$ stability limit is generally accepted as the most reliable criterion for all kinds of reasonable equations of state (see the excellent review by McClintock [63]).

On the basis of this argument we currently have three classic X-ray binaries which are considered as plausible black holes in terms of precise measurements of their mass functions. These are Cyg X-1, LMC X-3 and A0620-00[8] and their main properties are presented in Table 3 (adapted from [63]).

---

[8] I have omitted the HMXB system LMC X-1 since the radial velocity fits and other arguments, presented by Hutchings et al. [64] [65] for taking a 0.14 $M_\odot$ mass function to the reported 4 $M_\odot$ as lower limit of M_X, are rather speculative. In particular, the fits performed to the emission line radial velocities, used to derive $q = M_C/M_X > 2$, are hardly believable. Further work is clearly needed.

Table 3. Properties of Three Black Hole Candidates

	Cyg X-1	LMC X-3	A0620-00
L_X (erg s^{-1})	2 x 10^{37}	3 x 10^{38}	1 x 10^{38}
MK type	09.7Iab	B3V	K5V
D (kpc)	2.5	55	1
m_v	9	17	18
v_c sin i	76	235	457
P (days)	5.6	1.7	0.32
f (M/M$_\odot$)	0.25	2.3	3.18
M_X/M$_\odot^\dagger$	16	9	13

† These are the likely masses from [66], [67] and [68], respectively.

Due to their HMXB condition, the mass function of the first two systems
is highly dependent on the companion star mass and hence does not provide a
strong constraint on its own. In order to do so it is necessary to invoke additional
arguments related to the inclination (absence of eclipses) and the likely mass of
the optical star as derived from its spectral type. This yields lower limits of 5
M$_\odot$ and 7 M$_\odot$ for Cyg X-1 [69] and LMC X-3 [67] respectively. Unfortunately
it has been found that the OB companions of X-ray binaries are frequently
undermassive for their spectral types [70] which introduces a serious handicap to
the previous discussion. Paczynski [71] proposed a way to sort out the problem
by including the dereddened energy distribution of the companion star. This
provides a robust lower limit of M_X, only dependent on the distance to the
source. Distance estimates for both objects yield lower limits of 3.4 M$_\odot$ and 6
M$_\odot$ respectively. However, Mazeh et al. [72] warned about the possiblity that
important systematic errors were present in the velocity data of LMC X-3 as
a result of the non-negligible contribution of emission from the accretion disc.
These authors finally lowered the M_X limit to 2.5 M$_\odot$.
On the other hand, the limit derived on the LMXB A0620-00 is far more
solid than the previous candidates since it is derived from the mass function
alone without any assumption of the secondary star type or inclination. Furthermore McClintock & Remillard [68] discuss the action of possible contamination
effects on the radial velocity curve (e.g. X-ray heating or partial filling of absorption lines by emission from the accretion disc) and demonstrate that they
are absolutely negligible.
The recent claim by McClintock & Remillard [73] of a 3.1 ± 0.5 M$_\odot$ mass
function for the newly discovered X-ray transient Nova Muscae 1991 (GS1124-
683) seems to add another LMXB black hole system to the previous list. However,
as pointed out by McClintock [63], the mass function of all these systems are
"perilously close to the causality limit (3 M$_\odot$)" and so "the holy grail in the
search for black holes (...) is a system with a mass function that is plainly 5

$M_\odot$ or greater". As we will discover in the following sections this is what makes V404 Cyg (GS2023+338) a very important reference point in the universe of black hole candidates.

5 V404 Cyg in Outburst and Decline

On May 1989 the Japanese satellite *GINGA* discovered a new X-ray source, called GS2023+338 [74]. Its striking X-ray properties were unique and are summarized below [45] [50] [75]:

- The X-ray flux at the peak of the outburst reached 21 Crabs, being the brightest X-ray source ever detected after A0620-00.
- The X-ray flux possibly reached saturation at 21 Crabs (1-37 keV). This corresponds to $L_X \sim 1.5 \ 10^{38}$ (D/1 kpc)^2 erg s^{-1} which easily exceeds the Eddington limit for a 1 $M_\odot$ compact object.
- It showed a strong and chaotic variability (e.g. it once varied by a factor of 200 in matter of minutes).
- On short timescales it showed intense flickering, mimicking the behaviour of Cyg X-1.
- It exhibited a very hard X-ray spectrum, consistent with a single power-law of slope 1.3-1.5. An ultrasoft component, expected from an optically thick accretion disc, was absent.

The *optical counterpart* was found to be an old Nova which outburst in 1938: V404 Cygni [76]. Other outbursts in 1956 and possibly 1979 were also detected following an inspection of photographic plates [77]. The X-ray properties coupled with the dramatic amplitude of the optical outburst (~ 8 magnitudes in B, from 20 to 12 mag.) strongly suggested that V404 Cyg is an X-ray transient LMXB [78].

The first optical spectra [79] [42] showed emission lines (HI, HeI, HeII, FeII, Bowen) superposed on a red continuum (see Fig.6). From the analysis of absorption features the interstellar extinction was estimated to be ~ 3 magnitudes, implying a minimum distance of 1.5 kpc [42]. P-Cygni profiles were also detected in a number of Balmer and HeI lines from June 1-5, suggesting the expansion of a gas shell early in the outburst [80]. In the same work we report a form of S-wave variability in the double-peaked HeII $\lambda 4686$ line both on short ($\sim$ hours) and long ($\sim$ days) timescales. In particular, a detailed analysis of the asymmetric profile observed on the night of June 8 led us to constrain any possible periodicity to $\geq 3^{\text{hr}}$.

Determining the orbital period for a newly discovered X-ray transient is the first step towards understanding the system. Several periods have been proposed in the literature for the outburst and decline phases of V404 Cyg. They range from 10^{min} to 5^{hr} and are summarized below.

- Wagner et al. [79] announced a 10 min. photometric modulation.

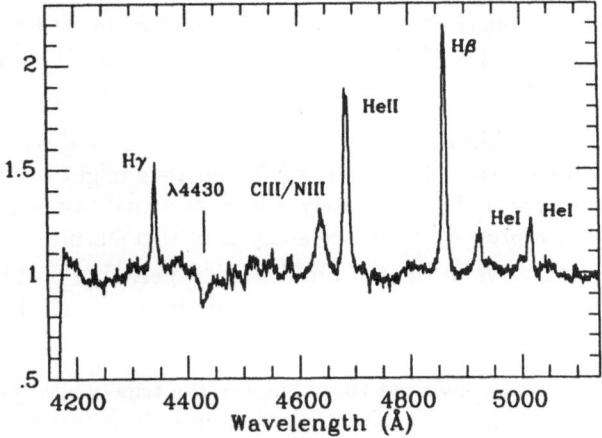

Fig. 6. Sum of 48 high-resolution spectra of V404 Cyg covering the maximum of the 1989 outburst (June 1-11). Main features, as broad emission lines and the interstellar absorption band at $\lambda 4430$, are indicated. From Ref.80.

- Haswell & Shafter [81] reported a 5.0±0.4 hour modulation in the radial velocity of the Na D_1 absorption line.
- Gotthelf et al. [82] performed high-speed photometry and did not detect any stable periodicity. Instead they found *transient modulations* of 3-10 mins and *low-frequency flickering*.
- Leibowitz et al. [83] and Udalsky & Kaluzny [84] claimed a $\sim$ 3 hour R band photometric periodicity.
- Gotthelf et al. [85] observed ≥ 3 hour trends during the radial velocity analysis of the HeII $\lambda 4686$ emission line.

However, such periodicities are rather common amongst X-ray transients during their active phases (e.g. Aql X-1 [86], A0620-00 [87]) and they have subsequently been found to be spurious and unrelated to the true orbital period. Only when the system has returned to quiescence will the secondary star not be masked by the high luminosity of the accretion and, hence, it can then offer the best opportunity for an unambiguous determination of the system parameters.

6 V404 Cyg in Quiescence

Since July 1990 V404 Cyg has settled down in quiescence at R$\sim$16.5 mag [88] with an X-ray flux below or close to the detection threshold of X-ray satellites[9].

---

[9] Very recently though Mineshige et al. [89] have reported the possible detection of V404 Cyg with Ginga during quiescence at $L_X \leq 5 \times 10^{34} \, (D/3\mathrm{kpc})^2$ erg s^{-1}.

In July 1990 we performed high resolution Hα spectroscopy with the 2.5m Isaac Newton Telescope at the Observatorio del Roque de los Muchachos. These were the first optical spectra of V404 Cyg in quiescence and their analysis provided the following results:

- Detection of a possible absorption feature at λ6495 in the summed spectrum. This is a hallmark of G-K stars [90] and thus might be the detection of the companion star. Unfortunately, the poor signal-to-noise ratio of the individual spectra prevented further analysis of this feature.
- Detection of an S-wave in the Hα profile with a period of 0.24^d. This modulation was also observed in the centroid of the line and the continuum flux.

At this point we were convinced that this was the true orbital period on the basis of two supporting facts. Firstly, it agrees perfectly with the maximum in the distribution of LMXB orbital periods [29]. Secondly, it is a stable modulation, both photometrically and spectroscopically. Kato & Hirata [91] reported the same periodicity in I-band photometry conducted just two weeks after our observations. On the other hand, it was possible to isolate a 0.24^d modulation in the Hβ radial velocities from the outburst epoch (see Fig.7). In this latter analysis, we deliberately excluded the first 6 nights after the outburst peak (1-6 June) due to peculiar velocity shifts produced by the P-Cygni profiles (see Casares & Charles [92] for further details).

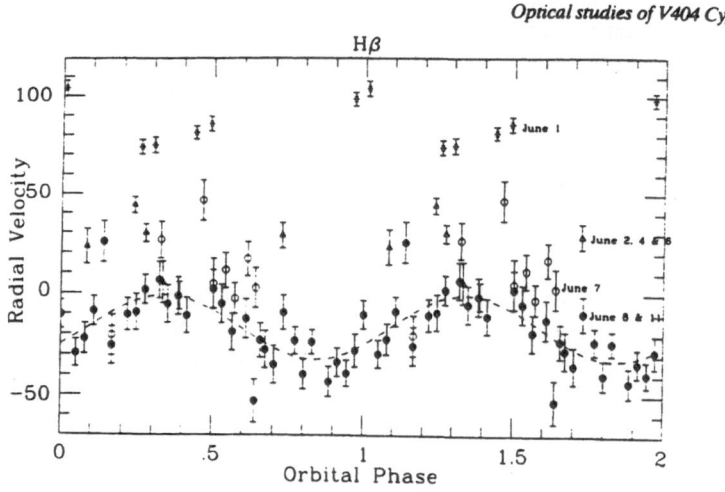

Optical studies of V404 Cyg

Fig. 7. Radial velocities of Hβ through the outburst maximum, folded on the 0.24^d period. The effect of contamination by the P-Cygni profiles is clearly observed for the early days (June 1-6). From Ref.92.

A more complete dataset was acquired one year later with the powerful 4.2m William Herschel Telescope, which also operates at the Observatorio del Roque de los Muchachos. The new database consists of 12 blue ($\lambda\lambda 3500$-5185) and 55 high-resolution red ($\lambda\lambda 6050$-6990) spectra, whose averages are shown in Fig.8.

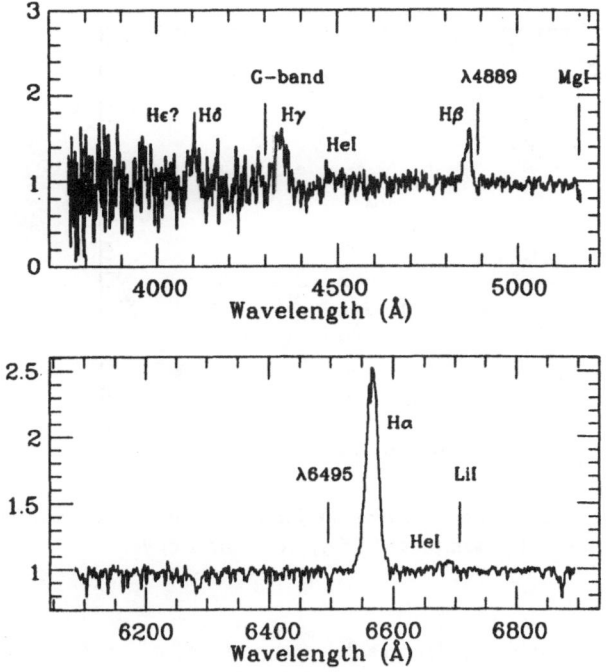

Fig. 8. Overview of V404 Cyg spectral features at quiescence in the blue (top) and red (bottom) optical ranges. Averages were performed in the reference frame of the secondary star (e.g. correcting every single spectrum from its corresponding dopplershift prior to summing) in order to allow the sharp late-type features to be more clearly visible. From Ref.93.

We detect Balmer lines down to Hδ and a number of HeI lines at $\lambda 6678$ and $\lambda 4471$. However, the high excitation lines, HeII $\lambda 4686$ and the Bowen blend, have faded completely, as is to be expected if X-ray emission has switched off. Although some absorption features corresponding to the optical star can be faintly seen in the blue, (i.e. the G-band or the MgI triplet), absorption lines are much more evident in the vicinity of Hα due to the much stronger signal in the red. Every single spectrum was cross-correlated in the range $\lambda\lambda 6350$-6530 with a template star of similar spectral type (61 Cyg A) to derive their relative radial velocities. The subsequent power-spectrum analysis provided a period of 6.473 $\pm$ 0.001 days, 27 times longer than the S-wave periodicity previously reported ! Figure 9 shows the resulting radial velocity curve folded on the 6.5^d period with the least-squares sine fit overlayed. This fit gives a semi-amplitude of 211 $\pm$ 4 km s^{-1}. The implied mass function (a firm lower limit to M_X) is 6.3 $\pm$ 0.3 M$_\odot$, which makes V404 Cyg the strongest black hole candidate yet found since it is almost twice that of Cyg X-1 and A0620-00 (details of this work will be found in [94]).

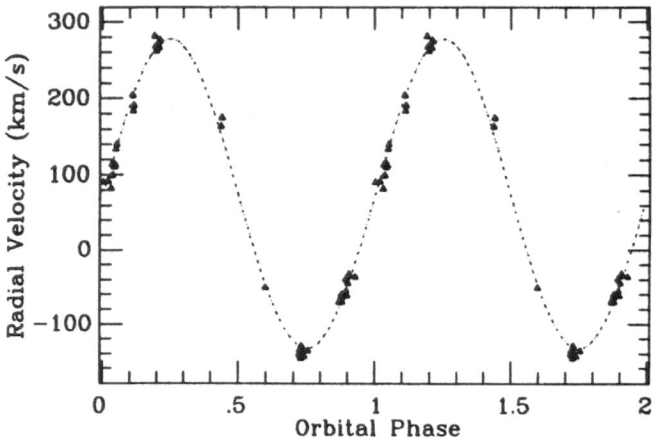

Fig. 9. Radial velocity curve and best sine fit for the secondary of V404 Cyg with respect to the rest frame of the template 61 Cyg A. From Ref.94.

The System Parameters In order to solve completely for the V404 Cyg system parameters, we must first estimate M_c and i. The former parameter can be estimated from the spectral type and luminosity class of the companion star. A quantitative comparison of selected ratios of close line pairs with respect to those listed in a catalogue of main sequence (V) and giant (III) stars provides a likely classification in the range G9V ($\pm$ 1) to K0III ($\pm$ 4). However, the first possibility can be ruled out on the basis of:

- The detection of an ellipsoidal modulation in the light-curve with the 6.5^d orbital period [95]. This implies that the optical star must fill (or nearly fill) its Roche lobe, whereas a G9V would only fill up to 14 percent of the lobe.
- The level of the interstellar extinction found ($A_v \sim 4$ mag., [93]) yields a distance of only 0.5 kpc for a G9V star. This is inconsistent with the reddening studies of Neckel & Klare [96] and Forbes [97], which predict a negligible extinction for distances <0.8 kpc.

Therefore, the optical companion must be a Roche lobe filling K0 III or IV. We favour the latter class since, with a distance of 2.7 kpc, it would be consistent with an Eddington limited L_x at the peak of the outburst (for a 10 $M_\odot$ black hole; see [9]). The estimated mass for the subgiant companion would thus be $M_c \sim 1.2\ M_\odot$ [98].

On the other hand, a solid constraint to the inclination comes from the absence of X-ray eclipses through outburst [45], which implies $i \leq 80°$. Furthermore, a sensible estimate for i can be obtained from the *rotational broadening* (V_{rot}

sin i) of the absorption lines. Our analysis gives $V_{rot} \sin i \leq 35$ km s^{-1} which, combined with the expected value of 55 km s^{-1} for a Roche-lobe filling and corotating K0IV star, implies $i \leq 40°$. The rotational broadening also provides an estimate of the *mass ratio* of the system, $q = M_c/M_X$, through the following expression [99]

$$V_{rot} \ sin \ i \ \sim \ 0.462 \ K_c \ q^{1/3} \ (1 + q)^{2/3} \tag{8}$$

which makes use of (3). Hence we obtain $q \leq 0.04$, which implies an extraordinary ≥ 28 M$_\odot$ black hole for the case of a 1.2 M$_\odot$ K0IV companion.

Further support for such a low inclination comes from the thinness of the Hα emission line when compared with that of the other black-hole LMXB A0620-00. In addition, the extrapolation of Bochkarev et al.'s ellipsoidal models [100], applied to a K0IV companion and $i \sim 40°$, gives an amplitude of ~ 0.2 mag for the ellipsoidal modulation, in perfect agreement with Wagner et al.'s light-curve. We therefore support an intermediate inclination ($i \leq 40°$) for V404 Cyg. The finally proposed parameters are listed in Table 4 (further details will be found in [93]).

Table 4. System Parameters

Period (days)	K_c (km s^{-1})	$V_{rot} \sin i$ (km s^{-1})	q (=M_c/M_X)	i	M_c (M$_\odot$)	M_X (M$_\odot$)	R_c (R$_\odot$)	a (R$_\odot$)
6.473	211	≤ 35	≤ 0.04	$\leq 40°$	1.2	≥ 28	~ 7	~ 45

Final Remarks I will end this paper by briefly referring to two more discoveries which may have important implications.

- The discovery of the LiI $\lambda 6708$ line in the summed spectrum (see Fig.6) is totally unexpected for a late-type star in a supposedly old-population system [101]. This strongly suggests that either LiI is created through spallation reactions during the X-ray outbursts or the system is very young.
- The 0.24^d modulation is also present in photometry and Hα radial velocities of the new database [102] [93]. A possible explanation could involve an inner companion, in which case V404 Cyg would be a triple system. However, this possibility opens the almost insurmountable question of how such a triple system could have evolved to its present state. On the other hand, the anti-phasing between the 0.24^d photometric and radial velocity curves is very puzzling and is currently unexplained.

References

1.   Giaconni, R., Gursky, H., Paoli, F., & Rossi, B., 1962. *Phys. Rev. Lett.*, **9**, 439.
2.   Gursky, H., Giaconni, R., Gorenstein, P., Waters, J.R., Oda, M., Bradt, H.V., Garmire, G. & Sreekantan, B.V., 1966. *Ap.J.*, **144**, 1249.
3.   La Dous, C., 1992. in *Cataclysmic Variables and Related Objects*, ed. M. Hack, NASA/CNRS Monograph Series on Non-Thermal Phenomena in Stellar Atmospheres.
4.   Bath, G.T., 1985. *Rep. Prog. Phys.*, **48**, 483.
5.   Paczynski, B., 1971. *Ann. Rev. Astron. Astrophys.*, **9**, 183.
6.   Eggleton, P.P., 1983., *Ap. J.*, **268**, 368.
7.   Pringle, J.E., 1981. *Ann. Rev. Astron. Astrophys.*, **19**, 137.
8.   Geyer, F., Herold, H. & Ruder, H., 1990. in *Proc. 11th North American Workshop on CVs and LMXBs*. Ed. C.W. Mauche (CUP), p. 307.
9.   Shakura, N.I. & Sunyaev, R.A., 1973. *A.&A.*, **24**, 337.
10.   Smak, J., 1969. *Acta Astr.*, **19**, 155.
11.   Smak, J., 1981. *Acta Astr.*, **31**, 395.
12.   Osaki, Y., 1974. *P.A.S.J.*, **26**, 429.
13.   Meyer, F. & Meyer-Hofmeister, E., 1981. *A.&A.*, **104**, L10.
14.   Bath, G.T., 1973. *Nat. Phys. Sci.*, **246**, 84.
15.   Bath, G.T., Evans, W.D., Papaloizou, J.C.B. & Pringle, J.E., 1974. *MNRAS*, **169**, 447.
16.   Bath, G.T. & Pringle, J.E., 1981. *M.N.R.A.S.*, **194**, 967.
17.   Livio, M. & Verbunt, F., 1988. *M.N.R.A.S.*, **232**, 1p.
18.   Bath, G.T., 1975. *M.N.R.A.S.*, **171**, 311.
19.   van Paradijs, J., 1983. in *Accretion Driven Stellar X-ray Sources*, ed. W.H.G. Lewin and E.P.J. van den Heuvel (Cambridge: CUP), p. 189.
20.   Bradt, H.V.D. & McClintock, J.E., 1983. *Ann. Rev. Astron. Astrophys.*, **21**, 13.
21.   Joss, P.C. & Rappaport, S.A., 1984. *Ann. Rev. Astron. Astrophys.*, **22**, 537.
22.   Ilovaisky, S.A., 1988. Mem. Soc. Astron. It., **59**, 229.
23.   van den Heuvel, E.P.J., 1992. in *X-ray Binaries and the Formation of Binary and Millisecond Radio Pulsars*, eds. E.P.J. van den Heuvel & S.A. Rappaport (Dordrecht), *in press*.
24.   Rappaport, S. & van den Heuvel, E.P.J., 1982. in *Be Stars*, ed. M. Jaschek and H. Groth (Dordrecht: Reidel), p. 327.
25.   van den Heuvel, E.P.J. & Rappaport, S., 1987. in *Physics of Be Stars*, IAU Coll. **92**, p.291. CUP.
26.   Cowley, A.P., Hutchings, J.B., Crampton, D. & Hartwick, F.D.A., 1987. *Ap. J.*, **320**, 296.
27.   Milgrom, M., 1978. *A.& A.*, **67**, L25.
28.   Mason, K.O., 1989. in *Proc. 23rd ESLAB Symp. on Two-Topics in X-Ray Astronomy*, ed. J. Hunt & B. Battrick (ESA Publ. Division), p. 113.
29.   Parmar, A.N. & White, N.E., 1988. *Mem. Soc. Astron. It.*, **59**, 147.
30.   Mason, K.O., Parmar, A.N. & White, N.E., 1985. *M.N.R.A.S*, **216**, 1033.
31.   Motch, C., Pedersen, H., Beuermann,K., Pakull, M.W. & Courvoisier, T.J.-L., 1987. *Ap. J.*, **313**, 792.
32.   Parmar, A.N., White, N.E., Giommi, P. & Gottwald, M., 1986. *Ap. J.*, **308**, 199.
33.   Hellier, C. & Mason, K.O., 1989. *M.N.R.A.S*, **239**, 715.

34. Mason, K.O., 1986. in *The Physics of Accretion onto Compact Objects*, eds. K.O. Mason, M.G. Watson & N.E. White (Springer-Verlag).

35. Hoffman, J.A., Marshall, H.L. & Lewin, W.H.G., 1978. *Nature*, **271**, 630.

36. Lewin, W.H.G., van Paradijs, J. & van der Klis, M., 1988. *Sp. Sci. Rev.*, **46**, 273.

37. Penninx, W.H., 1991. *PhD thesis*, Univ. of Amsterdam, Holland.

38. Alpar, M.A. & Shaham, J., 1985. *Nature*, **316**, 239.

39. van den Heuvel, E.P.J. & Taam, R.E., 1984. *Nature*, **309**, 235.

40. van Paradijs, J. & Lewin, W.H.G., 1986. in *The Evolution of Galactic X-ray Binaries*, eds. J. Truemper et al. (Reidel), p. 187.

41. Grindlay, J.E., 1988. *Adv. Space Res.*, **8**, (2)539.

42. Charles, P.A., 1989. in *Proc. 23rd ESLAB Symp. on Two-Topics in X-Ray Astronomy*, ed. J. Hunt & B. Battrick (ESA Publ. Division), SP-296, p. 129.

43. Callanan, P. & Charles, P.A., 1989. in *Proc. 23rd ESLAB Symp. on Two-Topics in X-Ray Astronomy*, ed. J. Hunt & B. Battrick (ESA Publ. Division), SP-296, p. 139.

44. Liang, E.P. & Nolan, P.L., 1984. *Sp. Sci. Rev.*, **38**, 353.

45. Tanaka, Y., 1989. in *Proc. 23rd ESLAB Symp. on Two-Topics in X-Ray Astronomy*, ed. J. Hunt & B. Battrick (ESA Publ. Division), p. 1.

46. Makishima, K., 1984. in *X-Ray Astronomy '84*. Ed. M. Oda & R. Giacconi. Pub. Institute of Space and Astronautical Science.

47. Thorne, K.S. & Price, R.H., 1975. *Ap. J. Lett.*, **195**, L101.

48. Guilbert, P.W. & Fabian, A.C., 1982. *Nature*, **296**, 226.

49. Inoue, H., 1990. *Adv. Space Res.*, **10**, (2)153.

50. Kitamoto, S., 1989. in *Proc. 23rd ESLAB Symp. on Two-Topics in X-Ray Astronomy*, ed. J. Hunt & B. Battrick (ESA Publ. Division), p. 43.

51. van Paradijs, J., Penninx, W. & Lewin, W.H.G., 1988. *M.N.R.A.S.*, **233**, 437.

52. Lamb, F.K., 1988. *Adv. Space Res.*, **8**, (2)421.

53. Stella, L. et al., 1985. Ap. J. Lett., **288**, L45.

54. Tennant, A. et al., 1986. *M.N.R.A.S.*, **221**, 27p.

55. Honey, W.B., 1989. *PhD thesis*, Univ. of Oxford, UK.

56. Miyamoto, S. & Kitamoto, S., 1985. in *Japan-US Seminar on Galactic and Extragalactic Compact X-ray Sources*. Ed. Y. Tanaka & W.H.G. Lewin, ISAS, p. 187.

57. Cheng, F.H., Horne, K., Panagia, N., Shrader, C.R., Gilmozzi, R., Paresce, F. & Lund, N., 1992. *Ap. J.*, **397**, 664.

58. van den Heuvel, E.P.J., 1992. in *Proceedings of the ISY Conference, "Space Science with particular emphasis on High Energy Astrophysics" Symposium*, Munich, ESA Publ. Division, p. 29.

59. Shapiro, S.L. & Teukolsky, S.A., 1983. *Black Holes, White Dwarfs and Neutron Stars* (New York: Wiley).

60. Rhoades, C.E. & Ruffini, R., 1974. *Phys. Rev. Lett.*, **32**, 324.

61. Friedman, J.L. & Ipser, J.R., 1987. *Ap. J.*, **314**, 594.

62. Bahcall, S., Lynn, B.W. & Selipsky, S.B., 1990. *Ap. J.*, **362**, 251.

63. McClintock, J.E., 1986. in *The Physics of Accretion onto Compact Objects*, eds. K.O. Mason, M.G. Watson & N.E. White (Springer-Verlag), p. 211.

64. Hutchings, J.B., Crampton, D. & Cowley, A.P., 1983. *Ap. J. Lett.*, **275**, L43.

65. Hutchings, J.B., Crampton, D., Cowley, A.P., Bianchi, L. & Thompson, I.B., 1987. *A. J.*, **94**, 340.

66. Gies, D.R. & Bolton, C.T., 1986. *Ap. J.*, **304**, 371.

67. Cowley, A.P., Crampton, D., Hutchings, J.B., Remillard, R. & Penfold, J.E., 1983. *Ap. J.*, **272**, 118.
68. McClintock, J.E., & Remillard, R.A., 1986. *Ap.J.*, **308**, 110.
69. Oda, M., 1977. *Sp. Sci. Rev.*, **20**, 757.
70. van den Heuvel, E.P.J., 1983. in *Accretion Driven Stellar X-ray Sources*, eds. W.H.G. Lewin & E.P.J. van den Heuvel (Cambridge: CUP), p. 303.
71. Paczynski, B., 1974. *A. & A.*, **34**, 161.
72. Mazeh, T., van Paradijs, J., van den Heuvel, E.P.J. & Savonije, G.J., 1986. *A. & A.*, **157**, 113.
73. McClintock, J.E., & Remillard, R.A., 1992. *IAUC* 5499.
74. Makino, F. & the Ginga Team, 1989. *IAUC* 4782.
75. Sunyaev, R.A. et al., 1991. *Sov. Astron. Lett.*, **17**, 123.
76. Marsden, B.G., 1989. *IAUC* 4783.
77. Richter, G.A., 1989. *Inf. Bull. Var. Stars.*, No 3362.
78. Charles, P.A., Hassall, B., Sahu, K., Broadhurst, T., Lawrence, A., Taylor, A., Hacking, P., Jones, D.H.P. & Carter, D., 1989. *IAUC* 4794.
79. Wagner, R.M. et al, 1990. in *Proc. of the 11th North American Workshop on CVs and LMXRBs*, ed. C. Mauche, C.W. CUP, p. 29.
80. Casares, J., Charles, P.A., Jones, D.H.P., Rutten, R.G.M. & Callanan, P.J., 1991. *M.N.R.A.S.*, **250**, 712.
81. Haswell, C.A. & Shafter, A.W., 1990. *IAUC* 5074.
82. Gotthelf, E., Patterson, J. & Stover, R.J., 1991. *Ap.J.*, **374**, 340.
83. Leibowitz, E.M., Ney, A., Drissen, L., Grandchamps, A. & Moffat, A.F.J., 1991. *M.N.R.A.S.*, **250**, 385.
84. Udalsky, A. & Kaluzny, J., 1991. *P.A.S.P*, **103**, 198.
85. Gotthelf, E., Halpern, J.P., Patterson, J. & Stover, R.J., 1992. *A.J.*, **103**, 219.
86. Watson, M.G., 1976. *M.N.R.A.S.*, **176**, 19P.
87. Matilsky, T. et al., 1976. *Ap. J.*, **210**, L127.
88. Wagner, R.M., 1992. *Bull. Am. Astron. Soc.*, **24**, 770.
89. Mineshige, S., Ebisawa, K., Takizawa, M., Tanaka, Y., Hayashida, K., Kitamoto, S., Miyamoto, S. & Terada, K., 1992., *P.A.S.J.*, **44**, 117.
90. Horne, K., Wade, R.A. & Szkody, P., 1986. *M.N.R.A.S.*, **219**, 791.
91. Kato, T. & Hirata, R., 1990. *Inf. Bull. Var. Stars*, 3529.
92. Casares, J. & Charles, P.A., 1992. *M.N.R.A.S.*, **255**, 7.
93. Casares, J., Charles, P.A., Naylor, T. & Pavlenko, E.P., 1993. *M.N.R.A.S.*, submitted.
94. Casares, J., Charles, P.A. & Naylor, T., 1992. *Nature*, **355**, 614.
95. Wagner, R.M., Kreidl, T.J., Howell, S.B. & Starrfield, S.G., 1992. *Nature*, submitted.
96. Neckel, Th. & Klare, G., 1980. *A.&A.*, **42**, 251.
97. Forbes, D., 1985. *A.J.*, **90**, 301.
98. Iben, I.Jr., 1967. *Ap.J.*, **147**, 624.
99. Wade, R.A. & Horne, K., 1988. *Ap.J.*, **324**, 411.
100. Bochkarev, N.G., Karitskaya, E.A. & Shakura, N.I., 1979, *Soviet Astr.*, **23**, 8.
101. Martín, E. L., Rebolo, R., Casares, J. & Charles, P.A., 1992. *Nature*, **358**, 129.
102. Casares, J., Charles, P.A. & Pavlenko, E.P., 1992. in *Proceedings of the ISY Conference, "Space Science with particular emphasis on High Energy Astrophysics" Symposium*, Munich, ESA Publ. Division, p. 147.

General Relativistic Stationary Axisymmetric Rotating Systems

S.Bonazzola, E.Gourgoulhon, M.Salgado

Département d'Astrophysique Relativiste et Cosmologie
Observatoire de Paris, Section de Meudon
F-92195 Meudon Cedex, France

Abstract: Following the 3+1 formalism of General Relativity we write the Einstein equations for a stationary rotating neutron star specially suited to be solved numerically with spectral methods. We also provide the equations of motion to study the normal modes once the star is perturbed from the stationary axisymmetric equilibrium configuration.

1 Introduction

One of the crucial steps for solving a physical problem is the choice of a coordinate system. This is specially important in General Relativity and even more in Numerical Relativity. For the present case of axisymmetric symmetry the natural choice is the polar spherical-type coordinates. Unfortunate ly this natural choice presents natural singularities. These mathematical singularities can be handled by imposing regularity conditions on the physical quantities [1,3]. We will not enter into details at this respect, but it can be proved [1] that the regularity conditions on the shift vector lead to two possible gauges for the metric. Namely the *radial gauge* and the *isothermal gauge*. Some people prefer to impose the radial gauge with a *polar slicing* condition for several reasons [1]. The main one is perhaps because in this gauge the Einstein equations for the shift vector and some metric variables are parabolic equations instead of elliptic equations obtained when the isothermal gauge is chosen. It is well known that these latter are more complicated to solve with the standard numerical methods than the parabolic equations. The problem with the radial gauge and the polar slicing condition is that the lapse function is not regular at the origin $r = 0$. To avoid this trouble some mixed condition has to be adopted and all the procedure become a little bit messy.

As the *spectral methods* have proved to be an excellent recipe to solve elliptic equations [3,4] we have imposed the isothermal gauge to the metric and the *maximal slicing* condition on the extrinsic curvature tensor rather than the other gauge. This approach is more advantageous and clearer than the first one. Actually, as we will see later on, we have made linear combinations on the Einstein equations in order to obtain only elliptic equations for the field variables.

We are also planning to study the normal modes of oscillation of the neutron star when it is perturbed from its equilibrium configuration. For doing so we have written the equations of motion for the small perturbations. This has been done in terms of physical quantities in such a way that the Newtonian and the flat-space limit could be recovered immediatly and the physical interpretation of some of the terms appearing in the equations could be done without difficulty. Furthermore the form of the equations is also convenient to use spectral methods in the numerical solution.

2 IGMS coordinates

The IGMS cordinates (*Isotropic Gauge - Maximal Slicing*), (t, r, θ, ϕ) are defined by the following conditions:

— *isotropic gauge*: the choice of three spatial coordinates (r, θ, ϕ) makes possible to write the spatial 3-metric h_{ij} in terms of three independent components by imposing the isothermal conditions $h_{r\theta} = h_{r\phi} = 0$, $h_{\theta\theta} = r^2 h_{rr}$.

— *maximal slicing*: the space-time foliation by $t = $ const. hypersurfaces, Σ_t, is such that the extrinsic curvature tensor of these hypersurfaces satisfies $\mathrm{Tr}\,\mathbf{K} = K_i{}^i = 0$.

The stationary axisymmetry is defined by the existence of two Killing vectors which commute: $\mathbf{e}_t = \frac{\partial}{\partial t}$ and $\mathbf{e}_\phi = \frac{\partial}{\partial \phi}$. In this case the shift vector, N^i, and the spatial 3-metric $\mathbf{h}$ verify [6]: $N^r = N^\theta = h_{\theta\phi} = 0$.

Consequently the 4-metric $\mathbf{g}$ reads:

$$g_{\alpha\beta} dx^\alpha dx^\beta = -N^2 dt^2 + A^4 B^2 r^2 \sin^2 \theta (d\phi - N^\phi dt)^2 + \frac{A^4}{B^2}(dr^2 + r^2 d\theta^2) \quad (1.1)$$

where N, N^ϕ, A, B are only functions of r, θ.

3 The field equations

We shall consider the neutron star matter as a perfect fluid. Then the energy-momentum tensor will be $\mathbf{T} = (e + p)\mathbf{u} \otimes \mathbf{u} + p\mathbf{g}$, being e, p the energy density and the presure of the fluid in the comoving frame respectively and $\mathbf{u}$ the fluid's 4-velocity.

In the stationary case $\mathbf{u}$ is a linear combination of the two Killing vectors. This means that: $u^r = 0$ and $u^\theta = 0$.

We also introduce the physical velocity in the ϕ-direction as follows: $U^\phi = A^2 B r \sin\theta \left(\Omega - N^\phi\right)/N$, where Ω is the coordinate angular velocity defined as $\Omega := u^\phi/u^t$. Finally we define a physical energy density as $E = \Gamma^2(e + p) - p$, being $\Gamma := \left(1 - U^{\phi 2}\right)^{-\frac{1}{2}}$, the Lorentz factor.

We will not give the curbersome details to arrive to the Einstein equations for the field variables; these could be seen in a paper to appear in the future. For instance we will just give the final form of the equations which are well suited for using our spectral methods library to find their numerical solution.

Let us define the following useful variables: $\xi(r, \theta) := \ln N(r, \theta)$, $\alpha(r, \theta) := \ln A(r, \theta)$, $\beta(r, \theta) := \ln B(r, \theta)$

3.1 Momentum constraint equation

From the three momentum constraint equations there is just one non-trivial equation for the shift component N^ϕ

$$\triangle \tilde{N}^\phi - \frac{1}{r^2 \sin^2\theta} \tilde{N}^\phi = -16\pi \frac{A^2 N}{B^3}(E + p)U^\phi - r\sin\theta \nabla N^\phi \cdot \nabla(6\alpha + 3\beta - \xi) \quad (3.1)$$

where $\tilde{N}^\phi := r\sin\theta\, N^\phi$. We can see on the left-hand side of equation (3.1) a pure linear expression where we can appreciate the flat-space Laplacian operator, $\triangle$ (to be taken in spherical coordinates). On the right-hand side there are two non-linear terms (the "source"). The first couples the field with the fluid and the second is just a scalar product in the flat space between two flat-space gradients. Outside the star the term coupling the fluid and the field vanishes.

3.2 Dynamic Einstein Equations

After making a convenient linear combination on the Einstein equations we obtain three elliptic equations for the three field variables left . The first one is the equation for the lapse function

$$\Delta \xi = \frac{A^4}{B^2} \left\{ 4\pi \left[E + 3p + (U^\phi)^2 (E + p) \right] + 2(k_1^2 + k_2^2) \right\} - \nabla \xi \cdot \nabla (\xi + 2\alpha + \beta) \quad (3.2)$$

where $k_1 = -\frac{B^2 r \sin\theta}{2N} \frac{\partial N^\phi}{\partial r}$, $k_2 = -\frac{B^2 \sin\theta}{2N} \frac{\partial N^\phi}{\partial \theta}$. In terms of the extrinsic curvature components $k_1 = \frac{K_{r\phi}}{A^4 r \sin\theta} = \frac{K_\phi{}^r}{B^2 r \sin\theta}$, $k_2 = \frac{K_{\theta\phi}}{A^4 r^2 \sin\theta} = \frac{K_\phi{}^\theta}{B^2 \sin\theta}$. Equation (3.2) has the same structure as equation (3.1). On the left-hand side we see the flat-space Laplacian operator whereas the non-linear terms appear in the right-hand side. There are the fluid-field coupling term and a high non-linear term coupling the conformal factor A and the B variable with the only non-zero components of the extrinsic curvature tensor k_1 , k_2. Finally a quadratic term is the flat-space scalar product between the flat-space gradient of ξ and the gradient of a li near combination of the metric functions.

The simplest field equation is the following one

$$\frac{\partial^2 \tilde{Q}}{\partial r^2} + \frac{1}{r} \frac{\partial \tilde{Q}}{\partial r} + \frac{1}{r^2} \frac{\partial^2 \tilde{Q}}{\partial \theta^2} = \frac{A^4}{B^2} 16\pi p \tilde{Q} \ . \quad (3.3)$$

where $\tilde{Q} := r \sin\theta N A^2 B$. On the left-hand side of equation (3.3) we recognize the flat-space lapacian operator in two dimensions. The "source" is just a fluid-field coupling term which vanishes outside the star. In vacuum the solution of the resulting homogeneous linear equation is well know.

The last field equation reads as follows:

$$\frac{\partial^2 H}{\partial r^2} + \frac{1}{r} \frac{\partial H}{\partial r} + \frac{1}{r^2} \frac{\partial^2 H}{\partial \theta^2} = \mathcal{S}$$

where $H := \xi + 2\alpha - \beta$ and $\mathcal{S} = \frac{A^4}{B^2} \left[8\pi \left(p + (E + p)(U^\phi)^2 \right) + 3(k_1^2 + k_2^2) \right] - (\nabla \xi)^2$.

As we have pointed out we are planning to solve numerically these set of field equations in addition to the fluid equations of motion (see next section) by using spectral methods [4]. In this reference it could be found a detailed description of how to use these numerical methods to invert the Laplacian operators. At infinity we will impose to the metric the flat space-time condition (the field variables A, B, N must go to 1 and the shift N^ϕ vanishes).

4 Equations of motion

4.1 Rigidly rotating model

When the coordinate angular velocity Ω is constant, the equations of motion $(T^{\mu\nu}{}_{;\nu} = 0)$ and the first law of thermodynamics at zero temperature lead us to a first integral $H - \frac{1}{2}\ln\left[N^2 - A^4B^2r^2\sin^2\theta\left(\Omega - N^\phi\right)^2\right] = \text{constant}$, where the pseudo-enthalpy H is defined in terms of the heat function f. This equation can be recognized as equation (6) of Bonazzola and Schneider [2].

4.2 Differentially rotating model

If the coordinate angular velocity is a function of r, θ the differential rotating model is specified by giving the *law* of rotation $F(\Omega)$ into the equation $F(\Omega) + \frac{A^4B^2r^2\sin^2\theta(\Omega-N^\phi)}{N^2-A^4B^2r^2\sin^2\theta(\Omega-N^\phi)^2} = 0$. The value of Ω is obtained by solving the last equation in Ω. The equations of motion and the first law of thermodynamics then reduce to $\frac{Nf}{T}\exp\left(\int_{\Omega_0}^{\Omega} F(\Omega')d\Omega'\right) = \text{const.}$

One of this rotating models within the field equations and the equation of state describe completely the axisymmetric rotating neutron star and the shape of the surrounding space-time. This configuration will be considered as the neutron star equilibrium configuration. Once the whole set of equilibrium equations have been solved we can study the small oscillations of the star around the equilibrium position and consequently the normal modes of the star.

5 Normal modes of oscillation

To obtain the equations of motion for the small oscillations of the star we have used the same perfect-fluid energy-momentum tensor with non-zero $r-$ and θ- velocity components and without considering any symmetry. By this we mean that all but the metric variables have been considered as functions of r, θ, ϕ, and t. Then we have linearized the equations around the equilibrium in which the equilibrium variables have of course only an r and θ dependence while the perturbations are those variables possessing no symmetry, i.e., they depend on time and the three spatial coordinates. This approximation means physically that the perturbations are small enough such the space-time is not modified by the tiny oscillations of the star. The limit of validity of this approach breaks down when the perturbations become as large as the equilibrium quantities.

Here we just give the non-linearized version of the equations, the linearized form could be seen in a more detailed future paper [5].

5.1 Equation of motion for the normal modes

$$
\frac{\partial U^i}{\partial t} + (\mathbf{V} \cdot \nabla)U^i = -\frac{1}{(\mathcal{E}+p)}\left\{ U^i\frac{\partial p}{\partial t} + (Nq^{ji} + U^iN^j)\frac{\partial p}{\partial x^j}\right\} + \mathcal{F}
$$

where $\mathcal{F} := NU^aU^b\left[U^i\mathcal{O}^0_{ba} - \mathcal{O}^i_{ba}\right]$ are the "inertial forces" (centrifugal force, Coriolis' force and general relativistic effects) given in terms of the Ricci rotation coefficients $\mathcal{O}^c_{ba}$; $\mathbf{q}$ is the matrix transformation between the coordinate and the orthonormal basis $\mathbf{e}_\mu$ and $\mathbf{e}_{(a)}$ respectively [1,5]; $\mathbf{U}$ is the physical velocity defined as $\mathbf{U} := -\mathbf{u}\cdot\mathbf{e}_{(a)}/(\mathbf{u}\cdot\mathbf{e}_{(0)})$; the operator $\mathbf{V} \cdot \nabla$ means $V^i\frac{\partial}{\partial x^i}$, where $V^i := dx^i/dt$. Here the i index takes the values $1, 2, 3$, the other indexes go from zero to three.

In equilibrium all the conservation equations are satisfied identically. However when the star fluid is perturbed the thermodynamical variables evolve in time. So to close the system of equations the conservation equations for the perturbed thermodynamical variables such as energy, entropy, baryon number, fraction of particles and enthalpy have to be taken into account [5,7].

6 Conclusion

We have showed how the isothermal gauge and the maximal slicing condition lead to non-linear elliptic equations for all the field variables which are perfectly treatable with spectral methods. By using these we hope not only find the standard results concerning the star radius and mass limits for a given equation of state but also push for the first time all the numerical calculations from the center of the star to "infinity" where the asymptotic flat space-time condition has to be imposed. Having these results, it will not be too difficult to study the star normal modes of oscillation and the stability of this compact object.

References

1.   J.M. Bardeen, T. Piran: Phys. Rep. **96** 205 (1965)
2.   S. Bonazzola, J. Schneider: Astrophys. J. **191** 273 (1974)
3.   S. Bonazzola, J.A. Marck: Proceedings of the Frontiers in Numerical Relativity Conference (1988)
4.   S. Bonazzola, J.A. Marck: Journ.Comp.Phys. **87** 201 (1990)
5.   S. Bonazzola, E. Gourgoulhon, M.Salgado: to be published
6.   B. Carter: J. Math. Phys. **10** 70 (1969)
7.   E. Gourgoulhon: PhD thesis, Université Paris 7 (1992)

The Superposition of Two Kerr-Newman Solutions

A. Chamorro [1], V. S. Manko [1] and N. R. Sibgatullin [2]

[1] Dpto. Física Teórica, Universidad del País Vasco, Bilbao, Spain
[2] Department of Hydrodynamics, Moscow State University,
Moscow 119899, Russia

Abstract: An exact asymptotically flat 8-parameter solution of the Ernst equations which may represent the exterior gravitational field of two charged spinning masses located on the symmetry axis is obtained. Its particular cases would be the new double Kerr solution and the double Reissner-Nordstrom solution.

Recently the method [1] developed by one of the authors for the construction of exact solutions of the Ernst equations has been successfully applied to finding the first examples of asymptotically flat metrics able to describe the exterior field of a magnetized rotating source [2]. The aim of the present communication is to show how this method could be used to construct a solution corresponding to another astrophysically interesting situation. That solution might be interpreted as describing the field of two Kerr-Newman masses located on the symmetry axis. It should be mentioned that earlier this problem was considered only for some special cases of the rotating charged sources [3], and also for the sources restricted beyond the extreme case [4]. The solution which will be considered below contains eight independent parameters corresponding to the masses, angular momentum, electric charges and locations of the sources on the symmetry axis, and it may be helpful for the analysis of more general cases of two interacting Kerr-Newman sources.

According to the method which we are using, the complex potentials ε and Φ satisfying the Ernst equations [5] can be found from the integrals

$$\varepsilon = \frac{1}{\pi} \int_{-1}^{1} \frac{\mu(\sigma)e(\xi)d\sigma}{\sqrt{1-\sigma^2}}; \quad \Phi = \frac{1}{\pi} \int_{-1}^{1} \frac{\mu(\sigma)f(\xi)d\sigma}{\sqrt{1-\sigma^2}} \tag{1}$$

where the unknown function $\mu(\sigma)$ satisfies the integral equations

$$\int_{-1}^{1} \frac{[e(\xi) + \tilde{e}(\eta) + 2f(\xi)\tilde{f}(\eta)]\mu(\sigma)d\sigma}{\sqrt{1-\sigma^2}} = 0; \quad \int_{-1}^{1} \frac{\mu(\sigma)d\sigma}{\sqrt{1-\sigma^2}} = \pi \tag{2}$$

Here $\xi \equiv z + i\rho\sigma$, $\eta \equiv z + i\rho\tau$, $(\sigma, \tau \in [-1,1])$; $e(\xi)$ and $f(\xi)$ are the locally holomorphic continuations of the functions $e(z) \equiv \varepsilon(\rho = 0, z)$, $f(z) \equiv \Phi(\rho = 0, z)$

into the complex plane $z + i\rho$ ((ρ, z) are the Weyl canonical coordinates which enter into (1), (2) as parameters); $\tilde{e}(\eta) \equiv [e(\eta^*)]^*$, $\tilde{f}(\eta) \equiv [f(\eta^*)]^*$ and f denotes the principal value of the respective integral.

For arbitrary rational functions $e(z)$ and $f(z)$ the corresponding $\mu(\sigma)$ should be searched in the form

$$\mu(\sigma) = A_0 + \sum_{k=1}^{m_1} A_1^k(\xi - \xi_1)^{-k} + \ldots + \sum_{k=1}^{m_n} A_n^k(\xi - \xi_n)^{-k} \qquad (3)$$

where the coefficients A_0, A_i^k depending on ρ and z have to be found from (2), the ξ_i being the roots of the equation

$$e(\xi) + \tilde{e}(\xi) + 2f(\xi)\tilde{f}(\xi) = 0 \qquad (4)$$

with corresponding multiplicity $m_i(i = 1, \ldots, n)$.

Tentatively a plausible form for the Ernst potentials ε and Φ on the symmetry axis ($\rho = 0$) leading to the double Kerr-Newman solution is the following

$$e(z) = 1 - \frac{2m_1}{z + z_1 - ia_1} - \frac{2m_2}{z + z_2 - ia_2},$$

$$f(z) = \frac{q_1}{z + z_1 - ia_1} + \frac{q_2}{z + z_2 - ia_2}, \qquad (5)$$

where the parameters $m_{1,2}, a_{1,2}, q_{1,2}$ respectively describe the mass, angular momentum per unit mass and eletric charge of each source, and the parameters $z_{1,2}$ are related to the location of the sources on the symmetry axis (note, that in the case $m_2 = q_2 = 0$ [or $m_1 = q_1 = 0$] and the subsequent choice $z_1 = m_1$ [or $z_2 = m_2$], formulae (5) give the $e(z)$ and $f(z)$ of the Kerr-Newman solution [6]).

From (4) and (5) it then follows that the function $\mu(\sigma)$ should be looked for in the form

$$\mu(\sigma) = A_0 + \sum_{n=1}^{4} A_n(\xi - \alpha_n)^{-1}, \qquad (6)$$

where α_n are the roots of the algebraic equation of the fourth order

$$\xi^4 + 2(z_1 + z_2 - m_1 - m_2)\xi^3 + \left(a_1^2 + a_2^2 + (z_1 + z_2)^2 + 2z_1z_2 + (q_1 + q_2)^2\right.$$

$$-2\left(m_1(z_1 + 2z_2) + m_2(2z_1 + z_2)\right])\xi^2 + 2[(z_1 - m_1)(a_2^2 + z_2^2) + (z_2 - m_2)(a_1^2 + z_1^2)$$

$$-2z_1z_2(m_1 + m_2) + (q_1 + q_2)(z_1q_2 + z_2q_1))\xi + (a_1^2 + z_1^2)(a_2^2 + z_2^2)$$

$$-2[m_1z_1(a_2^2 + z_2^2) + m_2z_2(a_1^2 + z_1^2)] + (a_1q_2 + a_2q_1)^2 + (z_1q_2 + z_2q_1)^2 = 0. \qquad (7)$$

It should be mentioned that in (6) we have assumed that all α_n are different; however, in particular cases some of the roots α_n may become multiple. That may require the performance of limiting transformations in the final expressions for ε and Φ with the aid of l'Hospital rule. The coefficients A_0 and A_n may be obtained from (2), that after integration yields the following elegant system of five linear algebraic equations

$$A_0 + \sum_{n=1}^{4} \frac{A_n}{r_n} = 1, \quad A_0 - \sum_{n=1}^{4} \frac{A_n}{\alpha_n + z_1 - ia_1} = 0, \quad A_0 - \sum_{n=1}^{4} \frac{A_n}{\alpha_n + z_2 - ia_2} = 0,$$

$$\sum_{n=1}^{4} \left(-m_1 + \frac{q_1^2}{\alpha_n + z_1 - ia_1} + \frac{q_2 q_1}{\alpha_n + z_2 - ia_2} \right) \frac{A_n}{(\alpha_n + z_1 + ia_1)r_n} = 0, \quad (8)$$

$$\sum_{n=1}^{4} \left(-m_2 + \frac{q_2^2}{\alpha_n + z_2 - ia_2} + \frac{q_2 q_1}{\alpha_n + z_1 - ia_1} \right) \frac{A_n}{(\alpha_n + z_2 + ia_2)r_n} = 0,$$

where $r_n \equiv [\rho^2 + (z - \alpha_n)^2]^{\frac{1}{2}}$.

Finally, from (1) we find the form of the potentials ε and Φ in terms of A_n

$$\varepsilon = 1 - 2 \sum_{n=1}^{4} \left(\frac{m_1}{\alpha_n + z_1 - ia_1} + \frac{m_2}{\alpha_n + z_2 - ia_2} \right) \frac{A_n}{r_n},$$

$$\Phi = \sum_{n=1}^{4} \left(\frac{q_1}{\alpha_n + z_1 - ia_1} + \frac{q_2}{\alpha_n + z_2 - ia_2} \right) \frac{A_n}{r_n}. \quad (9)$$

Formulae (9) together with system (8) and (7) fully determine the new stationary electrovacuum solution which may describe a superposition of two Kerr-Newman sources. The first three Simon's relativistic multipole moments [7] of this solution calculated with the aid of the Hoenselaers-Perjés procedure [8] are found to have the form (M_i, J_i, E_i and B_i respectively describe the distributions of mass, angular momentum, electric charge and magnetic moment)

$$M_0 = m_1 + m_2; \quad M_1 = (m_1 + m_2)^2 - m_1 z_1 - m_2 z_2;$$

$$M_2 = (m_1 + m_2)^3 - m_1(z_1^2 - a_1^2) + m_2(z_2^2 - a_2^2) - 2m_1 m_2(z_1 + z_2) - 2(m_1^2 z_1 + m_2^2 z_2);$$

$$J_0 = 0; \quad J_1 = m_1 a_1 + m_2 a_2;$$

$$J_2 = 2[(m_1 + m_2)(m_1 a_1 + m_2 a_2) - m_1 a_1 z_1 - m_2 a_2 z_2];$$

$$E_0 = q_1 + q_2; \quad E_1 = (m_1 + m_2)(q_1 + q_2) - q_1 z_1 - q_2 z_2; \quad (10)$$

$$E_2 = (q_1 + q_2)(m_1 + m_2)^2 - (z_1 + z_2)(m_1 q_2 + m_2 q_1) +$$
$$q_1(z_1^2 - a_1^2 - 2m_1 z_1) + q_2(z_2^2 - a_2^2 - 2m_2 z_2);$$

$$B_0 = 0; \quad B_1 = a_1 q_1 + a_2 q_2;$$

$$B_2 = (a_1 + a_2)(m_1 q_2 + m_2 q_1) + 2a_1 q_1(m_1 - z_1) + 2a_2 q_2(m_2 - z_2),$$

from where one can see that the solution obtained is asymptotically flat ($J_0 = 0$), and that the physical meaning of the parameters is the one previously mentioned. An interesting feature of the solution is that its total angular momentum defined by J_1 in (10) becomes zero when $a_2 = -a_1 \frac{m_1}{m_2}$ in which case the system apparently is endowed with differential rotation.

The main limiting cases of the solution are the following :

In the absence of electric charges ($q_1 = q_2 = 0$) it reduces to a new solution that would represent the superposition of two Kerr masses in which all parameters have a clear physical interpretation.

When the rotation parameters are equal to zero ($a_1 = a_2 = 0$), one comes to the solution describing the exterior field of two Reissner-Norsdtrom sources located on the symmetry axis.

The particular one black hole solutions could be obtained from the general formulae by performing the respective limits in α_n [(7) admits simple roots for these special cases].

A detailed analysis of the physical properties of our solution together with the explicit form of the coefficients A_0, A_n and of the corresponding metric functions will be published elsewhere [9].

Acknowledgements

This work has been supported by Contract No. 172.310-E132/91 from the Universidad del País Vasco (A. Ch.) and by Dirección General de Investigación Científica y Técnica of Spain (V.S.M.).

References

1.   N. R. Sibgatullin: "Oscillations and Waves in Strong Gravitational and Electromagnetic Fields" (Nauka, Moscow) 1984; English translation: Springer-Verlag, 1991
2.   V. S. Manko and N. R. Sibgatullin: Class. Quantum Grav. 9 L87 (1992); Phys. Lett. A (to be published).
3.   L. Parker, R. Ruffini and D. Wilkins: Phys. Rev. D 7 2874 (1973)
4.   G. A. Aleksejev: Abstracts GR-11, 227 (1986)
5.   F. J. Ernst: Phys. Rev 168 1415 (1968)
6.   E. T. Newman, E. Couch, K. Chinnapared, A. Exton, A. Prakash and R. Torrence: J. Math. Phys. 6 918 (1965)
7.   W. Simon: J. Math. Phys. 25 1053 (1984)
8.   C. Hoenselaers and Z. Perjés: Class. Quantum Grav. 7 1819 (1990)
9.   A. Chamorro, V. S. Manko and N. R. Sibgatullin: in preparation.

Stationary Black Holes Rotate Differentially

L. Herrera and V. S. Manko

Dpto. Física Teórica, Universidad del País Vasco, Bilbao, Spain

Abstract: It is shown through the discovery of the origin of relativistic rotational multipole moments that the sources corresponding to the stationary black hole solutions should exhibit differential rotation

In the paper [1] the first example of an asymptotically flat stationary vacuum metric possessing zero total angular momentum has been obtained with the aid of Sibgatullin's method [2] for the construction of exact solutions of the Ernst equation [3]. The investigation of the physical properties of that metric in the weak field and slow rotation approximation allowed us to make the conclusion that any possible source for that exterior solution should rotate differentially.

An important generalization of this result is that any axisymmetric solution with zero total angular momentum corresponding to the dipole rotational relativistic moment [4, 5], and nonzero higher rotational moments also exhibits differential rotation. For an arbitrary axisymmetric stationary solution possessing zero total angular momentum, the imaginary part of its Ernst potential ε [3] defining the angular momentum distribution can be expressed in the form (on the symmetry axis $\rho = 0$),

$$Im\varepsilon(\rho = 0) = \sum_{n=3}^{\infty} \alpha_n z^{-n}$$

where ρ and z are the Weyl-Papapetrou cylindrical coordinates, and α_n are arbitrary real constants. It follows then that the spinning mass solutions possessing only one rotational Geroch-Hansen multipole moment of order higher than the dipole one (the existence of solutions of this type follows from the Hauser-Ernst proof [6] of the well known Geroch conjecture [7]) are contained in the above equation as special cases corresponding to the particular choice of the parameters α_n. Since the massive sources for such solutions should be the differentially rotating ones, apparently becoming static in the absence of the rotational multipole moment, one immediately arrives at the conclusion about the general nature of the relativistic rotational multipole moments: they determine the differential rotation of a massive source.

The main consequence of this result is that the sources of the Kerr [8] and Kerr-Newman [9] black hole metrics should exhibit differential rotation due to the presence in these metrics of infinite sets of relativistic rotational moments.

The following conjecture may also be put forward [10] :

Any compact rigidly rotating massive source is characterised by the only dipole Geroch rotational relativistic moment corresponding to its total angular momentum. Other rotational multipole moments higher than the dipole one describe the deviations of the source from the rigidly rotating case.

References

1.   L. Herrera and V. S. Manko: Phys. Lett. A **167** 238 (1992)
2.   N. R. Sibgatullin: "Oscillations and Waves in Strong Gravitational and Electromagnetic Fields" (Nauka, Moscow) 1984; English translation: Springer-Verlag, 1991
3.   F. J. Ernst: Phys. Rev. **167** 1175 (1968)
4.   R. Geroch: J. Math. Phys. **11** 2580 (1970)
5.   R. O. Hansen: J. Math. Phys. **15** 46 (1974)
6.   I. Hauser and F. J. Ernst: J. Math. Phys. **22** 1051 (1981)
7.   R. Geroch: J. Math. Phys. **13** 394 (1972)
8.   R. P. Kerr: Phys. Rev. Lett. **11** 237 (1963)
9.   E. T. Newman, E. Couch, K. Chinnapared, A. Exton, A. Prakash and R. Torrence: J. Math. Phys. **6** 918 (1965)
10.   L. Herrera and V. S. Manko: Phys. Rev. Lett. (submitted).

Differentially Rotating Perfect Fluids

F.J. Chinea

Departamento de Física Teórica II, Facultad de Ciencias Físicas,
Universidad Complutense, 28040-Madrid, Spain

Abstract: Recent work on exact interior gravitational fields corresponding to stationary, axisymmetric perfect fluid bodies with differential rotation is described.

1 Differential Rotation

Due to its astrophysical implications, it is of interest to consider axisymmetric bodies of perfect fluid in stationary rotation, within the context of general relativity. If additional symmetries are excluded, very few interior exact solutions of this type are known (such extra symmetries are usually not acceptable if one is dealing with compact bodies, as they might imply that the body is cylinder- or slab-like); at present, there is no exact solution of such a restricted type describing both the interior of the fluid and its matching exterior vacuum field.

Among the very few known interior solutions of the type described, until recently all correspond to *rigidly rotating* bodies (*i.e.* masses of fluid with vanishing shear). On the other hand, the effect of differential rotation cannot be ignored if one is trying to construct realistic stellar models; in particular, differential rotation is expected to play a significant role in collapse and other dynamical processes. The interest of differential rotation in astrophysics has increased recently [1-4].

Interior fields corresponding to stationary, axisymmetric perfect fluids with differential rotation and no further isometries have been found quite recently [5-7]. In the following, I shall describe briefly the formalism used for constructing the solutions of references [5] and [6] (Sect. 2) and the main features of the solutions (Sect. 3).

2 The Equations

The orthonormal tetrad $\{u, \theta^1, \theta^2, \theta^3\}$ (where u is the 1-form corresponding to the fluid four-velocity u) is used. The symmetries correspond to the timelike Killing field ∂_t and the spacelike one ∂_ϕ; u is assumed to lie in the space generated by the Killing fields, so that the fluid motion is azimuthal. The equations to be solved are [8]:

$$du = a \wedge u + w \wedge \theta^1 \tag{2.1}$$

$$d\theta^1 = (b - a) \wedge \theta^1 + s \wedge u \tag{2.2}$$

$$d\theta^2 = -\nu \wedge \theta^3 \tag{2.3}$$

$$d\theta^3 = \nu \wedge \theta^2 \tag{2.4}$$

$$da = w \wedge s \tag{2.5}$$

$$db = 0 \tag{2.6}$$

$$dw = -(b - 2a) \wedge w \tag{2.7}$$

$$ds = (b - 2a) \wedge s \tag{2.8}$$

$$d * (w - s) + 2a \wedge *w + 2(a - b) \wedge *s = 0 \tag{2.9}$$

$$d * a + b \wedge *a + \frac{1}{2}w \wedge *w - \frac{1}{2}s \wedge *s = \frac{1}{2}(\mu + 3p)\theta^2 \wedge \theta^3 \tag{2.10}$$

$$d * b + b \wedge *b = 2p\theta^2 \wedge \theta^3 \tag{2.11}$$

$$d\nu + a \wedge *b - a \wedge *a + \frac{1}{4}(s - w) \wedge *(s - w) = \frac{1}{2}(\mu + p)\theta^2 \wedge \theta^3 \tag{2.12}$$

$$d\tilde{b} + b \wedge \tilde{b} - 2b \wedge \tilde{a} + 2a \wedge \tilde{a} - \frac{1}{2}(s - w) \wedge (\tilde{s} - \tilde{w}) + 2\nu \wedge *\tilde{b} = 0 \tag{2.13}$$

$$d * \tilde{b} + b \wedge *\tilde{b} - 2b \wedge *\tilde{a} + 2a \wedge *\tilde{a} - \frac{1}{2}(s - w) \wedge *(\tilde{s} - \tilde{w}) - 2\nu \wedge \tilde{b} = 0 \tag{2.14}$$

$$dp + (\mu + p)a = 0. \tag{2.15}$$

An explanation of the quantities and operators introduced in the above set of differential-form equations is in order. d and $\wedge$ have their usual meaning as exterior derivative and exterior product of forms. u and θ^1 are 1-forms in dt and $d\phi$, while all the other 1-forms that enter into the equations are linear superpositions of the differentials of the two remaining spacetime coordinates. The coefficients of all the forms, including u and θ^1, are functions of those two remaining coordinates. The asterisk denotes the Hodge dual in the θ^2-θ^3 subspace. It is defined by

$$*\theta^2 = \theta^3, \quad *\theta^3 = -\theta^2.$$

The tilde operation is also defined in the θ^2-θ^3 subspace, by

$$\tilde{\theta}^2 = \theta^2, \quad \tilde{\theta}^3 = -\theta^3,$$

and extended by linearity. While the Hodge dual $*$ is intrinsically defined, the definition given for $\tilde{}$ depends on the basis chosen in the θ^2-θ^3 subspace; as θ^2 and θ^3 are determined only up to a gauge rotation among themselves, it does not seem immediately clear that the equations as written above are independent of the choice of θ^2 and θ^3 (and the corresponding $\tilde{}$). However, a simple calculation shows that the system of equations written above is *invariant* under gauge rotations (some equations are transformed into themselves, while others are transformed into a linear superposition with constant coefficients of some equations in the set).

The 1-form a is the acceleration of the fluid, while the *roticity* w and the *deformity* s are related to the standard shear (σ) and vorticity (ω_T) tensors of the fluid by the following relations:

$$\sigma = \theta^1 \otimes_s s, \quad \omega_T = \theta^1 \wedge w; \tag{2.16}$$

the relation among w and the vorticity 1-form ω (*i.e.* the vorticity *vector* expressed in its covariant rather than in its contravariant form) is quite simple:

$$w = *\omega. \tag{2.17}$$

The scalar quantities p and μ are, respectively, the pressure and the energy density of the fluid. The 1-form b generalizes to the the non-vacuum case the gradient of the logarithm of the Weyl cylindrical radial coordinate: In vacuum ($p = 0$, $\mu = 0$) one has $b = \rho^{-1}d\rho$, where ρ is the mentioned cylindrical coordinate. Notice, however, that although b still defines a certain coordinate by integrating (2.6), the corresponding coordinate is no longer harmonic in the fluid interior. Finally, ν is the connection 1-form in the θ^2 -θ^3 subspace, as witnessed by (2.3-4).

The origin of the equations is the following: Equations (2.1-4) are Cartan's first structure equations, expressing here the fact that the torsion vanishes. Equations (2.5-8) are the Bianchi first identities, *i.e.* the integrability conditions for the previous equations (2.1-4). Equations(2.9-14) are Einstein's field equations for the present case. Finally, the contracted second Bianchi identities reduce to the Euler equation (2.15) for the fluid.

An important technical point is that (2.13-14) can be solved algebraically for ν, which can then be substituded into (2.3-4). By using the gauge freedom corresponding to rotations among θ^2 and θ^3, one can align θ^2 in the direction of b. The resulting equations are given in detail in [8].

The set of equations (2.1-15) include the general case of differential rotation. The rigidly rotating case corresponds to $s = 0$. The existence of a barotropic equation of state (a functional relation among p and μ) is not assumed; when its existence is assumed, one can conclude by imposing the integrability condition of (2.15) ($da = 0$) and taking into account (2.5) that w and s are collinear.

3 Solutions with Differential Rotation

3.1 An Irrotational Fluid

It has been mentioned that the rigid rotation case is characterized by $s = 0$. One may want to look for a fluid rotating in such a way that its motion is *irrotational*, corresponding to the opposite case $w = 0$. A solution of this type was found in [5]. As an Ansatz, the condition

$$b \wedge s = 0 \qquad (3.1.1)$$

was assumed; the meaning of this choice is explained in [8]. In order to simplify the resulting equations, the further restriction

$$\frac{\mu + 3p}{p} b \wedge a + a \wedge \tilde{a} = 0 \qquad (3.1.2)$$

was imposed.[1] An immediate consequence of these two Ansätze is that the equation of state must necessarily be

$$\mu = p. \qquad (3.1.3)$$

If we write $b = \rho^{-1} d\rho$, then by taking p and ρ as the non-ignorable coordinates, the metric can be expressed in terms of explicit functions of p and ρ and a function $T(\rho)$ [5]. The function $T(\rho)$ has to satisfy the differential equation

$$T_{\rho\rho} + (\frac{1}{\rho} + \frac{k\rho}{T^2})T_\rho = 0, \qquad (3.1.4)$$

where k is a positive constant. The angular velocity of the fluid at one of its points, $\Omega = u^\phi / u^t$, is given by

$$\Omega = \lambda \int_{\rho_0}^{\rho} \rho' T^{-1}(\rho') d\rho' + \Omega_0, \qquad (3.1.5)$$

where λ, ρ_0, and Ω_0 are constants.

Equation (3.1.4) can be reduced to a first-order one; this results in an Abel equation of the first kind [5]. No explicit closed-form solution in terms of elementary functions has been found. In spite of this, it has been shown that the bounding two-surface of the fluid at any instant of time (the surface $p = 0$) is compact. It appears that the fluid body is multiply connected, possibly toroidal in shape.

---

[1] The reasons for this condition are not immediately obvious from what has been explained here so far. It appears under a different guise in [8], where its origin becomes clear.

3.2 A Solution with Shear and Vorticity

Recently, a solution with non-vanishing shear and vorticity has been found [6]. The Ansatz

$$b = 2a, \quad s = \text{constant} \times w \qquad (3.2.1)$$

was imposed. This immediately determines the solution; in particular, the equation of state has to be

$$\mu = 2p, \qquad (3.2.2)$$

while the proportionality between s and w becomes fixed:

$$s = -\frac{1}{7}w. \qquad (3.2.3)$$

The metric is explicitly given in terms of elementary functions, except for the quadrature

$$\int \frac{1}{\sqrt{1 - \gamma \sigma^{-2} e^{\sigma^2}}} d\sigma, \qquad (3.2.4)$$

where γ is a constant [6]. The surface $p = 0$ has infinite area; the solution seems to correspond to a cosmological fluid or to an accretion disk of infinite extent.

Acknowledgments

The present work has been partially supported by DGICYT (Dirección General de Investigación Científica y Técnica, Spain) Project PB89-0142.

References

1.   R.A. Donahue, S.L. Baliunas: *Astrophys. J.* **393** L63 (1992)
2.   V.S. Geroyannis, A.A. Hadjopoulos: *Astrophys. J. Suppl. Series* **81** 377 (1992)
3.   P. Ghosh, M.A. Abramowicz: *Astrophys. J.* **366** 221 (1991)
4.   M.H. Pinsonneault, C.P. Deliyannis, P. Demarque: *Astrophys. J.* **367** 239 (1991)
5.   F.J. Chinea, L.M. González-Romero: *Class. Quantum Grav.* **7** L99 (1990)
6.   F.J. Chinea: *A differentially rotating perfect fluid*, submitted to *Class. Quantum Grav.*
7.   J.M.M. Senovilla: *Stationary and Axisymmetric Perfect-Fluid Solutions to Einstein's Equations*, in this volume
8.   F.J. Chinea, L.M. González-Romero: *Class. Quantum Grav.* **9** 1271 (1992)

Rotating Barotropes

L.M. González-Romero

Departamento de Física Teórica II, Facultad de Ciencias Físicas,
Universidad Complutense, 28040-Madrid, Spain

1 Introduction

Recently, a new approach to rotating perfect fluids (rigidly and differentially rotating) in General Relativity has been presented [1]; a summary of this approach appears in [2]. In this contribution, we discuss some results for barotropic perfect fluids (fluids with a one-parameter EOS –Equation Of State–). We can find this kind of fluids in astrophysical objects:

- When all the matter is cold at the endpoint of the thermonuclear evolution of a star.
- When the changes in the state of a star are adiabatic (convective equilibrium).

The main results in the contribution are:

- A classification of the type of motions avaibles for barotropes, as well as, some general properties for all cases, generalizations of Newtonian ones.
- An Ernst-like formalism for these fluids.
- A new family of solutions with examples of different types of motion.

We use the same definitions as in [1], [2]; some references to equations appearing in [2] wil be used so we take for them the notation "(II-eq. num.)".

2 Barotropes

A fluid with a barotropic EOS (barotrope), satisfies:

$$\mu = \mu(p), \quad \mu = \text{density of energy}, \quad p = \text{pressure},$$

from (II-2.15) we get [1]:

$$da = 0 \iff w \wedge s = 0 \Rightarrow w = \alpha s.$$

---

[1] we use the same definition as in [2] for a, b, w, and s

The integration of equations (II-2.5-8) for barotropes gives:

$$a = du$$
$$b = dv$$
$$w = e^{2u-v}d\psi$$
$$s = e^{v-2u}d\chi$$

A better parametrization for the ratio between w and s is:

$$(\lambda - 1)w = (\lambda + 1)s, \quad \lambda = -\frac{e^{v-2u}\chi' + e^{2u-v}\psi'}{e^{v-2u}\chi' - e^{2u-v}\psi'}$$

χ' and ψ' denote the derivatives with respect to the common variable, say x.

From these forms for a, b, w, and s arises a classification for the types of motion avaible for barotropes, because if we parametrize u, θ^1 as follows:

$$u = e^v[Adt + Bd\phi]$$
$$\theta^1 = e^{v-u}[Cdt + Dd\phi]$$

when the Killing vector fields are $\xi = \partial_t$ and $\eta = \partial_\phi$, the equations (II-2.1-2), can be written in the following matricial form:

$$dS\,S^{-1} = \begin{pmatrix} & d\psi \\ d\chi & \end{pmatrix}, \quad S = \begin{pmatrix} A & B \\ C & D \end{pmatrix} \tag{1}$$

Hence, as the velocity vector of the fluid u can be written:

$$u = -\frac{D}{AD - BC}e^{-u}\left[\partial_t - \frac{C}{D}\partial_\phi\right],$$

the angular velocity is $\Omega = -\frac{C}{D}$ and different forms for λ, as a consequence of different forms for w and s, give different angular velocities $\Rightarrow$ diferent kinds of motion.

Some general properties can be deduced from (1):

- A, B, C, and D, are functions of ψ (or χ), and $\det(S) = \text{const} \Rightarrow$ $S = \text{const.} \times SL(2,R)$ matrix
- $\Omega = \Omega(\psi)$ and[2] [3] $u = F(\Omega)G(p)[\partial_t + \Omega\partial_\phi]$

---

[2] Generalization of the Poincaré-Wavre theorem [3]
[3] the explicit form for $G(p)$ depends of the EOS

The classification of barotropes we propose (corresponding to the classification of the matrix S) reads as follows:

- $\lambda \neq const.$
    $w \uparrow s \downarrow$: Type G_-
    $w \uparrow s \uparrow$: Type G_+
- $\lambda = const.$
    $\lambda = 1$: RR (rigid rotation)
    $\lambda = -1$: IM (irrotational motion)
    $\psi = \psi(v - 2u)$
        $w \uparrow s \downarrow$: Type $D1_-$
        $w \uparrow s \uparrow$: Type $D1_+$
    $v - 2u = const.$
        $w \uparrow s \downarrow$: Type $D2_-$
        $w \uparrow s \uparrow$: Type $D2_+$

3 Ernst-like Formalism for Barotropes

For barotropes a formalism that reduces to the well-known Ernst formalism, for vacuum axisymmetric spacetimes, can be obtained using the approach presented in [1].

We define complex 1-forms:

$$\eta = b + i * b$$
$$\gamma = (1 - \lambda)b + 2\lambda a + i * (w - s)$$

All the equations can be reduced to[4]

$$d\eta + \frac{1}{2}\eta \wedge \bar{\eta} = pe^{2Q}\bar{\eta} \wedge \eta$$

$$d\gamma + \frac{1}{2}\bar{\gamma} \wedge \gamma = d \, ln|\lambda| \wedge Re(\gamma - \eta)$$

$$d * \gamma + Re\eta \wedge *\gamma = i\gamma_\lambda \wedge *Im\gamma + \mu_\lambda e^{2Q}\bar{\eta} \wedge *\eta + d \, ln|\lambda| \wedge *Re(\gamma - \eta)$$

$$dp = -\frac{1}{2}[Re\eta + \frac{1}{\lambda}Re(\gamma - \eta)](\mu + p)$$

$$dQ \wedge \eta = \ldots$$

where $\gamma_\lambda = Re\gamma + i\lambda^2 Im\gamma$ and $\mu_p = \frac{1}{2}\lambda[\mu + (2\lambda + 1)p]$.

If $d \, ln|\lambda| \wedge Re(\gamma - \eta) = 0$, an Ernst-like potencial $\mathcal{E}$ can be defined:

$$\gamma = 2\frac{d\mathcal{E}}{\mathcal{E} + \bar{\mathcal{E}}}$$

and the equations can be rewritten in terms of it.

---

[4] Q is defined by the following identity: $\theta^2 = e^Q b$; more details are given in [4]

For vacuum ($\mu = p = 0$) we can choose $\lambda = 1$ or $\lambda = -1$, without loss of generality, to get the Ernst equation [1][5]:

$$\eta = \frac{1}{\rho}(d\rho + idz)$$

$$d * d\mathcal{E} + \rho^{-1}d\rho \wedge *d\mathcal{E} = 2\frac{d\mathcal{E} \wedge *\mathcal{E}}{\mathcal{E} + \bar{\mathcal{E}}}.$$

4 A New Family of Solutions

To show how the Ernst-like formalism works we study a new family of solutions containing solutions of types G_-, G_+, RR, and IM[6]. We impose two simplifying ansatze:

$$u = u(v), \quad \text{and} \quad \Pi_\Omega = 0 \; (G_-, G_+, IM) \quad \text{or} \quad \Pi_\chi = 0 \; (RR)$$

where $\Pi \equiv pe^{2Q}$. When the ansätze are imposed on the equations of the previous section, all of them are reduced to an algebraic equation and a differential equation on Π:

$$\Pi_v = \Delta_3\Pi + \Delta_4\Pi^2$$

$$\Delta_2\Pi^2 + \Delta_1\Pi + \Delta_0 = 0$$

where $\Delta_i, i = 0, 1, 2, 3, 4$ are functions of u, derivatives of u, and $\mu = \mu(p)$. If we isolate Π from the algebraic equation and the result is introduced in the Bernoulli ordinary differential equation, only one ordinary differential equation for two functions $u, \frac{\mu}{p}$ is left.

Let us note that in principle we can fix the EOS beforehand ($\Leftrightarrow$ to fix μ as a function of p) and the problem is reduced to solve the equation for u (this equation in general is complicate, a fourth-order equation). If, instead of fixing the equation of state we fix an explicit form for the dependence of u on v, the equation to be solved determines the EOS, and in general is a second-order differential equation. A simple case appears if we impose $\frac{du}{dv} = const. \neq 1/2$, in which case we get an Abel equation [3].

Properties of the solutions

In this section we will present some basic properties of the solutions discussed above.

All of them have a three-dimensional algebra of isometries; more precisely, the Bianchi type of these algebras is:

- For G_-: Type VII$_0$

---

[5] The equation $dQ \wedge \eta = \dots$ decouples from the others
[6] Some other solutions with $s \neq 0$ and $w \neq 0$ were recently obtained [5, 6]

- For G_+: Type VI_0
- For RR: Type II
- For IM: Type II

Concentrating our study in the irrotational case we obtain the additional properties:

- The solution is not static.
- The isobaric surfaces of $p = const$ have the form of a surface of revolution in an Euclidean spacetime in cilindrical coordinates; Figure 1. represents this embedded surface, and the solid line is the trajectory of the additional Killing vector field projected on the isobaric surface.

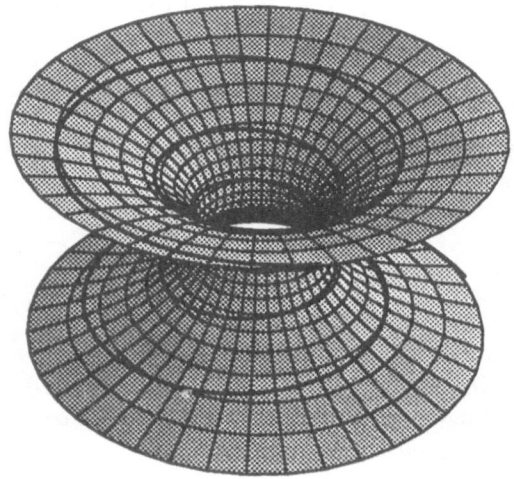

Fig. 1. Approximate form of an isobaric surface for the irrotational solution discussed in the text

Given the form of the surfaces and the others properties of the solution we think that it could be interpreted as the stationary support for the inner part of an accretion disk.

Acknowledgments

This work has been supported in part by DGICYT (Dirección General de Investigacion Científica y Técnica, Spain) Project PB89-0142.

References

1. F.J. Chinea, L.M. González-Romero: *Class. Quantum Grav.* **9** 1271 (1992)
2. F.J. Chinea: *Differentially rotating perfect fluids*, in this volume
3. J.L. Tassoul: *Theory of Rotating Stars*, Princeton University Press, Princeton, New Jersey (1978)
4. L.M. González-Romero: *Rotating Barotropes: An Ernst-like formalism*, in preparation
5. F.J. Chinea: *A differentially rotating perfect fluid*, submitted to *Class. Quantum Grav.*
6. J.M.M. Senovilla: *Stationary and Axisymmetric Perfect-Fluid Solutions to Einstein's Equations*, in this volume

Matching of Stationary Axisymmetric Space-times

M. Martín-Prats and José M. M. Senovilla

Departament de Física Fonamental, Universitat de Barcelona,
Diagonal 647, 08028 Barcelona
and
Laboratori de Física Matemàtica, Societat Catalana de Física, I.E.C.,
Barcelona, Spain.

Although the general solution of Einstein's equations for stationary and axisymmetric space-times is still unknown, these symmetries strongly restrict the form of the metric and fournish us with a helpful tool to face the problem of matching two such solutions. The aim of this work is trying to clarify the restrictions conveyed by the matching of interior and exterior solutions (otherwise called matter and vacuum solutions) on the nature of the allowed joining hypersurfaces and resulting exterior metrics, in order to offer, thereby, an interpretation in terms of the physical components of their energy-momentum tensors.

Setting of the problem

The work deals with the matching of two 4-dimensional manifolds endowed with Lorentzian metrics (V_4, g) representing respectively interior and exterior solutions of Einstein's equations with the above prescribed symmetries. The only restriction to be placed beforehand is continuity of the resulting global potentials on the linking hypersurface.

Both metrics can be reduced to a simple structure by defining two coordinates adapted to the Killing vectors, such that the metric functions depend only on the other two coordinates. In this way, the manifolds to match can be described by [1]:

Interior manifold (V_4, g_1) :

$$ds_1^2 = -e^{2u}(dt + Ad\varphi)^2 + e^{-2u}[e^{2k}(d\rho^2 + dz^2) + w^2 d\varphi^2]$$ (1)

where u, A, k and w are functions of ρ and z.

Exterior manifold (V_4, g_2) :

$$ds_2^2 = -e^{2v}(dT + Bd\phi)^2 + e^{-2v}[e^{2h}(dr^2 + d\theta^2) + r^2 d\phi^2] \qquad (2)$$

where v, B and h are functions depending on r and θ which obey a set of additional conditions to ensure $G_{\alpha\beta} = 0$, namely :

$$v_{rr} + v_{\theta\theta} + v_r/r = -e^{4v}(B_r^2 + B_\theta^2)/2r^2, \qquad (3)$$
$$B_{rr} + B_{\theta\theta} - B_r/r = -4(B_r v_r + B_\theta v_\theta), \qquad (4)$$
$$h_r = r(v_r^2 - v_\theta^2) - e^{4v}(B_r^2 - B_\theta^2)/4r, \qquad (5)$$
$$h_\theta = 2[r v_r v_\theta - e^{4v} B_r B_\theta/4r], \qquad (6)$$

where subindices denote partial derivation, and (3,4) are the integrability conditions for (5,6).

To preserve the Killing's adapted structure of the metric on the resulting global spacetime, it seems a logical choice for the joining hypersurface to identify $t = T$ and $\varphi = \phi$ and to parametrise the rest of coordinates with λ. The result is a unique hypersurface Σ with coordinates $\{\lambda, t, \varphi\}$.

Matching

Finding the conditions to correctly perform the matching is just a matter of calculation, the particular expression of Darmois' junction conditions (continuity of first and second fundamental forms on the hypersurface) in this case being:

a) $[ds^2]_\Sigma = 0$ b) $[\mathbb{K}]_\Sigma = 0$

$$u \doteq v$$
$$A \doteq B$$
$$w^2 \doteq r^2$$
$$e^{2k}/(w_\rho^2 + w_z^2) \doteq e^{2h}$$

$$u_\rho \dot{z} - u_z \dot{\rho} \doteq v_r \dot{\theta} - v_\theta \dot{r} \qquad (7)$$
$$A_\rho \dot{z} - A_z \dot{\rho} \doteq B_r \dot{\theta} - B_\theta \dot{r} \qquad (8)$$
$$\dot{\theta} \doteq \epsilon(w_\rho \dot{z} - w_z \dot{\rho}) \qquad (9)$$
$$k_z \dot{\rho} - k_\rho \dot{z} - \frac{(\dot{w}_z w_\rho - \dot{w}_\rho w_z)}{(w_\rho^2 + w_z^2)} \doteq h_\theta \dot{r} - h_r \dot{\theta} . \qquad (10)$$

It should be noted that in all computations involving the vectors orthogonal to hypersurface, normalization to unity has been disregarded, preserving only the equality of moduli on both sides of the hypersurface. To clear up notation we should mention that $\doteq$ stands for equality on Σ and upper dots mean derivation with respect to λ. The ϵ appearing in (9.b) introduces a sign to be taken as positive if, as usual, r and w are to be related by increasing functions.

Some remarks may clarify the meaning of such a set of equalities at this point. First, (7.a,8.a) give a boundary condition on the functions involved in the second order differential equations (3,4). Moreover (7.b,8.b) show out equality of their normal derivatives on Σ. In addition to that, (9) relate the coordinates on both sides at the joining hypersurface.

Proceeding for further insight, equations (7.a-10.a) can be derived on the surface to obtain:

$$u_\rho \dot\rho + u_z \dot z \doteq v_r \dot r + v_\theta \dot\theta, \tag{11}$$

$$A_\rho \dot\rho + A_z \dot z \doteq B_r \dot r + B_\theta \dot\theta, \tag{12}$$

$$w_\rho \dot\rho + w_z \dot z \doteq \dot r, \tag{13}$$

$$(k_\rho \dot\rho + k_z \dot z) - (\dot w_z w_z + \dot w_\rho w_\rho)/(w_\rho^2 + w_z^2) \doteq h_\theta \dot\theta + h_r \dot r, \tag{14}$$

which enable us to single out both the exterior potentials and their partial derivatives in terms of the interior ones. The next step is extending this dependence to a certain neighbourhood of Σ.

Determination of the Hypersurface

In case we knew the metric functions for a certain given interior stationary axisymmetric solution, we can try to find the hypersurface where to fit the exterior solution. It is easy to see that the matching conditions, themselves, restrict the description of such a hypersurface in the following sense. First of all, (10.b) can be transformed into an equation linking $\dot\rho$ and $\dot z$ with coefficients depending only on interior potentials given schematically by $F(\rho, z)\,\dot\rho \doteq G(\rho, z)\,\dot z$. From this and provided that G/F fulfills Lipschitz's condition, it is possible to extract an intrinsic description of Σ in (V_4, g_1), namely $\rho = \rho(z)$. Using then (9), we can build up the expressions $r(z)$, $\theta(z)$ up to constants, which define the hypersurface Σ on the exterior (V_4, g_2).

Local Determination of the Exterior Solution

Once the hypersurface has been determined, we proceed further to the extension of potentials through Σ. Propagation equations for them can be found in the vacuum additional equations (3-6). The first couple in this set represent elliptic differential equations which, given the Cauchy conditions supplied by (7) and (8), determine v and B in a neigbourhood of Σ as the Cauchy-Kowalevsky theorem states [2] . Then, we can solve (5,6) by linear integration to determine $h(r, \theta)$ locally up to constants.

Global Exterior Solution. Static Case.

The restriction to the static case is achieved by setting $A = 0$ where needed. Under this assumption the number of additional conditions is reduced to:

$$v_{rr} + v_{\theta\theta} + v_r/r = 0, \tag{15}$$

$$h_r = r\,(v_r^2 - v_\theta^2), \tag{16}$$

$$h_\theta = 2r\,v_r v_\theta. \tag{17}$$

Closely analyzing (15) we observe that it is exactly the Laplacian Δv corresponding to the spatial metric $dl^2 = e^{2h}(dr^2 + d\theta^2) + r^2 d\varphi^2$ coming from the projection of ds_2^2 orthogonal to the timelike Killing multiplied by its norm. Similarity with Newtonian potential theory follows from the fact that Raychaudhuri's equation for the present interior metric (which is shear-, rotation- and expansion-free) states:

$$e^{-2k}[u_{\rho\rho} + u_{zz} + (u_\rho w_\rho + u_z w_z)/w] = e^{-2u}(\rho + 3p)/2 \tag{18}$$

or equivalently $\Delta u = e^{-2u}(\rho + 3p)/2$ which recalls the problem of uniquely the potential corresponding to a Poisson interior equation and a Laplacian exterior one, once its value and normal derivative on the closed compact linking surface are given.

One might expect troubles arising from the use of Cauchy's conditions, instead of Dirichlet or Neumann ones, as boundary conditions on the above equation. The difference amounts to avoiding the usual prescription of regularity at infinity. Even so, we can preserve this condition by demanding:

$$\int\int\int_V L(x,y) e^{-2u} \frac{\rho + 3p}{2}(y)\, dV_y = \int\int_S \left\{ L(x,y)\frac{du}{dn_y}(y) - u(y)\frac{dL(x,y)}{dn_y} \right\} dS_y \tag{19}$$

where $L(x,y)$ is the fundamental solution of Laplace's equation in this metric, $\frac{d}{dn}$ means normal derivative and V, S stand respectively for the volume enclosed and the joining surface in every particular instant of t This will work whenever no other singularity exists in the outer region of space-time

Interpretation

To make sense of the mathematically imposed conditions, and without loss of generality, we can decompose the energy-momentum tensor of the interior in the most general form:

$$T_{\alpha\beta} = (\rho + p)u_\alpha u_\beta + pg_{\alpha\beta} + q_\alpha u_\beta + q_\beta u_\alpha + \Pi_{\alpha\beta} \tag{20}$$

where $q_\alpha u^\alpha = 0, \Pi_{\alpha\beta}u^\beta = 0, \Pi_\alpha^\alpha = 0, \Pi_{\alpha\beta} = \Pi_{\beta\alpha}$ and u^α is a unit velocity vector supposed to be tangent to Σ at Σ.

Labelling coordinates $\{t, \varphi, \rho, z\}$ as $\{0, 1, 2, 3\}$ and for the general fluid velocity vector $u^\alpha = fe^{-u}(\delta_0^\alpha + \Omega\delta_1^\alpha)$ with f as a normalization factor, we have: $q_0 + \Omega q_1 = 0; \Pi_{0\beta} + \Omega\Pi_{1\beta} = 0; \Pi_i^i - \Omega\Pi_1^0 = 0$ with i=2,3; to these general conditions one must add those imposed the metric, whose Einstein tensor non vanishing components are separated in two boxes, G_{ab} for $a,b = \{0,1\}$ and G_{ij} for $i,j = \{2,3\}$. Altogheter, they determine $q_2 = q_3 = 0$ for the general stationary case and $q_\alpha = 0$ for the more restrictive static case.

Beyond these particularizations, as is well known, the matching conditions imply directly Israel's conditions, which state $[T_{\alpha\beta}n^\beta]_\Sigma = 0 \leftrightarrow [G_{\alpha\beta}n^\beta]_\Sigma = 0$

where n stands for the normal vector to Σ and $[]_\Sigma$ denotes the discontinuity of any function across Σ. Using a fluid-adapted orthogonal tetrad, given by n, u and two mutually orthogonal vectors spanning the rest of directions and orthogonal to n and u, $e_1^\alpha = X\delta_0^\alpha + Y\delta_1^\alpha$; $e_2^\alpha = \dot\rho\delta_2^\alpha \dot z \delta_3^\alpha$ we can write down these conditions explicitly:

$$G_{\alpha\beta}n^\alpha n^\beta \mid_\Sigma = 0 \leftrightarrow pn^\alpha n_\alpha + \Pi_{\alpha\beta}n^\alpha n^\beta \doteq 0, \tag{21}$$

$$G_{\alpha\beta}n^\alpha u^\beta \mid_\Sigma = 0 \leftrightarrow q_2 n^2 + q_3 n^3 \doteq 0, \tag{22}$$

$$G_{\alpha\beta}n^\alpha e_1^\beta \mid_\Sigma = 0 \leftrightarrow \Pi_{\alpha\beta}n^\alpha e_1^\beta \doteq 0, \tag{23}$$

$$G_{\alpha\beta}n^\alpha e_2^\beta \mid_\Sigma = 0 \leftrightarrow \Pi_{\alpha\beta}n^\alpha e_2^\beta \doteq 0, \tag{24}$$

Among these, the only physically relevant are (21,24), the others being automatically satisfied because of the structure of $T_{\alpha\beta}$ discussed above. From (21) it becomes clear that the matching conditions ensure the vanishing of normal pressure on the joining hypersurface, and there also appears condition (24) restricting the possible values on Σ of some anisotropic pressures.

Previous results on this subject can be found in the work by W. Roos [3,4] , who started from more restrictive hypothesis (like imposing a perfect-fluid model for $T_{\alpha\beta}$ all over the space-time) and required only vanishing of normal pressure on the hypersurface, which is what must be done as follows from our general results.

To make a summary, matching conditions allow the definition of the external metric in a neighbourhood of a certain hypersurface in the following form:

Equation (10.b) defines the hypersurface Σ_1 (first condition on Σ)

Equations (9) relate Σ_1 to Σ_2.

Equations (7,8) are boundary Cauchy conditions for the exterior problem.

Equation (10.a) adds a new restriction to be imposed a posteriori on the interior functions on Σ (second condition on Σ).

It can be shown that the first and second conditions imply the physical restrictions derived above by means of the Israel conditions. Therefore, given any hypersurface on which normal pressure vanishes and interior potential functions fulfill relation (10.a) matching is allowed and the exterior solution is locally determined.

References

1.   Kramer,D., Stephani,H., MacCallum,M. Herlt,E.: Exact solutions of Einstein's field equations. Cambridge University Press. Cambridge (1980)
2.   Courant,R. and Hilbert,D.: Methods of Mathematical Physics, Vol II. Insterscience. New York (1966)
3.   Roos,W.: On the existence of interior solutions in General Relativity. Gen. Rel. Grav. 7 (1976) 431–444
4.   Roos,W.: The matter-vacuum matching problem in General Relativity. General Methods and special cases. Gen. Rel. Grav. 8 (1977) 753–760

Axial Symmetry and Conformal Killings

Marc Mars and José M.M. Senovilla

Departament de Física Fonamental, Universitat de Barcelona,
Diagonal 647, 08028 Barcelona
and
Laboratori de Física Matemàtica, Societat Catalana de Física, I.E.C.,
Barcelona, Spain

1 Introduction

Conformal symmetries have a very important application in stationary and axially symmetric spacetimes due to their importance for the description of the
gravitational field of isolated objects. In fact, most *known* stationary and axisymmetric perfect-fluid solutions possess one non-isometric symmetry. Namely,
The Wahlquist solution [1] (and its limit cases found by Kramer [2], [3], [4])
have a non-trivial Killing tensor, while the solutions presented in Refs. [5] and
[6] have a proper conformal Killing vector. It seems therefore very interesting
to study general stationary and axisymmetric spacetimes with one proper (or
homothetic) conformal Killing vector. An important result due to Carter [9]
establishes that in stationary and axisymmetric spacetimes, the timelike and
axial Killing vectors commute or there is a larger isometry group of at least
four dimensions. Do results similar to this hold for conformal Killings in axially
symmetric spacetimes? The answer is yes, at is shown here. More precisely, we
are able to show that 1) In axially symmetric spacetimes with one and only one
conformal Killing vector, the axial Killing and the conformal Killing commute.
Therefore, in axially symmetric spacetimes with a conformal Killing, if they do
not commute then there is an at least three-dimensional conformal group. 2)
In axially symmetric spacetimes with a *timelike* conformal Killing vector, there
is no restriction in assuming that the axial Killing and the timelike conformal
Killing commute and also either there are only those two symmetries or else
there is a, at least, four-dimensional conformal group. 3) In axially symmetric
spacetimes with one more (and only one) Killing vector and one and only one
conformal Killing vector, the axial Killing commutes with the other two. Therefore, we also have: 4) In stationary and axially symmetric spacetimes with one
conformal Killing vector, the axial Killing commutes with the two others. Thus

we see that the three-dimensional conformal group of stationary and axisymmetric spacetimes with one conformal Killing cannot be arbitrary but rather it can only take one of the few forms in which the axial Killing commutes with the other two symmetries. It should be noticed that this is a general result and it does not depend on the particular form of the energy-momentum tensor or any other thing at all. Obviously, similar results also hold for cylindrically symmetric spacetimes with one conformal Killing vector. Of course, all these results are also valid for homothetic and true Killing vectors and thereby we reobtain, in this last case, the main results already known for stationary and axisymmetric spacetimes or cylindrically symmetric spacetimes. The plan of this paper is not to give proofs of these subjects and therefore we will restrict ourselves to explaining the meaning of these results and what consequences can be extracted from them. For further results which have not been presented here for the sake of brevity, as well as for complete proofs of the assertions stated in this text, see [10].

2 Axially Symmetric Spacetimes with Conformal Isometries

A space-time, V_4, has axial symmetry if there is an effective realization of the one-dimensional torus T into V_4 that is an isometry and such that its set of fixed points constitutes a two-dimensional surface. Mathematically, these two conditions are expressed by:

1. There is a map τ

$$\tau : T \times V_4 \to V_4$$

$$(\phi, x) \to \tau(\phi, x) \equiv \tau_\phi(x)$$

which is a realization of the Lie group T where each τ_ϕ is an isometry of V_4.
2. The set of fixed points of τ, that is to say

$$W_2 \equiv \{x \in V_4; \ \tau_\phi(x) = x \ \forall \phi \in T\}$$

is a two-dimensional surface in V_4. From now on, we shall often refer to W_2 as the *axis of symmetry.*

The Killing vector field that τ defines will be called σ throughout this paper. At any point $Q \in W_2$ we denote by L_Q the linear space tangent to the surface and by P_Q the linear subspace of $T_Q(V_4)$ orthogonal to L_Q.

In what follows, we are going to write down general results concerning the commutators of general vector fields with the axial Killing vector field. These general results will be essential for the main theorems below. To that end, let α be any vector field. The following can be shown:

Lemma 1. *For all vector fields α and every point $Q \in W_2$, the vector $[\alpha, \sigma]|_Q$ belongs to P_Q and is orthogonal to $\alpha|_Q$.*

Moreover, one can prove the following interesting theorem, which will be important for proving the main theorems of this paper.

Theorem 2. *Let α be a vector field in an axisymmetric space-time and $Q \in W_2$.*

1. *$\alpha|_Q$ is tangent to the axis at Q iff $[\alpha, \sigma]|_Q = 0$.*
2. *$\alpha|_Q(\neq 0)$ is normal to the axis at Q iff $\alpha|_Q$ and $[\alpha, \sigma]|_Q$ are linearly independent vectors and $[[\alpha, \sigma], \sigma]|_Q$ depends linearly on the previous.*
3. *α is neither tangent nor normal to the axis at Q iff $\alpha|_Q$, $[\alpha, \sigma]|_Q$ and $[[\alpha, \sigma], \sigma]|_Q$ are linearly independent vectors and $[[[\alpha, \sigma], \sigma], \sigma]|_Q$ depends linearly on the previous two.*

From now on, we will consider the case of axially symmetric spacetimes with conformal symmetries, which are the subject of this paper. First of all let us recall that a conformal Killing vector field λ is a vector field which satisfies $\mathcal{L}_\lambda g = \Psi g$ where $\mathcal{L}_\lambda$ is the Lie derivative with respect to λ, g is the metric tensor field in the spacetime and Ψ is a scalar function called scale factor of λ. Therefore, we have

$$\nabla_\alpha \lambda_\beta + \nabla_\beta \lambda_\alpha = \Psi g_{\alpha\beta} \ .$$

The first result we present related to this case is the following proposition.

Proposition 3. *In an axisymmetric spacetime, let λ be a conformal Killing vector field tangent to the axis for all $Q \in W_2$ and with associated scale factor Ψ. Then $[\lambda, \sigma] = 0$ if and only if $\sigma(\Psi) = 0$.*

Therefore, the necessary and sufficient condition such that a conformal Killing vector field tangent to the axis commutes with the axial Killing is that the scale factor be constant along the orbits of the axial Killing vector field. A trivial consequence is that all homothetic Killing vector fields (and also all Killing vector fields) tangent to the axis commute with the axial symmetry. Despite of this, it seems that, in principle, this is not true for general conformal Killing vector fields. We shall see, however, that this property does hold for general conformal Killings. In order to prove it, we first need the following fundamental result, which strengthens previous similar results [9] and states that an axial conformal Killing vector field and an axial Killing vector field with the same axis of symmetry in a given spacetime must coincide.

Theorem 4. *Let $\{\eta_\phi\}$ be an effective realization of the Lie group T that is a conformal isometry with a surface of fixed points $\widetilde{W_2}$ in an axisymmetric space-time. If $\widetilde{W_2} = W_2$, then $\{\eta_\phi\}$ is in fact an isometry and coincides with the axial isometry.*

With this theorem one can easily see that in axially symmetric spacetimes with a conformal Killing and no other conformal Killing (nor Killing) vector fields, the axial symmetry and the conformal symmetry commute. More precisely, we have the following theorem:

Theorem 5. *In an axisymmetric spacetime with a conformal Killing vector* λ, *if there is no more conformal symmetry then*

$$[\lambda, \sigma] = 0 \ .$$

Another immediate consequence of the previous results is that, in an axially symmetric spacetime, there is no restriction in assuming that a timelike conformal Killing commutes with the axial Killing (this was already known for Killing vector fields, see [9]). The precise statement is given in the Corollary following the next Proposition.

Proposition 6. *In an axisymmetric spacetime, let* λ *be a conformal Killing vector field which does not commute with* σ. *If at some point Q of the axis* $\lambda|_Q$ *is not normal to the surface of fixed points, then there always exists another conformal Killing vector field that commutes with the axial Killing vector field.*

Corollary *In an axisymmetric spacetime. if there is a timelike conformal Killing field, then there always exists a timelike conformal Killing field that commutes with the axial Killing field.*

Let us remark that all the results shown in this section for conformal Killings hold also for homothetic Killings and real Killings, as is obvious. Most of these results were known for Killing fields but, as far as we know, they were peviously unknown for general conformal Killing vector fields.

3 Axisymmetric Spacetimes with Another Symmetry and a Conformal Symmetry

Until now we have been considering an axisymmetric spacetime with a conformal Killing vector field. In General Relativity it has much interest the case of axisymmetric spacetimes with another symmetry which commutes with the axial symmetry, for example stationary and axisymmetric spacetimes or cylindrically symmetric spacetimes. It is obvious that all we have done in the case of conformal Killing fields applies for Killing fields as well, and so we can recover the main results proved by Carter in the early seventies. Moreover, it has been recently found [7] that a class of stationary and axisymmetric exact solutions [5] possesses a conformal Killing vector. Due to the importance that this type of metrics may have in describing the gravitational field of isolated objects, as explained in the Introduction, some papers have recently appeared in the literature considering the case of stationary and axisymmetric exact solutions with a third proper conformal Killing vector field and studying the different Bianchi types that these three vector fields can adopt.

In that direction the previous lemmas and theorems allow us to show the following result.

Theorem 7. *In an axisymmetric spacetime with another Killing vector field ξ and a conformal Killing vector field λ, if there is no more conformal symmetry then*

$$[\sigma, \xi] = 0 ,$$

$$[\sigma, \lambda] = 0 .$$

We see, therefore, that in in stationary and axisymmetric spacetimes , if there is one (and only one) proper (or homothetic) conformal Killing vector field, then it must commute with the axial symmetry. Of course, the same happens in cylindrically symmetric spacetimes. This is a very interesting result and, in fact, it simplifies largely the Bianchi types one has to study in these cases. Thus, for instance, it has been recently considered the case of stationary and axisymmetric perfect-fluid spacetimes in Ref.[8]. In this paper, Bianchi types with $[\sigma, \lambda] \neq 0$ have been studied with the result of the impossibility of getting solutions to the field equations for a perfect-fluid energy-momentum tensor. In fact, as we have shown, no matter what the energy-momentum is, there exists *no* spacetime with that property. Theorem 7 above should be taken into account for future work in spacetimes with two symmetries and one conformal symmetry, if one of the symmetries is required to be axial. Similarly, for cases with an axial symmetry and only one more conformal symmetry we have proven in the previous section that these two symmetries must commute. Therefore, if we want to study conformally stationary and axially symmetric spacetimes, we can assume without restriction that these two symmetries commute (analogously to what happens in stationary and axisymmetric manifolds), and set up the coordinate systems accordingly. Many other consequences can be extracted from the results herein shown, but as they are self-evident we do not believe necessary to explain them here further.

References

1.   H.D. Wahlquist: Interior solution for a finite rotating body of perfect fluid. *Phys. Rev.* **172** 1291-1296 (1968)

2.   D. Kramer: A new solution for a rotating perfect fluid in general relativity. *Class. Quantum Grav.* **1** L3-L7 (1984)

3.   D. Kramer: Rigidly rotating perfect fluids. *Astron. Nachr.* **307** 309-312 (1986)

4.   J.M.M. Senovilla: Stationary axisymmetric perfect-fluid metrics with $\rho + 3p =$ const. *Phys. Lett.* **A123** 211-214 (1987)

5.   J.M.M. Senovilla: On Petrov type-D stationary axisymmetric rigidly rotating perfect-fluid metrics. *Class. Quantum Grav.* **4** L115-L119 (1987)

6. J.M.M. Senovilla: New family of stationary and axisymmetric perfect-fluid solutions. Accepted for publication in *Class. Quantum Grav.*
7. D. Kramer: Perfect fluids with conformal motion. *Gen. Rel. Grav.* **22** 1157-1162 (1990)
8. D. Kramer, J. Carot: Conformal symmetry of perfect fluids in general relativity. *J. Math. Phys.* **32** 1857-1860 (1991)
9. B. Carter: The commutation property of a sationary axisymmetric system. *Commun. Math. Phys.* **17** 233-238 (1970)
10. M. Mars, J.M.M. Senovilla: Axial symmetry and conformal Killings. Submitted to *Class. Quantum Grav.* (1992)

II

Other Topics

Numerical Relativistic Hydrodynamics

José Mª. Ibáñez

Departament de Física Teòrica, Universitat de València,
46100-Burjassot, València, Spain

Abstract: Some *high-resolution shock-capturing methods* have been designed recently to solve nonlinear hyperbolic systems of conservation laws. We have extended them to the relativistic hydrodynamic system of equations *via a local characteristic approach* . We are presenting tests of our procedure in the ultrarelativistic case. We have studied the gravitational collapse of spherically symmetric configurations. Finally, preliminary results in multidimensional applications are displayed.

1 Introduction

In the present lecture I am going to report recent preliminary results in a research project developing a fully relativistic and multidimensional hydrodynamical code. This work is currently being carried out in collaboration with José A. Font, A. Marquina, José Mª. Martí, Juan A. Miralles and José V. Romero, at the University of València.

The term *relativistic hydrodynamics* refers to that part of Physics devoted to the study of the dynamics of both those flows in which the bulk Lorentz factor $W \equiv (1 - v^2)^{-1/2}$ exceeds one in more than a few percent (v is the flow velocity in units of the speed of light) or those where the effects of the background gravitational field -or that generated by the matter itself- are so important that a description in terms of the Einstein theory of gravity must be taken into account.

Relativistic hydrodynamics plays a major role in the realm of Astrophysics. High-velocity outflows can be found in galactic jets (see, for example [2]), in the stellar collapse of iron cores into massive stars which precedes the Supernovae II explosions [9], or in the material accreting onto a compact object [59]. In galactic jets the fluid material reaches the ultrarelativistic regime (i.e., $W \geq 2$). The existence of strong gravitational fields in some of the above astrophysical scenarios complicates the problem and a fully general-relativistic description is necessary.

Our main aim is to correctly model the formation and propagation of strong shocks. Strong relativistic shocks are a very important feature in several problems arising not only in Astrophysics (see above) but also in Cosmology [46] and Plasma Physics [1].

A multidimensional description is necessary in order to understand the complex structures of these shocks when interacting with matter of the interstellar or intergalactic medium or even more interesting when dynamical instabilities of different kinds are developed (Rayleigh-Taylor, Kelvin-Helmholtz,...) at the interfaces between two fluids. Finally, a multidimensional analysis will be necessary if we are interested in describing the release -even in a quasi-Newtonian description - of gravitational radiation coming from the gravitational collapse of cores in evolved massive stars or during the collision of two compact objects.

From the numerical point of view the correct modelling of shocks has attracted the attention of many researchers in Astrophysics and in Computational Fluid Dynamics. A numerical scheme in conservation form allows for *shock-capturing*, i.e., it guarantees the correct jump conditions across discontinuities. Traditionally, shock-capturing methods introduced *artificial viscosity* terms in the scheme in order to damp the oscillations and instabilities associated with the numerical computation of discontinuities. Historically, researchers working in relativistic -both special and general- hydrodynamics (see references [56], [61]; the *Kyoto group* : [50], [49], [51]; *Wilson's school* : [10], [16], [32]), following Wilson's pioneering work ([68], [69]), have used a combination of artificial viscosity and upwind techniques in order to get numerical solutions of the relativistic hydrodynamic equations.

Wilson wrote the system as a set of advection equations. In order to do this he has to treat terms containing derivatives (in space or time) of the pressure as source terms. This procedure breaks -physically and numerically- an important property of the relativistic hydrodynamic system of equations: its *conservative* character (see below).

In recent years a number of new shock-capturing finite difference approximations have been constructed and found to be very useful in shock calculations (see, e.g., [71] and references cited therein). In addition to conservation form, these schemes are usually constructed to have the following properties:

a) Stable and sharp discrete shock profiles.

b) High accuracy in smooth regions of the flow

Schemes with these characteristics are usually known as *high-resolution schemes*. They avoid the use of artificial viscosity terms when treating discontinuities and, after extensive experimentation, they appear to be a solid alternative to classical methods with artificial viscosity. As a sample, the Piecewise Parabolic Method described in [11] has become quite popular among people interested in Newtonian astrophysical simulations.

We have recently proposed an extension of these *modern high-resolution shock-capturing methods* specifically designed to solve *nonlinear hyperbolic systems of conservation laws*. This has been applied to the relativistic hydrodynamic system of equations by Martí *et al.* in [44], which, as it is well-known, have the important property of being the expression of *local conservation laws*. This is a crucial point in our approach.

To end this section let me set out some alternative techniques which are currently being developed in order to solve the equations of relativistic hydrodynamics numerically: *Spectral methods* and *Smooth particle hydrodynamics*.

Spectral methods:

The mathematical development of spectral methods can be found in Gottlieb and Orszag [25]. Basically, they are an extension of Fourier Analysis. Each unknown function is expanded in some characteristic polynomials (Legendre, Chebychev...) according to the boundary conditions of the problem. The main advantage of the spectral methods is their accuracy: the global error on the solution decreases exponentially with the number of degrees of freedom. The handling of shock waves with spectral methods, which is one of the most severe problems concerning these techniques - due to the Gibbs phenomenon- , is currently being studied by Bonazzola and Marck (BM) in the Relativistic Astrophysics Group at Meudon (see [4], [5], [6] and [7]). By combining moving grids and shock tracking techniques BM have obtained promising results for the 1D case in [6]. BM have developed a Newtonian pseudo-spectral 3D hydro-code and studied the gravitational collapse - infall epoch- of a rotating stellar core embedded in an external tidal potential. Preliminary results corresponding to the epoch after bounce are reported by Bonazzola and Marck in [7].

Smooth particle hydrodynamics (SPH):

Derived by Gingold and Monaghan (see, e.g., [22] and [47]) and by Lucy in [38]. Interested readers can address to the recent review by Monaghan in [48].

The classic SPH approximates the density of a Newtonian fluid with the expression

$$\rho(\mathbf{x}, t) \approx \sum_a m_a \mathcal{W}(|\, \mathbf{x} - \mathbf{x}_a \,|, h_a) \tag{1}$$

where m_a is the mass of a fluid "particle" a, $\mathbf{x}_a$ is its position, the function $\mathcal{W}$ is the so-called kernel, a function strongly peaked around $|\, \mathbf{x} - \mathbf{x}_a \,| = 0$ which smooths the particle over some typical "smoothing length" h_a. The properties of the kernel $\mathcal{W}$ and its different expressions can be found in the above references. The SPH formulation of the hydrodynamic equations involves the smoothed estimates of physical quantities. The smoothed estimate of any physical quantity $A(\mathbf{x})$ is:

$$< A(\mathbf{x}) > \approx \sum_a m_a \frac{A(\mathbf{x}_a)}{\rho(\mathbf{x}_a)} \mathcal{W}(|\, \mathbf{x} - \mathbf{x}_a \,|, h_a) \tag{2}$$

The main advantage of the method -as it has been emphasized by Steinmetz and Müller in [62]- is that it does not require a computational grid, making it suitable for multidimensional applications. In reference [62] a critical discussion of the capabilities and limits of SPH has been presented via the computation of a set of problems: one-dimensional standard shock tube tests, three-dimensional simulation of the adiabatic collapse of an initially isothermal gas sphere, the encounter of a star (polytrope) and a massive black hole. Some of the conclusions of the work of Steinmetz and Müller in [62] are: i) SPH is able to get accurate results for problems involving strong shocks. ii) To obtain reliable three-dimensional results SPH requires large particle numbers of up to several tens of thousands. iii) SPH and finite difference methods should be looked upon as complementary methods.

Some preliminary attempts in the extension of SPH to find numerical solutions of relativistic hydrodynamics have been carried out by Mann in [39].

In the next section (§2) I will set out some basic notions of the theory of hyperbolic systems. Section §3 reviews some historical finite-difference schemes pointing out the fundamental differences among them. Section §4 is devoted to Godunov-type methods with which we are mainly concerned. Due to its importance the particular Riemann solver derived by Roe is presented in section §5. The theoretical ingredients recorded so far are applied to the relativistic hydrodynamic system of equations in section §6. The two particular algorithms used in our calculations are explained in section §7. The two final sections §8 and §9 focus on our one-dimensional and multi-dimensional calculations. The last section §10 summarizes our results.

2 Some definitions and reminders

Let me first give some definitions and several fundamental ideas concerning the solutions of the initial-value problem (IVP) for hyperbolic systems of conservation laws.

A one dimensional *hyperbolic system of conservation laws* is:

$$\frac{\partial \mathbf{u}}{\partial t} + \frac{\partial \mathbf{f(u)}}{\partial x} = 0 \tag{3}$$

where $\mathbf{u}$ is the N-dimensional vector of unknowns and $\mathbf{f(u)}$ are N-vector-valued functions called *fluxes* . The above system (3) is said to be *strictly hyperbolic* if the Jacobian matrix

$$\mathbf{A} = \frac{\partial \mathbf{f(u)}}{\partial \mathbf{u}} \tag{4}$$

has real and distinct eigenvalues $\{\lambda_\alpha(\mathbf{u})\}_{\alpha=1,..N}$ and the set of eigenvectors is complete in $\mathcal{R}^N$. If some of the eigenvalues are equal the system is a non-strictly hyperbolic one. We will assume that the eigenvalues are arranged in increasing order.

The equation

$$dx/dt = \lambda_\alpha(\mathbf{u}) \tag{5}$$

defines the αth *characteristic field* .

Lax [36] has shown that the above IVP has at most one C^1 solution (*classical solution*) in the small. In the large, discontinuous solutions (*weak solutions*) are admitted. A weak solution is one that satisfy (3) in the sense of distribution theory, i.e.,

$$\int_0^\infty \int_{-\infty}^\infty \left(\frac{\partial \omega}{\partial t} \mathbf{u} + \frac{\partial \omega}{\partial x} \mathbf{f(u)} \right) dx dt + \int_{-\infty}^\infty \omega(x,0) \mathbf{u}_0(x) dx = 0 \tag{6}$$

for all C^∞ test functions $\omega(x, t)$ that vanish for $| x | + t$ large. This is equivalent to requiring that the relation obtained by integrating (3) over the rectangle $(a, b) \times (t_1, t_2)$ should hold:

$$\int_a^b \mathbf{u}(x, t_2)dx - \int_a^b \mathbf{u}(x, t_1)dx + \int_{t_1}^{t_2} \mathbf{f}(\mathbf{u}(b, t))dt - \int_{t_1}^{t_2} \mathbf{f}(\mathbf{u}(a, t))dt = 0 \quad (7)$$

A piecewise-smooth weak solution of (3) satisfies (3) pointwise in each smooth region; across each curve of discontinuity the *Rankine-Hugoniot relation* (R-H)

$$\mathbf{f}(\mathbf{u}_R) - \mathbf{f}(\mathbf{u}_L) = s(\mathbf{u}_R - \mathbf{u}_L) \tag{8}$$

holds, where s is the propagation speed of discontinuity, and $\mathbf{u}_R$ and $\mathbf{u}_L$ are, respectively, the states on the right and on the left of the discontinuity.

The class of all weak solutions is too wide in the sense that there is no uniqueness for the IVP, and an additional principle is needed for determining the physically relevant solution. Usually this principle -sometimes called *viscosity principle*- identifies the physically relevant solutions, defined as those that are the limit as $\epsilon \to 0$ of solutions $\mathbf{u}(\epsilon)$ of the viscous equations.

$$\frac{\partial \mathbf{u}}{\partial t} + \frac{\partial \mathbf{f}(\mathbf{u})}{\partial x} = \epsilon \frac{\partial^2 \mathbf{u}}{\partial x^2} \quad , \quad \epsilon > 0 \tag{9}$$

For the scalar case, Oleinik in [54] has shown that discontinuities of such admissible solutions can be characterized by the following condition:

$$\frac{f(u) - f(u_L)}{u - u_L} \geq s \geq \frac{f(u) - f(u_R)}{u - u_R} \tag{10}$$

for all values of u between u_L and u_R (*entropy condition*). A discontinuity is called a *shock* if the above inequalities are strict. A discontinuity is called a *contact* discontinuity if equalities hold identically. If f is convex the above characterization is equivalent to

$$u_L \geq u_R \tag{11}$$

or

$$f'(u_L) \geq s \geq f'(u_R) \tag{12}$$

where primes stand for derivative with respect to the argument. This last relation has a geometrical interpretation: the characteristic curves at each side must converge to the discontinuity curve.

For systems of conservation laws, Lax has given the corresponding characterization of the admissible solutions to (3). For a given α, it is

$$\lambda_\alpha(\mathbf{u}_L) \geq s \geq \lambda_\alpha(\mathbf{u}_R) \tag{13}$$

$$\lambda_{\alpha-1}(\mathbf{u}_L) < s < \lambda_{\alpha+1}(\mathbf{u}_R) \tag{14}$$

These relations guarantee that α fields of characteristics converge to the discontinuity curve from the right and $N - \alpha + 1$ from the left. The information carried by these $N+1$ characteristic curves and the $N-1$ relations that we can derive from the R-H conditions, when s is eliminated, allows to know the $2N$ values of $\mathbf{u}$ at each side of the discontinuity.

In the following we will focus on numerical approximations to weak solutions of (3), $v(x,t)$, that are obtained by explicit schemes in *conservation form* :

$$v_j^{n+1} = v_j^n - \lambda(\widehat{f}_{j+1/2} - \widehat{f}_{j-1/2}) \tag{15}$$

where $\lambda = \frac{\Delta t}{\Delta x}$, $v_j^n = v(n\Delta t, j\Delta x)$, and $\widehat{f}_{j+1/2} = \widehat{f}(v_j^n, v_{j+1}^n)$ is a *numerical flux* which must verify the consistency relation $\widehat{f}(u,u) = f(u)$.

Lax and Wendroff proved in [37] that if a finite-difference scheme in conservation form converges to some function $u(x,t)$ as the grid is refined, then this function will in fact be a weak solution of the conservation law. Harten *et al.* showed in [28] that, in the scalar case, *monotonic schemes* in conservation form always converge to the physically relevant solution. Finally, a practical advantage of writing a finite-difference scheme in conservation form is that the quantities which ought to be conserved, according to the differential equation, are exactly conserved in the difference form.

3 Some historical finite-difference schemes

Let me pay attention to three classical finite-difference schemes proposed by several authors in the fifties for solving the scalar hyperbolic equation

$$\frac{\partial u}{\partial t} + \frac{\partial f(u)}{\partial x} = 0 \tag{16}$$

$$a = df/du \tag{17}$$

each of them having some particular feature that is worthwhile to notice.

UPWIND (Courant *et al.* [13]):

It was the first in pointing out the role of characteristics in the numerical solution. Let us consider a be a constant, the upwind algorithm reads:

$$v_j^{n+1} = v_j^n - \lambda a(v_{j+1}^n - v_j^n) \tag{18}$$

for $a < 0$, or

$$v_j^{n+1} = v_j^n - \lambda a(v_j^n - v_{j-1}^n) \tag{19}$$

for $a > 0$.

By introducing the notation

$$a^- \equiv min(a,0) = \frac{1}{2}(a - \mid a \mid) \tag{20}$$

$$a^+ \equiv max(a,0) = \frac{1}{2}(a + \mid a \mid) \tag{21}$$

we can rewrite the upwind algorithm in several forms

$$v_j^{n+1} = v_j^n - \lambda \left[(a^+(v_j^n - v_{j-1}^n) + a^-(v_{j+1}^n - v_j^n) \right] \qquad (22)$$

or,

$$v_j^{n+1} = \lambda a^+ v_{j-1}^n + (1 - \lambda \mid a \mid)v_j^n - \lambda a^- v_{j+1}^n \qquad (23)$$

or,

$$v_j^{n+1} = v_j^n - \frac{\lambda}{2} a(v_{j+1}^n - v_{j-1}^n) + \frac{\lambda}{2} \mid a \mid (v_{j+1}^n - 2v_j^n + v_{j-1}^n) \qquad (24)$$

The last equation displays the conservative form of the scheme by defining the numerical flux:

$$\hat{f}_{j+1/2}^{U} = \frac{1}{2} \left[a(v_{j+1}^n + v_j^n) - \mid a \mid (v_{j+1}^n - v_j^n) \right] \qquad (25)$$

If a is not a constant, then, a generalization of the upwind scheme is

$$\hat{f}_{j+1/2}^{U} = \frac{1}{2} \left[f_{j+1} + f_j - \mid a_{j+1/2} \mid (v_{j+1}^n - v_j^n) \right] \qquad (26)$$

Equation (23) shows the monotone character of the scheme provided that the condition $\sigma \equiv \lambda \mid a \mid \leq 1$ is satisfied. This is the Courant-Friedrich-Lewy condition (CFL) which expresses the fact that the dependence domain of the solution of the finite difference equation must include the dependence domain of the solution of the partial differential equation at all grid points.

GODUNOV [24]:

As the upwind scheme, the Godunov method -in its original formulation- is first order accurate. We will pay particular attention to this scheme in section §4, which, in a generic conservative form, can be written as

$$v_j^{n+1} = v_j^n - \lambda(\hat{f}_{j+1/2}^{G} - \hat{f}_{j-1/2}^{G}) \qquad (27)$$

TWO-STEP LAX-WENDROFF [37]:

In the nonlinear case, the two-step Lax-Wendroff (LW) scheme can be written in a conservative form

$$v_j^{n+1} = v_j^n - \lambda(\hat{f}_{j+1/2}^{LW} - \hat{f}_{j-1/2}^{LW}) \qquad (28)$$

where the LW-numerical flux is defined by

$$\hat{f}_{j+1/2}^{LW} = f(v_{j+1/2}^{n+1/2}) \qquad (29)$$

$$v_{j+1/2}^{n+1/2} = \frac{1}{2}(v_{j+1}^n + v_j^n) - \frac{\lambda}{2} \left(f(v_{j+1}^n) - f(v_j^n) \right) \qquad (30)$$

The LW scheme is second-order accurate, both in space and time, provided that the CFL condition be satisfied.

Let us comment on some peculiarities which can arise, in a numerical application, if we, naively, choose the LW scheme guided only by its high accuracy. In [28] an example is given of a solution to the above scalar hyperbolic conservation law with $f(u) = u - 3\sqrt{3}\,u^2(u-1)^2$ and initial condition $u(x,0) = 1$ if $x \leq 0.5$ or $u(x,0) = 0$ if $x > 0.5$. The numerical results indicated that the approximate solution converges to a weak solution (satisfies the R-H condition at the discontinuities) which violates the entropy condition. The nonmonotonicity of the LW method is responsible for this erroneous solution; but, also, it is responsible for the development of oscillations. The presence of spurious oscillations is another peculiarity related to the poor representation of shock waves by a numerical method. It is a reminder of the well-known, in Fourier analysis, Gibbs phenomenon.

Von Neumann and Richtmyer (VNR) developed in [67] an *artificial viscosity* term (**Q**) which was introduced in the Lagrangian form of the equations of gas dynamics. The goal of the artificial viscosity was to reduce the oscillations while allowing the shock transition to occupy only a few grid points and having negligible effect in smooth regions. VNR explicited the following requirements to be satisfied by **Q** : i) The system must have solutions without discontinuities. ii) The thickness of the shock layers, δ_s, must be everywhere of the same order as the interval length Δx, independently of the strength of the shock. iii) The effect of the terms containing **Q** must be negligible outside the shock layers. iv) The R-H relations must hold when all other dimensions characterizing the flow are large compared to δ_s. VNR showed that an expression which meets the above prescriptions is:

$$\mathbf{Q}(\mathrm{x,t}) = \begin{cases} -\alpha \partial v/\partial x & \text{if } \partial v/\partial x < 0 \quad \text{or } \partial \rho/\partial t > 0 \\ 0 & \text{otherwise} \end{cases}$$

A particular form, suggested by VNR, for the coefficient α is

$$\alpha = \rho(a\Delta x)^2 \mid \frac{\partial v}{\partial x} \mid \qquad (31)$$

where v is the velocity of the fluid, ρ its density, and a is an adjustable parameter with which the user must experiment in order to spread out the shock into a small number of zones and damp down spurious oscillations behind the shock.

This technique has been largely used and improved. One of the main advantages is its low cost, since it is easy to implement -in explicit codes- once the free parameters have been fixed and does not require much CPU time.

Recently, Noh in [52] has pointed out several errors induced by the artificial viscosity: excess **Q** heating, **Q** errors when shocks are propagated over a nonuniform mesh, and **Q** errors in propagating shocks in spherical symmetry. These errors are intrinsic to the method, in the sense that they are contained in the exact solution of the differential equation with **Q**. Versions more refined of this method, which employ an artificial heat flux or a tensorial formulation (see [52], and references cited therein), give good results for the standard test problems, especially when they are combined with adaptive mesh techniques (Norman and Winkler in [53]).

Several disadvantages of the *artificial viscosity* technique must be noted: i) It needs, in each particular application, a previous numerical experimentation to fit the free parameters to the particular problem to be solved in order to find the appropriate balance between the dissipation necessary to preserve monotonicity without causing unnecessary smearing. ii) As Freistühler and Pitman have noticed in [20], not only is the convergence of the viscosity method for general strictly hyperbolic systems still an open question, but this approach is, in general, not well-behaved when the system is a non-strictly hyperbolic one.

4 Godunov-type methods

In 1959 Godunov described in [24] an ingenious method for one-dimensional fluid dynamic problems with shocks. Let $j - 1/2$ and $j + 1/2$ be the lower and the upper interfaces, respectively, of the numerical cell j. Godunov's original idea was to consider that there is a real discontinuity at the interface of the variables of our problem, and, consequently, he proposed using the exact solution of *local Riemann problems* .

A Riemann problem is an *initial value problem* for (3) with initial data:

$$\mathbf{u}(x,0) = \begin{cases} \mathbf{u}_L & \text{if } x < x_{shell} \\ \mathbf{u}_R & \text{if } x > x_{shell} \end{cases}$$

where $\mathbf{u}_{L,R}$ are constant states left and right of a given discontinuity at $x = x_{shell}$.

In general, the solution of that Riemann problem depends only on the states $\mathbf{u}_L$, $\mathbf{u}_R$ and the ratio x/t; it will be denoted by $\mathbf{u}\left((x - x_{shell})/t; \mathbf{u}_L, \mathbf{u}_R\right)$.

The main features of the *Godunov's algorithm* are (see [43], for details)

1) It is written in *conservation form*

$$\mathbf{u}_j^{n+1} = \mathbf{u}_j^n - \frac{\Delta t}{\Delta x}(\widehat{\mathbf{f}}_{j+1/2} - \widehat{\mathbf{f}}_{j-1/2}) \tag{32}$$

where $\widehat{\mathbf{f}}$ is the *numerical flux* (see below), $\mathbf{u}_j^{n+1}$ is

$$\mathbf{u}_j^{n+1} = \frac{1}{\Delta x} \int_{I_j} \mathbf{u}_n(x, t_{n+1}) dx \tag{33}$$

The above integration is carried out into the numerical cell $I_j = [(j-1/2)\Delta x, (j+1/2)\Delta x]$ and $\mathbf{u}_n(x,t)$ can be expressed exactly in terms of the solution of local Riemann problems:

$$\mathbf{u}_n(x,t) = \begin{cases} \mathbf{u}\left(\frac{x-(j+1/2)\Delta x}{t-t_n}; \mathbf{u}_j^n, \mathbf{u}_{j+1}^n\right) & \text{if } j\Delta x < x < (j+1/2)\Delta x \\ \mathbf{u}\left(\frac{x-(j-1/2)\Delta x}{t-t_n}; \mathbf{u}_{j-1}^n, \mathbf{u}_j^n\right) & \text{if } (j-1/2)\Delta x < x < j\Delta x \end{cases}$$

with $t_n \le t \le t_{n+1}$

The theorem of Lax and Wendroff [37] -see at the end of section §2- assures that a weak solution can be found. Harten *et al.* in [30] showed that the exact solution $\mathbf{u}_n(x,t)$ satisfies the entropy condition.

2) The *numerical flux*, $\widehat{\mathbf{f}}$, is defined by

$$\widehat{\mathbf{f}}_{j+1/2} = \mathbf{f}(\widehat{\mathbf{u}}_{j+1/2})$$

where
$$\widehat{\mathbf{u}}_{j+1/2} = \mathbf{u}(0; \mathbf{u}_j^n, \mathbf{u}_{j+1}^n)$$
is the exact solution of the Riemann problem at the interface $x = (j + 1/2)\Delta x$.

The exact solution of this problem for the dynamics of ideal gases can be found, for example, in reference [57] and the corresponding algorithm will be called *Godunov's Riemann solver* .

When $\mathbf{u}_n(x,t)$ is some approximation to the exact solution of the Riemann problem, then, the scheme is called a *Godunov-type* method (see, e.g., references [30], [65], [14], [21]).

The original Godunov method is only first order accurate. The renewed interest in these methods arises after the works of Van Leer ([63], [64]) which focus on the improvements of the spatial accuracy by cell-reconstruction techniques.

Given the computational cost involved in solving exactly the Riemann problem for general nonlinear hyperbolic systems of conservation laws or for materials with a general equation of state (EOS), several *approximate Riemann solvers* have been derived in recent years ([58], [12], [14] [23]).

5 Roe's Riemann solver

Given the importance of this seminal technique we are going to point out the main steps leading to the so-called *Roe's numerical flux*. The following discussion of Roe's Riemann solver is taken from [58] (see, also [43] for details).

Roe looks for approximate solutions of the above Riemann problem which are exact solutions to the approximate problem resulting when the original Jacobian matrix $\mathbf{A}$ of the system (3) is replaced by a new matrix $\widetilde{\mathbf{A}}$ such that: i) $\widetilde{\mathbf{A}}$ is a constant matrix that depends only on the initial data. ii) As $\mathbf{u}_L$ and $\mathbf{u}_R \rightarrow \mathbf{u}$ is

$$\widetilde{\mathbf{A}}(\mathbf{u}_L, \mathbf{u}_R) \rightarrow \mathbf{A}(\mathbf{u})$$

this is a *consistency relation*. iii) For any $\mathbf{u}_L$ and $\mathbf{u}_R$ is

$$\widetilde{\mathbf{A}}(\mathbf{u}_L, \mathbf{u}_R).(\mathbf{u}_L - \mathbf{u}_R) = \mathbf{f}_L - \mathbf{f}_R$$

iv) The eigenvectors of $\widetilde{\mathbf{A}}$ are *linearly independent*.

The conditions iii) and iv) are necessary and sufficient for the algorithm to recognize a shock wave.

In this way, the numerical fluxes can be written

$$\widehat{\mathbf{f}}(\mathbf{u}_L, \mathbf{u}_R) = \frac{1}{2}(\mathbf{f}_L + \mathbf{f}_R - \sum_{\alpha=1}^{N} |\widetilde{\lambda}_\alpha| \Delta \widetilde{\omega}_\alpha \widetilde{\mathbf{e}}_\alpha) \qquad (34)$$

where $\widetilde{\lambda}_\alpha$ and $\widetilde{\mathbf{e}}_\alpha$ ($\alpha = 1, ..., N$) are the eigenvalues (*characteristic speeds*) and the eigenvectors of the Jacobian $\widetilde{\mathbf{A}}$, respectively. The quantities $\Delta \widetilde{\omega}_\alpha$ are the jumps of the *local characteristic variables* across discontinuities and are obtained from:

$$\mathbf{u_R} - \mathbf{u_L} = \sum_{\alpha=1}^{N} \Delta \widetilde{\omega}_\alpha \widetilde{\mathbf{e}}_\alpha \tag{35}$$

Some comments are in order:

1. Since the scheme is based on a linear decomposition of the characteristic fields, one can apply different numerical schemes for each field.

2. The scheme is non-oscillatory in the sense that no new extrema are created for the linear systems or a single nonlinear conservation law. No spurious numerical oscillations contaminate the calculations.

3. The above construction gives a difference approximation of first-order accuracy and may admit entropy violating shocks. Transonic rarefactions may generate unphysical solutions; they are prevented by adding some artificial viscosity only at the sonic points (see, for example, [29]).

4. By construction Roe's Riemann solver is exact for constant coefficient linear system of conservation laws.

Further details can be found in [58].

6 The equations of Relativistic Hydrodynamics as a system of conservation laws

The fundamental equations of motion describing a relativistic fluid flow are governed by *local conservation laws*: the local conservation of baryon number density

$$\nabla_\mu J^\mu = 0 \tag{36}$$

and the local conservation of energy-momentum

$$\nabla_\nu T^{\mu\nu} = 0 \tag{37}$$

where the current J^μ and the energy-momentum tensor $T^{\mu\nu}$ are

$$J^\mu = \rho u^\mu \tag{38}$$

$$T_{\mu\nu} = \rho h u_\mu u_\nu + p g_{\mu\nu} \tag{39}$$

($\alpha = 0, 1, 2, 3$ and latin indices run from 1 to 3). In the above equations ρ is the rest-mass density, p is the pressure, h is the specific enthalpy, $h = 1 + \epsilon + p/\rho$, ϵ is the specific internal energy, u^μ is the four-velocity of the fluid and $g_{\mu\nu}$ defines the space-time $\mathcal{M}$ where the fluid evolves.

We have constrained the energy-momentum tensor to be a perfect fluid, neglecting heat conduction, viscous interactions and magnetic fields. The Minkowski metric is used throughout, that is, gravitational effects of any nearby bodies and of the fluid itself have been neglected.

In Cartesian coordinates the above system of equations (36, 37) can be written in a more explicit way

$$\frac{\partial \mathbf{F}^\alpha(\mathbf{w})}{\partial x^\alpha} = 0 \tag{40}$$

where the 5-vector of unknowns is

$$\mathbf{w} = (\rho, v_i, \epsilon) \qquad (41)$$

and the quantities $\mathbf{F}^\alpha$ are

$$\mathbf{F}^0(\mathbf{u}) = (\rho W, \rho h W^2 v_j, \rho h W^2 - p) \qquad (42)$$

$$\mathbf{F}^i(\mathbf{u}) = (\rho W v_i, \rho h W^2 v^i v_j + p\delta^i_j, \rho h W^2 v^i) \qquad (43)$$

In the above expressions is $x^\alpha = (t, x, y, z)$, the three-velocity $v^i \equiv u^i/u^0$ has the components $v^i = (v^x, v^y, v^z)$ and the Lorentz factor defined by $W \equiv u^0$ satisfies the familiar relation $W = (1 - v^2)^{-1/2}$ ($v^2 = \delta_{ij} v^i v^j$). The components of $\mathbf{F}^0(\mathbf{w})$ are, respectively, the relativistic rest-mass density, the relativistic momentum density and the total energy density.

An equation of state $p = p(\rho, \epsilon)$ closes, as usual, the system. A very important quantity derived from the equation of state is the local sound velocity c_s:

$$h c_s^2 = \chi + (p/\rho^2)\kappa \qquad (44)$$

with $\chi = \partial p/\partial \rho$ and $\kappa = \partial p/\partial \epsilon$.

From the computational point of view, the last equation of the system (40) is not very useful since, in the Newtonian limit ($p \simeq \rho\epsilon \ll \rho$, $v \ll 1$) and for all practical purposes, this equation is identical to the first one. In practice, the last equation has been substituted by the one which results when subtracting the first one from it.

The set of equations (40) constitutes a particular quasi-linear system of 5 first order partial differential equations for the unknown field $\mathbf{w}$, for which the (5×5)-matrices $\mathcal{A}^\alpha(\mathbf{w})$ are the Jacobian matrices associated to the 5-vector $\mathbf{F}^\alpha(\mathbf{w})$, the *flux* in the α-direction. The explicit expressions of the matrices $\mathcal{A}^\alpha$ can be found in reference [19]. System (40) can be rewritten as

$$\mathcal{A}^\alpha(\mathbf{w})\frac{\partial \mathbf{w}}{\partial x^\alpha} = 0 \qquad (45)$$

The hyperbolic character of system (45) has been exhaustively studied by Anile and collaborators (see [1] and references cited therein) for a general space-time in which the fluid evolves (*test fluid approximation*). In particular, they have derived the spectral decomposition of the above Jacobian matrices in the particular reference frame for which matter is at rest. In [19] we have generalized this analysis for an arbitrary reference frame filling, as far as we know, a gap in the scientific literature.

According to Anile [1] the above system (45) will be hyperbolic in the time-direction defined by the vector field ξ with $\xi_\alpha \xi^\alpha = -1$, if the following two conditions hold:

i) $\det(\mathcal{A}^\alpha \xi_\alpha) = 0$
ii) for any ζ such that $\zeta_\alpha \xi^\alpha = 0$, $\zeta_\alpha \zeta^\alpha = 1$, the eigenvalue problem

$$\mathcal{A}^\alpha(\zeta_\alpha - \lambda\xi_\alpha)\mathbf{r} = 0 \qquad (46)$$

has only real eigenvalues λ and N linearly independent eigenvectors $\mathbf{r}$.

These conditions have a particular interest from a technical point of view. We will return to this point at the end of this section.

Equations (40) can be written as a system of *conservation laws in the sense of Lax* [36] allowing to apply specific numerical techniques for solving them numerically. With this aim let us define the vector

$$\mathbf{u} = \mathbf{F}^0(\mathbf{w}) \qquad (47)$$

and let us introduce the three 5-vectors $\mathbf{f}^i$ defined by

$$\mathbf{f}^i = \mathbf{F}^i \circ (\mathbf{F}^0)^{-1} \qquad (48)$$

where $\circ$ means composition of functions.

With the above definitions, system (40) reads as a system of conservation laws in the sense of Lax [36] for the new vector of unknowns $\mathbf{u}$

$$\frac{\partial \mathbf{u}}{\partial x^0} + \frac{\partial \mathbf{f}^i(\mathbf{u})}{\partial x^i} = 0 \qquad (49)$$

In the above system (49) we can define (5×5)-Jacobian matrices $\mathcal{B}^i(\mathbf{u})$, the Jacobian matrices associated to the 5-vector $\mathbf{f}^i(\mathbf{u})$, the so-called *flux* in the j-direction of the system (49) as:

$$\mathcal{B}^i = \frac{\partial \mathbf{f}^i(\mathbf{u})}{\partial \mathbf{u}} \qquad (50)$$

Let $\{\lambda_\alpha, \mathbf{r}_\alpha\}$ be, respectively, the eigenvalues and righteigenvectors of matrices $\mathcal{A}^i$ solution of the eigenvalue problem (46). Then the following result is easy to verify: the spectral decomposition of the Jacobian matrices $\mathcal{B}^i$ is given by the set $\{\lambda_\alpha, \mathcal{A}^0 \mathbf{r}_\alpha\}$. All these quantities are known in terms of the original variables $\mathbf{w}$ (see reference [19] for more details).

The above conclusion sets up the technical ingredients for extending modern high-resolution shock-capturing methods to multidimensional relativistic hydrodynamics.

7 Two high-resolution shock-capturing algorithms: MUSCL and PHM

Let

$$\frac{\partial \mathbf{u}}{\partial t} + \frac{\partial \mathbf{f}(\mathbf{u})}{\partial x} = \mathbf{s}(\mathbf{u}) \qquad (51)$$

a one dimensional *hyperbolic system of conservation laws* with *source terms* $\mathbf{s}(\mathbf{u})$. Strictly speaking a conservation law implies that the source term is zero. In practice, these terms may have an important influence in the calculations as a source of numerical difficulties. Hence, I will consider, in the next, hyperbolic systems like (51).

Two modern high-resolution shock-capturing algorithms have been implemented into our relativistic hydro-code: i) Our version of MUSCL (from Monotonic Upstream Schemes for Conservation Laws [64]), which is globally second order accurate, and ii) PHM (from Piecewise Hyperbolic Method, designed by Marquina in [40]).

Let us summarize the main features of both algorithms in the next subsections.

7.1 Our MUSCL-version

The main ingredients of our MUSCL algorithm - an approximate Riemann solver and cell reconstruction techniques- were presented in [44]. Let me describe them in some detail.

1. At each time level, the data are the *cell averages* of the conserved quantities

$$v_j^n = \frac{1}{\Delta x_j} \int_{x_{j-1/2}}^{x_{j+1/2}} u(x, t^n) dx \tag{52}$$

2. *Reconstruction procedure* of the solution from its cell averages:
   A monotonicity preserving linear reconstruction of the original variables using the 'minmod' function as a 'slope limiter' (see reference [64]).
3. Evaluation of the *numerical fluxes* at the cell interfaces:
   The ith component of the numerical flux is computed as follows

$$\widehat{f}_{j+\frac{1}{2}}^{(i)} = \frac{1}{2} \left(f^{(i)}(u_{j+\frac{1}{2}}^L) + f^{(i)}(u_{j+\frac{1}{2}}^R) - \sum_{\alpha=1}^{3} | \widetilde{\lambda}_\alpha | \Delta \widetilde{\omega}_\alpha \widetilde{R}_\alpha^{(i)} \right) \tag{53}$$

where L and R stand for the left and right states at a given interface $(j + \frac{1}{2})$, $\widetilde{\lambda}_\alpha$ and $\widetilde{R}_\alpha^{(i)}$ ($\alpha = 1, 2, 3$) are, respectively, the eigenvalues (i.e., the *characteristic speeds*) and the ith-component of the α-righteigenvector of the Jacobian matrix

$$A_{j+\frac{1}{2}} = \left(\frac{\partial f(u)}{\partial u} \right)_{u=(u_{j+\frac{1}{2}}^L + u_{j+\frac{1}{2}}^R)/2} \tag{54}$$

and the quantities $\Delta \widetilde{\omega}_\alpha$ - the jumps in the local characteristic variables across each cell interface- are obtained from:

$$u_R^{(i)} - u_L^{(i)} = \sum_{\alpha=1}^{3} \Delta \widetilde{\omega}_\alpha \widetilde{R}_\alpha^{(i)} \tag{55}$$

$\widetilde{\lambda}_\alpha$, $\widetilde{R}_\alpha^{(i)}$ and $\Delta \widetilde{\omega}_\alpha$, as functions of u, are evaluated at each interface and, therefore, they depend on the particular values u_L and u_R.

4. *Time evolution* :
   The 'method of lines' version of the scheme is:

$$\frac{d\mathbf{u}_j(t)}{dt} = -\frac{\hat{\mathbf{f}}_{j+\frac{1}{2}} - \hat{\mathbf{f}}_{j-\frac{1}{2}}}{\Delta x} + \mathbf{s}_j \tag{56}$$

where

$$\hat{\mathbf{f}}_{j+\frac{1}{2}} = \widetilde{\mathbf{f}}(\mathbf{u}_{j-k}, \mathbf{u}_{j-k+1}, ..., \mathbf{u}_{j+k}) \tag{57}$$

is a consistent numerical flux vector, i.e., $\widetilde{\mathbf{f}}(\mathbf{u}, ..., \mathbf{u}) = \mathbf{f}(\mathbf{u})$. Once the procedure to evaluate $\hat{\mathbf{f}}_{j+\frac{1}{2}}$ is known, the system (56) can be integrated in time by using a suitable ODE (ordinary differential equation) solver.

MUSCL advances in time by a standard predictor-corrector method. This version of MUSCL is, then, globally second order accurate. In our multidimensional applications, MUSCL have used a third order Runge-Kutta method that preserves the conservation form of the scheme and does not increase the total variation of the solution at each time substep (see Shu and Osher in [60]).

7.2 PHM

The numerical results obtained in our tests (see [44]) made our updated MUSCL method reliable for the exploration of the ultrarelativistic regime. When the Lorentz factor increased we observed that our numerical scheme became less resolutive, and, consequently, we started looking for shock-capturing methods of higher resolution power.

With this aim, we approximated the solution to the equations of special-relativistic hydrodynamics in conservation form (1D planar case), by means of a third order shock-capturing method designed by Marquina [40] for 1D and 2D scalar conservation laws. We have used a *local characteristic approach* formulated by Harten *et al.* in [27] as a procedure for extending to systems Marquina's method.

Let me describe some of the technical components of PHM

1. At each time level, the data are the *point values* of the conserved quantities

$$\mathbf{v}_j^n = \mathbf{u}(\mathbf{x}_j, t^n) \tag{58}$$

2. Evaluation of the *numerical fluxes* at the cell interfaces:
   1) Obtain the fluxes $\mathbf{g}$ of the local characteristic variables at the center of each cell as

$$\mathbf{g}_j = \mathbf{R}_j^{-1}\mathbf{f}_j \tag{59}$$

where $\mathbf{R}_j$ is a matrix whose columns are the right eigenvectors of the Jacobian matrix

$$\mathbf{A}_j = \left(\frac{\partial \mathbf{f}(\mathbf{u})}{\partial \mathbf{u}}\right)_j \tag{60}$$

   2) With the data $\mathbf{g}_{j-1}, \mathbf{g}_j$ and $\mathbf{g}_{j+1}$, construct, componentwise, a hyperbola in $I_j = [x_{j-\frac{1}{2}}, x_{j+\frac{1}{2}}]$ as in [40].
   3) Evaluate $\widehat{\mathbf{g}}_{j+\frac{1}{2}}$, the numerical flux of the local characteristic variables at

the cell interface as in reference [40].

4) Compute $\widehat{\mathbf{f}}_{j+\frac{1}{2}}$ as follows

$$\widehat{\mathbf{f}}_{j+\frac{1}{2}} = \overset{*}{\mathbf{R}}_{j+\frac{1}{2}} \ \widehat{\mathbf{g}}_{j+\frac{1}{2}} \tag{61}$$

where each column in $\overset{*}{\mathbf{R}}_{j+\frac{1}{2}}$ is computed from hyperbolic reconstructions of the corresponding right eigenvectors of $\mathbf{A}_j$ or $\mathbf{A}_{j+1}$ according to the direction of each local characteristic wind as determined from the sign of the eigenvalues of $\mathbf{A}_j$ and $\mathbf{A}_{j+1}$.

3. *Time evolution*

PHM uses the third order Runge-Kutta method of Shu and Osher [60]. PHM is globally third order accurate.

7.3 General comments

Both algorithms are written in conservation form, thus the Rankine-Hugoniot conditions are automatically satisfied for each discontinuity, according to the Lax and Wendroff theorem cited above (section §2).

The *source terms* $\mathbf{s}_j$ can be calculated, to linear accuracy, from the values of the variables at the zone centres and at the previous time step. Nevertheless, particular caution must be taken with the source terms in the case of stiff problems. Several *time-splitting* algorithms allow to gain the accuracy required. An exact second order algorithm is the *Strang splitting* (see reference [71]) which can be written in a compact way

$$\mathbf{u}^{n+1} = \mathcal{L}_s^{\Delta t/2} \mathcal{L}_f^{\Delta t} \mathcal{L}_s^{\Delta t/2} \mathbf{u}^n \tag{62}$$

In the above equation $\mathcal{L}_f$ is the operator in finite differences which solves the homogeneous part of system (3). On the other hand $\mathcal{L}_s$ is the one that solves a system of ordinary differential equations of the form

$$\frac{\partial \mathbf{u}}{\partial t} = \mathbf{s}(\mathbf{u})$$

The local characteristic approach used to extend Marquina's method to systems has proven to be extremely fruitful (see next section and reference [41]). Although both procedures require knowledge of the spectral decomposition of the Jacobian matrix, reconstructing the fluxes of the local characteristic variables (as in PHM) eliminates the "noise" around the points where the discontinuities of $\mathbf{u}(x,t)$ interact (or are very close).

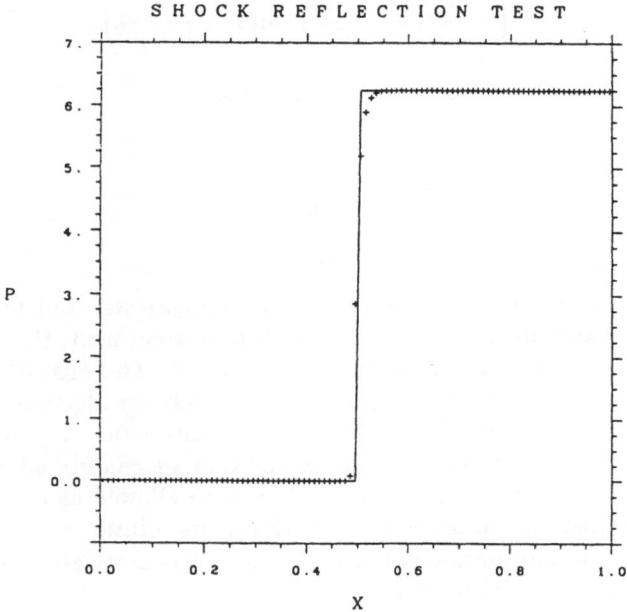

Fig. 1. Pressure in the *relativistic shock wall* test. Initial velocity is 0.9, in units of the speed of light

8 1D Relativistic Hydrodynamics: tests and applications.

In this section we will see how our hydro-code works in one-dimensional applications. In the first subsection I will display some results in the realm of the ultrarelativistic regime. In the second subsection I will present some preliminary general-relativistic stellar collapse calculations obtained with the general-relativistic version of our MUSCL-code.

8.1 Special-Relativistic Hydrodynamics: Ultrarelativistic Regime

The particularization of the relativistic hydrodynamic equations of the section §6 -system (49)- to the one-dimensional and planar case can be written

$$\frac{\partial \mathbf{u}}{\partial t} + \frac{\partial \mathbf{f(u)}}{\partial x} = 0 \tag{63}$$

being the 3-dimensional vector of unknowns $\mathbf{u}$

$$\mathbf{u} = (r, m, e)^{\mathrm{T}} \tag{64}$$

and the *flux* vector $\mathbf{f(u)}$

$$f = \left(\frac{rm}{e+p}, \frac{m^2}{e+p} + p, m\left(1 - \frac{r}{e+p}\right)\right)^T \tag{65}$$

where we have introduced the following variables:

$$r \equiv \rho W \tag{66}$$

$$m \equiv \rho h W^2 v \tag{67}$$

$$e \equiv \rho h W^2 - p \tag{68}$$

From the set of conserved variables $\aleph = \{r, m, e\}$ we need to recover the set of physical variables $\wp = \{\rho, v, \epsilon\}$ at each time step; next, the pressure will be evaluated from the EOS and the loop is closed. The algorithm for going through the $\aleph$ variables to the $\wp$ variables is merely an algebraic equation in p for which powerful methods can be applied. This is the only small price to pay for having a system written in such a way that we can use *ad hoc* methods. Writing the system in the above *conservation form* allowed us to efficiently use high-resolution methods in order to approximate its solutions.

The spectral decomposition of the Jacobian matrix associated to the flux can be found in references [44] or [19].

Our relativistic hydro-code has overcome several severe standard tests: *the relativistic Sod's test, the relativistic blast wave* and *the relativistic shock reflection*. Let me focus on the last one in this talk. Interested readers can be address to the references [44] and [41].

We have computed the solution to the *relativistic shock reflection* problem, that is, the shock thermalization of a cold, relativistically moving gas hitting a wall. Initial conditions are: $\rho = \rho_1 = 1$, $v = v_1$ and $\epsilon = 0$. An ideal gas law of adiabatic index $\Gamma = 5/3$ has been assumed.

The gas behind the shock is at rest ($v_2 = 0$) but its density increases with the initial velocity according to the following compression ratio σ ($\equiv \rho_2/\rho_1$)

$$\sigma = \frac{\Gamma + 1}{\Gamma - 1} + \frac{\Gamma}{\Gamma - 1}\epsilon_2 \tag{69}$$

where $\epsilon_2 = W_1 - 1$ and subscripts 1 and 2 stand for the states of the gas ahead and behind, respectively, of the shock. As it is well-known, in the Newtonian limit the compression ratio is independent of the initial velocity. In the ultrarelativistic regime the density of the gas behind the shock can grow without any limit ($\sigma \sim W_1$).

We run our code for several values of v_1 so as to cover all regimes, from the Newtonian to the ultrarelativistic one. An Eulerian grid of 100 points has been used.

Figures 1, 2 and 3 (taken from [41]) show the pressure of the gas at some instant when the shock is well formed, and for $v_1 = 0.9, 0.99, 0.999$.

For the sake of comparison we have displayed the analytical solution (continuous line). The mean relative error in our calculation of σ is 0.1%. This should be compared with the 5.6% reported by Centrella and Wilson in [10], for $v_1 = 0.9$

$(W_1 = 2.29)$ -the highest value of v_1 for which Centrella and Wilson have published results-. In [44] we have arrived, with our MUSCL version, at $W_1 \approx 23$. Indeed, as we noticed in [44], the resolution of the shock is poorer - and needs eight or nine points- at $W_1 \gg 1$, that is, when the difference between the light velocity and the initial velocity verifies $1 - v_1 \approx O(\Delta x^p)$, where p stands for the global accuracy of the algorithm. With PHM we have obtained very sharp shock profiles -see the pressure profile in figure 4 (taken from [41])- even when $v_1 = 0.9999$, i.e., $W_1 \approx 70$. The results are better than the ones reported by other authors (see, e.g., [32] or [53]) and, as far as we know, give the best resolution for fixed grids.

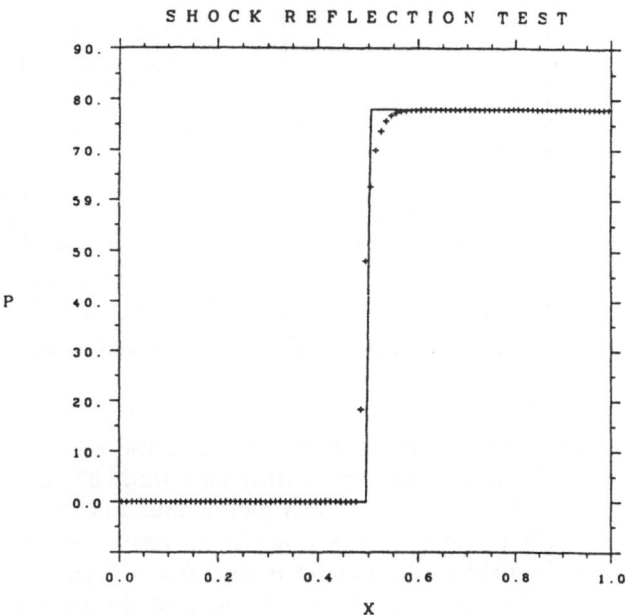

Fig. 2. Pressure in the *relativistic shock wall* test. Initial velocity is 0.99, in units of the speed of light

8.2 General-Relativistic Stellar Collapse

In a previous paper [42] Martí *et al.* focussed on the shock formation and propagation such as it appears in the standard scenario of the so-called *prompt mechanism* of type II Supernovae. In this reference [42] we have made a sample of Newtonian stellar collapse calculations with two codes: i) A standard finite-difference scheme which uses an artificial viscosity technique. ii) A Godunov-type method which uses a linearized Riemann solver. The initial model and the equation of state was kept fixed in order to be able to compare both methods directly. Differences in the behaviour of the global energetics of the collapse

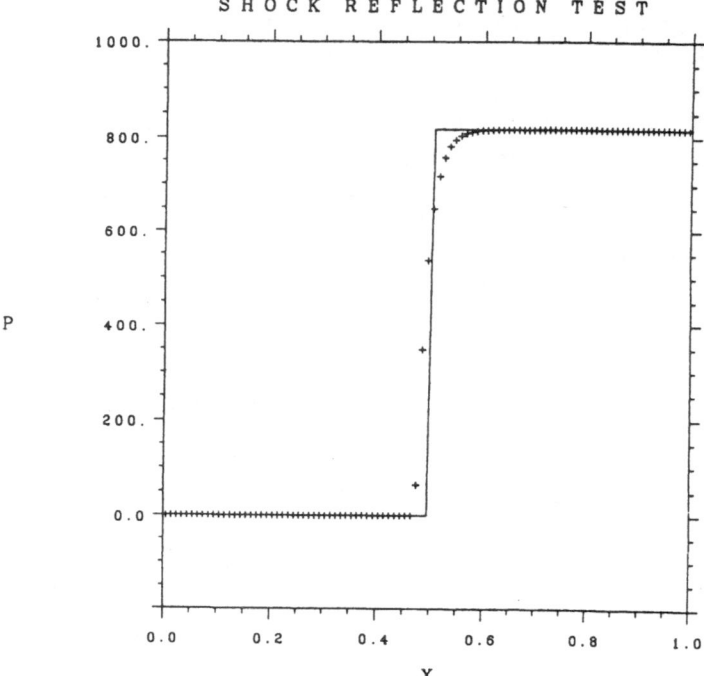

Fig. 3. Pressure in the *relativistic shock wall* test. Initial velocity is 0.999, in units of the speed of light

were found to depend on the particular way in which the artificial viscosity is implemented. Although the grid used was rather poor (only 50 nodes) and more sophisticated forms for the artificial viscosity can be found in the literature (see the paper of Noh in [52], referenced in section §3) our results in [42] might be of interest not only in the field of the *prompt mechanism* of Type II Supernovae, but even in the correct estimation of the efficiency of the energy released in form of gravitational radiation in non-spherical collapse. Indeed, as Bonazzola and Marck (in [4]) have emphasized, the gravitational power decreases with viscosity, ν, as $1/\nu^6$; hence, viscosity of a numerical code -explicit or intrinsic- may have dramatic effects.

The equations of general-relativistic hydrodynamics are the expression of the local laws of conservation of baryon number density and energy-momentum in a space-time $\mathcal{M}$, described by the four dimensional metric tensor $g_{\mu\nu}$. In our procedure, the metric is split into the objects α and γ_{ij}, keeping the line element in the form:

$$ds^2 = -\alpha^2 dt^2 + \gamma_{ij} dx^i dx^j \tag{70}$$

By defining the following set of variables

$$D \equiv \rho W$$

$$S \equiv \alpha T^{0r} = \rho h W^2 v$$

$$\tau \equiv \alpha^2 T^{00} = \rho h W^2 - p$$

the general-relativistic hydrodynamic equations in the one dimensional case (for example, spherical symmetry) can be written as a system of conservation laws in the sense of Lax [36]:

$$\frac{\partial \mathbf{u}}{\partial t} + \frac{\partial \mathbf{f(u)}}{\partial r} = \mathbf{s(u)}, \tag{71}$$

where

$$\mathbf{u} = (D, S, \tau)^{\mathrm{T}}$$

is the 3-dimensional vector of unknowns which defines the state of the system, the fluxes are

$$\mathbf{f} = \frac{\alpha}{\sqrt{\gamma_{rr}}} \left(\frac{DS}{\tau + p}, \frac{S^2}{\tau + p} + p, S \right)^{\mathrm{T}},$$

and the source terms are free of derivatives of hydrodynamic quantities and read:

$$\mathbf{s(u)} = \left(-D\frac{\partial ln\sqrt{\gamma}}{\partial t} - \frac{\alpha}{\sqrt{\gamma_{rr}}}\frac{DS}{\tau + p}\frac{\partial ln\gamma}{\partial r}, \right.$$
$$- S\frac{\partial ln(\sqrt{\gamma\gamma_{rr}})}{\partial t} - \frac{\alpha}{\sqrt{\gamma_{rr}}}\left[(\frac{S^2}{\tau + p})\frac{\partial ln\sqrt{\gamma}}{\partial r} + \tau\frac{\partial ln\alpha}{\partial r} + p\frac{\partial ln\sqrt{\gamma_{rr}}}{\partial r} \right],$$
$$\left. - \tau\frac{\partial ln\sqrt{\gamma}}{\partial t} - \frac{\alpha}{\sqrt{\gamma_{rr}}}S\frac{\partial ln\alpha\sqrt{\gamma}}{\partial r} - \frac{S^2}{\tau + p}\frac{\partial ln\sqrt{\gamma_{rr}}}{\partial t} - p\frac{\partial ln\sqrt{\gamma}}{\partial t} \right)^{\mathrm{T}}$$

In the above expressions, quantity γ is the determinant of the matrix γ_{ij}, v is defined by $v \equiv \sqrt{\gamma_{rr}}u^r/\alpha u^0$ (indexes 0 and r stands for the temporal and radial components, respectively) and represents the fluid velocity relative to an inertial observer at rest in the coordinate frame. The Lorentz-like factor is defined by $W \equiv \alpha u^0$.

From the *conserved* quantities $\aleph = \{D, S, \tau\}$, as in the special-relativistic case, we must obtain the set of quantities $\wp = \{\rho, v, \epsilon\}$ at each time step, by solving an implicit equation for pressure. In the Newtonian limit, the set of new variables $\aleph = \{D, S, \tau - D\}$ tends to the set $\{\rho, \rho v, \rho\epsilon + (1/2)\rho v^2\}$.

The hyperbolic character of the above system of equations as well as the spectral decomposition of the Jacobian matrix associated to the flux has been discussed in references [44], [43] and [34].

We have tested our one-dimensional general-relativistic code to reproduce some of the stationary solutions of the *spherical accretion onto a black hole* in two cases: i) *dust* accreting onto a Schwarzschild black hole [32] and ii) an *ideal gas* accreting onto a Schwarzschild black hole [45]. Details of these numerical experiments can be found in [43].

The initial model we have taken in our stellar collapse calculations is a white dwarf-like configuration having a central density 2.5×10^{10} g/cm^3. This is an equilibrium model for a particular EOS (Chandrasekhar's EOS with coulombian corrections) corresponding to the maximum of the "mass-radius" curve (see [33]). The numerical grid has been built up in such a way that the radius of the initial

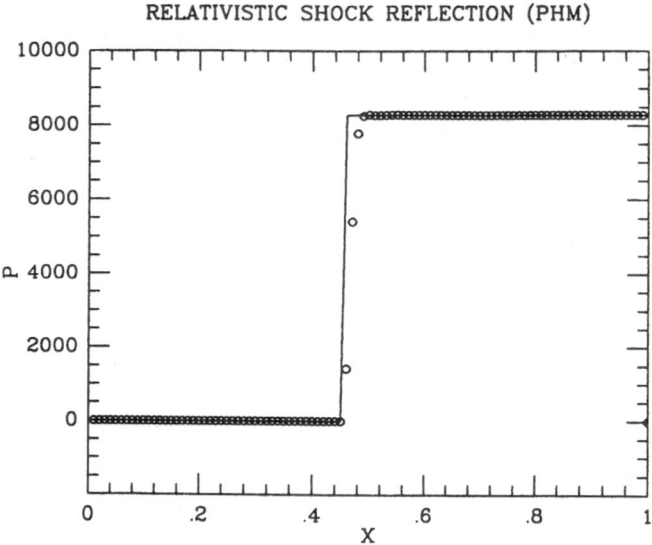

Fig. 4. Pressure in the *relativistic shock wall* test. Initial velocity is 0.9999, in units of the speed of light. PHM algorithm

model is partitioned into 200 zones distributed in geometric progression in order to have a finer resolution near the centre.

The EOS we have used is a Γ-law such that Γ varies with density according to Van Riper's prescription [66]:

$$\Gamma = \Gamma_{min} + \eta(\log \rho - \log \rho_b) \tag{72}$$

with: $\eta = 0$ if $\rho < \rho_b$ and $\eta > 0$ otherwise.

The parameters Γ_{min}, η and ρ_b are, typically: 4/3, 1, and 2.7×10^{14} gcm^{-3}, respectively, but we have also considered other values for Γ_{min} and η .

In these proceedings (see [3] in this volume) we present some results obtained using a harmonic time coordinate due to its singularity avoidance properties.

If we choose Schwarzschild-type coordinates (see, e.g., [8] or [26]) the full one-dimensional general-relativistic equations can be reduced to a merely hydrodynamical problem. In these coordinates the 3-metric reads

$$\gamma_{ij} = diag(X^2, r^2, r^2 sin^2\theta)$$

being

$$X = (1 - 2m/r)^{-1/2} \quad , \quad m = m(r,t)$$

The source terms have been explicitly written in reference [34].

At each time step functions m and α are integrated along the radius according to

$$\frac{\partial m}{\partial r} = 4\pi r^2 (\tau + D)$$

$$\frac{\partial \ln\alpha}{\partial r} = X^2 \left(\frac{m}{r^2} + 4\pi r(p + Sv)\right)$$

Let me notice the behaviour of the velocity field shown in figure 5 (taken from [34]). Shock is sharply solved in two zones and is free of spurious oscillations. The minimum of velocity -at the infall epoch- is about -0.40, that is $\approx 25\%$ greater than the value reported by [42], in the Newtonian case, for the same initial model, same EOS and a lagrangian version of our code which uses the same Riemann solver. The difference is entirely due to general-relativistic effects, although this point requires verification since a different grid has been used.

9 Multidimensional Problems

In two space dimensions a system of conservation laws takes the form

$$\frac{\partial \mathbf{u}}{\partial t} + \frac{\partial \mathbf{f}(\mathbf{u})}{\partial x} + \frac{\partial \mathbf{g}(\mathbf{u})}{\partial y} = 0 \tag{73}$$

where $\mathbf{u} = \mathbf{u}(x, y, t)$

The multidimensional character of the above system has been taken into account numerically by considering standard *operator splitting* techniques ([71]), similar to the ones used for treating the source terms, or the corresponding extension of the *method of lines* already discussed in section §7.

The splitting proceeds in two steps and the algorithm can be written -in a 2D case- as follows:

$$\mathbf{u}^{n+1} = \mathcal{L}_g^{\Delta t} \mathcal{L}_f^{\Delta t} \mathbf{u}^n \tag{74}$$

where $\mathcal{L}_f$ and $\mathcal{L}_g$ stand for the operators in finite differences associated, respectively, to the 1D systems

$$\frac{\partial \mathbf{u}}{\partial t} + \frac{\partial \mathbf{f}(\mathbf{u})}{\partial x} = 0$$

and

$$\frac{\partial \mathbf{u}}{\partial t} + \frac{\partial \mathbf{g}(\mathbf{u})}{\partial y} = 0$$

In this section I am going to discuss some multidimensional problems that we have used as bed-tests in our progress towards the construction of a multidimensional relativistic hydro-code. First, in subsection (§9.1) I will show the solution of some particular general-relativistic flows which can be described by a wave equation in a curved space-time. The next subsection (§9.2) is devoted to our preliminary results with relativistic multidimensional tests.

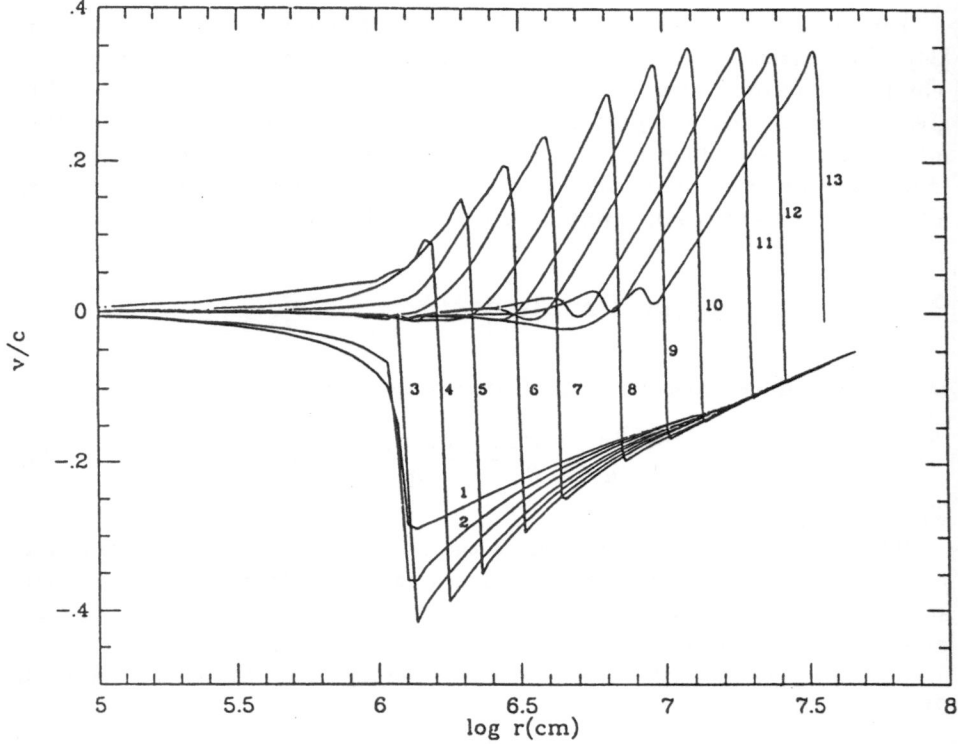

Fig. 5. General-relativistic stellar collapse. Snapshots of the velocity, in units of the speed of light, versus radial coordinate, in logarithmic scale and units of cm. Each curve is labeled by a number which establishes the temporal sequence into the interval $90.17 \leq t(msec.) \leq 93.34$

9.1 Potential flow passing over a black hole

In 1988 Petrich *et al.* (see [55]) derived, based on some assumptions, the analytical solution for the steady-state, subsonic accretion of a gas onto a Schwarzschild black hole. As a consequence of their assumptions the equations of general-relativistic hydrodynamics are simplified and lead to a wave equation for a potential ψ in a curved space-time:

$$\psi^{;\mu}_{,\mu} = 0 \tag{75}$$

where ; and , stand, respectively, for the usual covariant and ordinary derivatives. From this potential ψ the velocity field can be derived:

$$hu_\mu = \frac{\partial \psi}{\partial x^\mu} \tag{76}$$

In the Schwarzschild gravitational field generated by a source of mass M this equation reads

$$\frac{1}{r^2} \frac{\partial}{\partial r} \left[\left(1 - \frac{2M}{r} \right) r^2 \frac{\partial \psi}{\partial r} \right] + \frac{1}{r^2} \left[\frac{1}{\sin\theta} \frac{\partial}{\partial \theta} \left(\sin\theta \frac{\partial \psi}{\partial \theta} \right) + \frac{1}{\sin^2\theta} \frac{\partial^2 \psi}{\partial \phi^2} \right] -$$
$$- \left(1 - \frac{2M}{r} \right)^{-1} \frac{\partial^2 \psi}{\partial t^2} = 0 \tag{77}$$

A stationary solution of this equation which satisfies the appropriate boundary conditions (see [55]) is

$$\psi = -u^0_\infty t - 2M u^0_\infty ln(1 - 2M/r) + u_\infty(r - M)cos\theta \tag{78}$$

for a black hole. Here u^0_∞ and u_∞ are the asymptotic values of the temporal component of the four-velocity and the modulus of the three-vector $\mathbf{u}$, respectively. Let us notice the following relations between u^μ_∞ and the three-velocity vector $\mathbf{v}_\infty$

$$u^\mu_\infty = (u^0_\infty, \mathbf{u}_\infty) = (1 - v^2_\infty)^{-1/2}(1, \mathbf{v}_\infty) \tag{79}$$

where v_∞ is the absolute value of the asymptotic fluid three-velocity.

Although the above particular potential flow problem does not allow the presence of shock waves we have used this problem as a test of our 2D-MUSCL algorithm in order to describe the non-linearities induced by the geometrical terms related with the strong gravitational fields. With this aim, the above wave equation (77) has been rewritten as a hyperbolic system of conservation laws -in the sense of Lax [36]- by splitting it into a system of three equations of first order.

To do this let us introduce the following auxiliary functions

$$a = \frac{\partial \psi}{\partial t} \qquad b = \frac{\partial \psi}{\partial r} \qquad c = \frac{\partial \psi}{\partial \theta} \tag{80}$$

The hyperbolic system of conservation laws equivalent to (77) is

$$\frac{\partial \mathbf{u}}{\partial t} + \frac{\partial \mathbf{f}(\mathbf{u})}{\partial r} + \frac{\partial \mathbf{g}(\mathbf{u})}{\partial \theta} = \mathbf{s}(\mathbf{u}) \tag{81}$$

where

$$\mathbf{u} = (a, b, c) \tag{82}$$

is the 3-dimensional vector of unknowns, and

$$\mathbf{f(u)} = \left(-(1 - \frac{2M}{r})^2 b, -a, 0 \right) \tag{83}$$

and

$$\mathbf{g(u)} = \left(-(1 - \frac{2M}{r}) \frac{1}{r^2} c, 0, -a \right) \tag{84}$$

are the 3-vector-valued functions defining the *fluxes* in the r and θ directions, respectively. Finally

$$\mathbf{s(u)} = \left((1 - \frac{2M}{r})(1 - \frac{3M}{r}) \frac{2}{r} b + (1 - \frac{2M}{r}) \frac{\cos\theta}{r^2 \sin\theta} c, 0, 0 \right)$$

are the *source* terms.

In this application the source terms have been treated by operator splitting techniques in such a way that the operator associated to the source sector is the one which solves the system

$$\frac{\partial \mathbf{u}}{\partial t} = \mathbf{s(u)}$$

in the following form

$$[\mathbf{I} - \frac{1}{4} \Delta t s'(\mathbf{u}_i^n)] \Delta \mathbf{u}_i^* = \frac{1}{2} \Delta t s(\mathbf{u}_i^n)$$
$$\mathbf{u}_i^* = \mathbf{u}_i^n + \Delta \mathbf{u}_i^*$$

where $s' = \frac{\partial s}{\partial u}$.

The spectral decomposition of the Jacobian matrices associated to the fluxes in each direction as well as the characteristics of the computational cell and the initial and boundary conditions of our application can be found in reference [17].

In figure 6 (taken from [17]) we can see the evolution towards the steady-state accretion reached in a time $t \approx 65M$ for an asymptotic velocity $v_\infty^2 = 0.7$. This sequence of figures shows the flow lines for the following values of the temporal coordinate (in units of M): 0, 3.5, 65 and 100. We have obtained a value of $3.42 \le r_A/M \le 3.95$ (this interval is constrained by the resolution of our grid) for the radius of the critical cylinder inside which material is ultimately captured by the black hole, being the exact value $3.404M$ for an asymptotic velocity $v_\infty^2 = 0.7$.

The main conclusion from these figures is the following: we have succeeded in obtaining the stationary solution by using a *modern high-resolution shock-capturing* scheme of second order.

BLACK HOLE STEADY–STATE

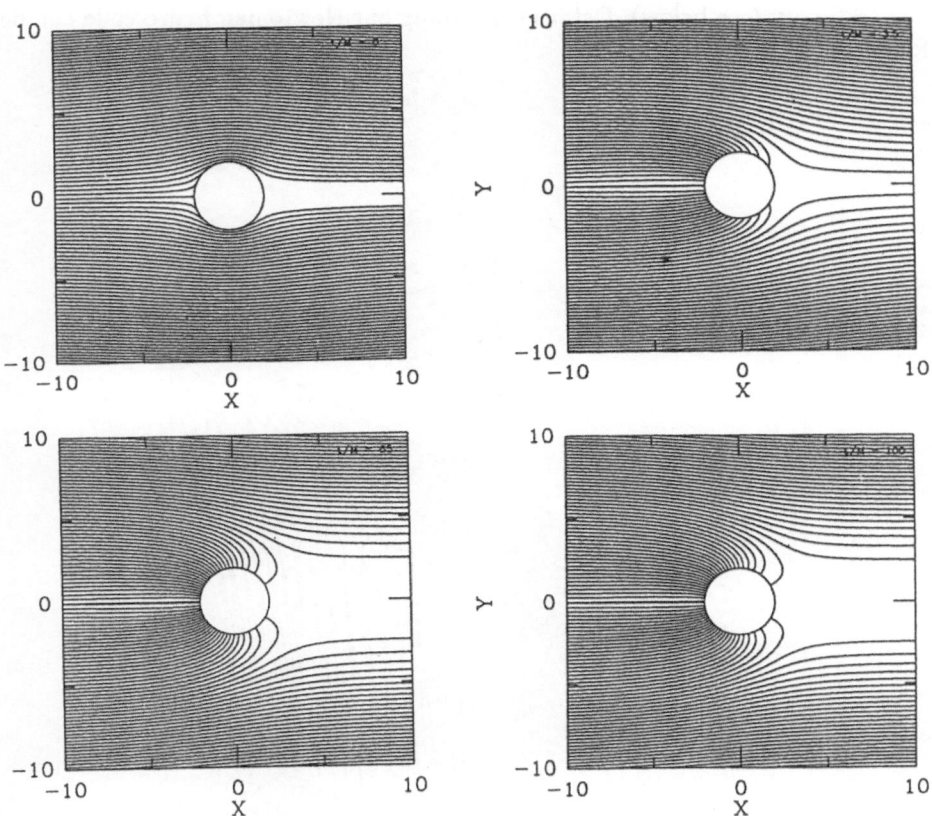

Fig. 6. Potential flow passing over a black hole. Evolution towards the steady-state accretion reached in a time $t \approx 65M$ for an asymptotic velocity $v_\infty^2 = 0.7$

9.2 2D Relativistic Hydrodynamics

As a first step to our objective of designing a multidimensional hydro-code to deal with the equations of relativistic hydrodynamics we built up a Newtonian two-dimensional hydro-code which has overcome the standard and severe test due to Emery (see below). Details concerning our Newtonian hydro-code can be found in [18].

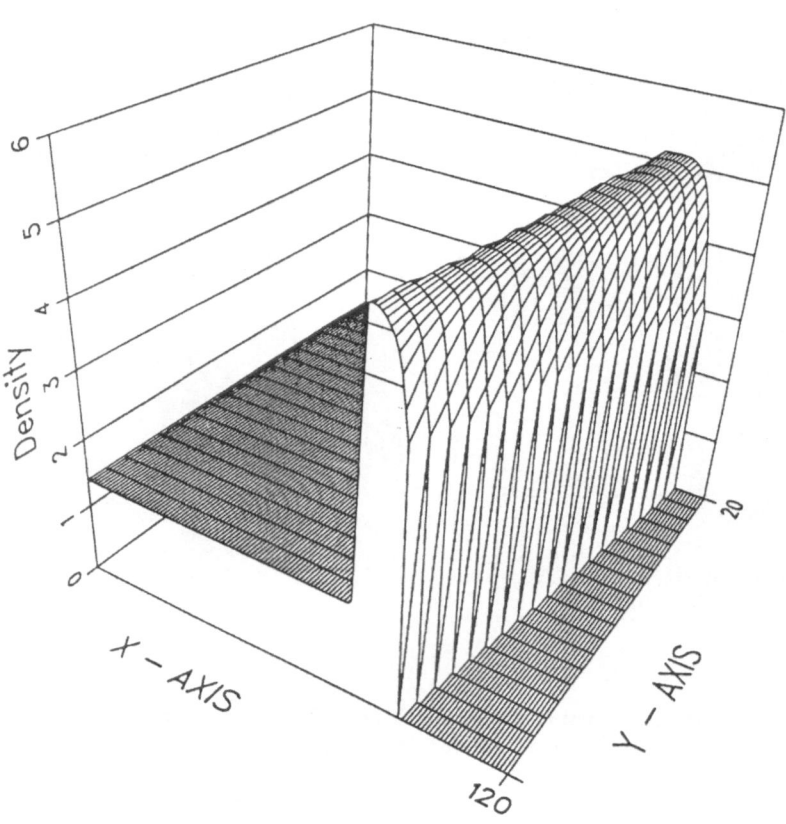

Fig. 7. Steady relativistic oblique shock. Initial inflow velocity 0.01 (in units of the speed of light). See text for details

Joining together the algebraic relations discussed in section §6 and the basic ingredients of our 2D MUSCL algorithm, once it has been tested with the above Newtonian numerical experiment, we are ready to show our preliminary relativistic results. These are -as far as we know- the first ones (in scientific literature) which have made use of modern high-resolution shock-capturing techniques in the realm of multidimensional relativistic hydrodynamics. Their interest goes further away of the special-relativistic case if we keep into account that, as it is well-known and we have shown in our 1D general-relativistic applications, the equivalence principle warranties the feasibility of extending our procedure to the study of test fluids evolving in strong gravitational fields.

The equations describing the two-dimensional flow of a relativistic ideal fluid are the trivial particularization of the ones displayed in section §6. The spectral decomposition of the Jacobian matrices associated to the fluxes in each direction have been derived in reference [19].

Let me display two numerical tests we have carried out with our relativistic 2D MUSCL code: i) the steady *relativistic oblique shock* wave, and ii) the relativistic version of the *Emery's step*.

Relativistic Oblique shock The basic algebraic relations which connect the two states at each side of a steady relativistic oblique shock have been derived by Königl (see reference [35]) for ideal gases with a constant adiabatic index Γ. A fundamental difference between the Newtonian and relativistic descriptions is the fact that the jump in density increases, in the relativistic case, with the upstream velocity and tends to infinity in the extreme-relativistic regime. Königl derives an algebraic equation which defines implicitly the jump in velocities in terms of the known upstream state. Once the jump in velocities is calculated the remaining unknowns are easily obtained (see [35], for details).

We have generated an oblique shock throwing an ideal gas with $\Gamma = 7/5$ through a corner (an oblique plane) with a wedge angle of 28°. The initial density is 1.4 and the pressure the one resulting for a Mach 3 flow (Newtonian definition). The initial velocity (in units of the speed of light) runs, in our experiments, from 0.01 to 0.95. Figures 7 and 8 (taken from [19]) show the density as a function of the spatial coordinates and at some instant in its evolution when the steady state has been reached.

The solution looks like the Newtonian. This can be explained if we take into account that as several authors have emphasized (see, e.g., [70] or [35]) the equations of steady special-relativistic gas dynamics and steady non-relativistic gas dynamics have a similar mathematical form, when expressed in some appropriate variables. This property has been used in order to find numerical solutions for relativistic steady flows.

As we can see in figures 7 and 8, the qualitative behaviour of density is the one described above, being the particular numbers agree with the Königl relations.

Relativistic Emery's step A severe test for two-dimensional flows in presence of shocks is the flat-faced step originally introduced by Emery (1968): a Mach 3 flow is injected into a tunnel containing a step. The tunnel is 3 units long and 1 unit wide. The step is 0.2 units high and is located 0.6 units from the left-hand end of the tunnel. Slab symmetry is assumed.

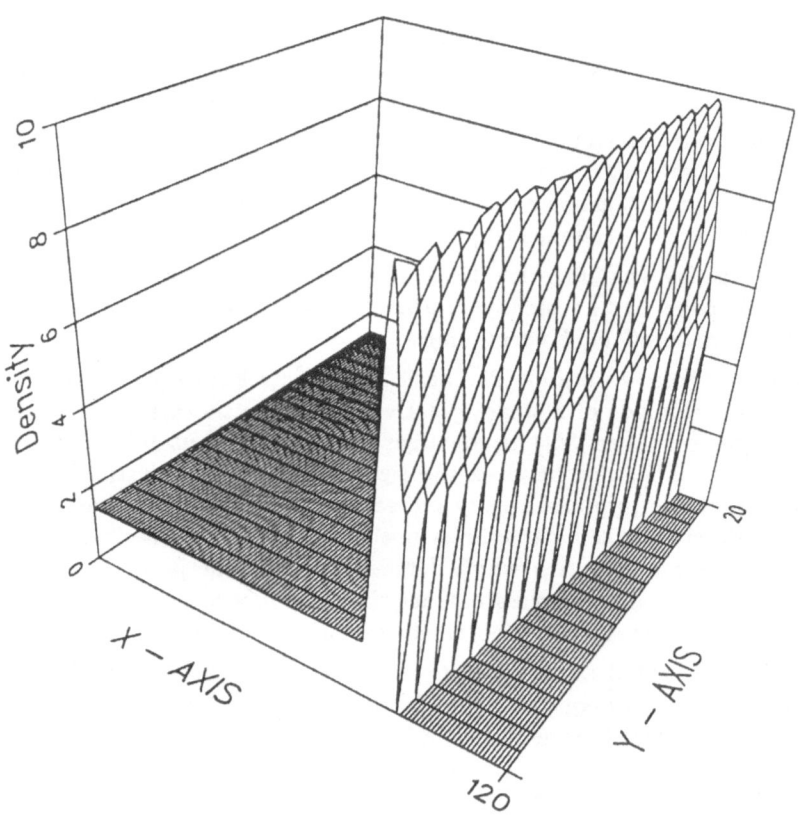

Fig. 8. Steady relativistic oblique shock. Initial inflow velocity 0.95 (in units of the speed of light). See text for details

The boundary conditions are: 1) Reflecting boundary conditions along the walls of the tunnel and at the left face and the bottom of the step. 2) On the right and the left sides of the tunnel, respectively, outflow and inflow boundary conditions are applied.

The initial conditions for the gas in the tunnel are given by

$$\rho(x, y, 0) = \rho_0 = 1.4$$

$$v^x(x, y, 0) = v_0^x$$

$$v^y(x, y, 0) = v_0^y = 0$$

for all x, y. The value of the initial pressure is derived from the other variables. The initial value of the x-component of the three-velocity v_0^x will be a free parameter.

The EOS considered is the one of an ideal gas with $\Gamma = 7/5$. Gas is continually fed in at the left-hand boundary with the flow variables given by their initial values. A rectangular grid of 120×40 has been used.

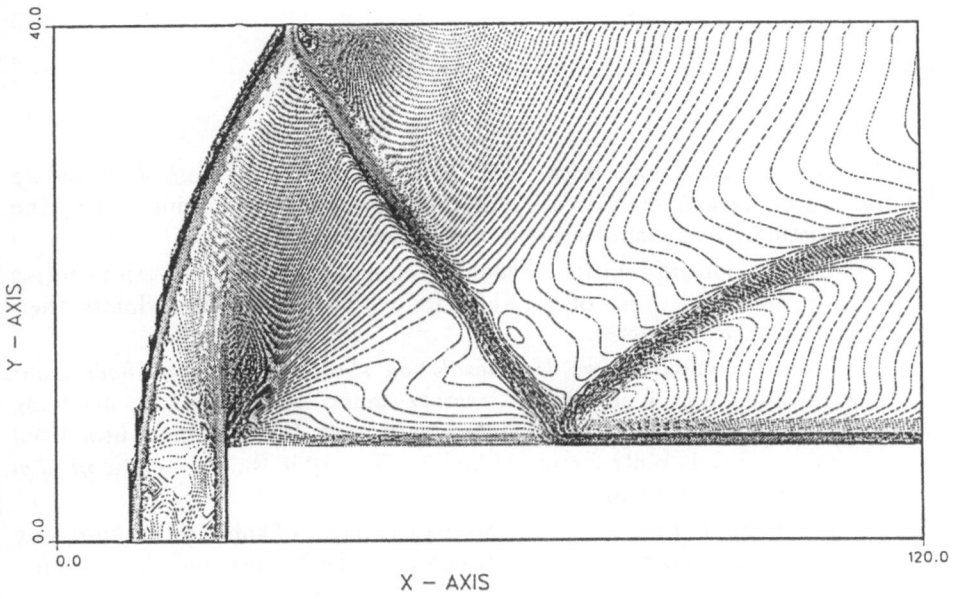

Fig. 9. Relativistic Emery's step. See text for details

As we have mentioned at the beginning of this section, an operator splitting technique in each spatial direction has been performed. We have also experimented with methods which avoid the splitting in spatial directions and which permit advancing in time with the third order Runge-Kutta method previously explained. In this last line of experimentation we have observed an important reduction of numerical noise and the best results have been obtained.

At the transonic rarefactions, where entropy violation may appear (and, in fact, it does), a local artificial viscosity according to the prescription of Harten and Hymann (1983) was incorporated.

Figure 9 (taken from [19]) shows the isodensity curves of the system at some instant of its evolution ($v_0^x = 0.9$). The main features of the solution are the Mach reflection of a *bow shock* at the upper wall, making the density distribution the most difficult to compute, and a *rarefaction fan* centered at the corner of the step. These general characteristics of the solution are similar to those found in the Newtonian case. Currently, we are experimenting with higher inflow velocities.

The severity of this test makes us confident of the feasibility of astrophysical applications as complex as the ones mentioned in the introduction of this paper.

10 Conclusions

1. We have extended some *high-resolution shock-capturing methods* (avoiding the use of an artificial viscosity in treating strong discontinuities) to the relativistic hydrodynamic system of equations.
   - The equations of relativistic hydrodynamics have been written in terms of a well-defined set of variables for which the system exhibits their conservative character.
   - We have compared two *high-resolution shock-capturing methods* : our version of MUSCL and PHM. Several standard and severe tests involving strong shocks have been successfully overcome. With PHM we have been able to reach Lorentz factors as high as $W \approx 70$ in the *relativistic normal reflection shock* problem.
2. We have computed the general-relativistic collapse of spherically symmetric configurations paying particular attention on the bounce and shock propagation.
3. We have succeeded in obtaining stationary solutions of general-relativistic potential flows in a Schwarzschild background.
4. By combining operator splitting techniques with the corresponding theoretical analysis of the spectral decomposition of the Jacobian matrices associated with each direction of the fluxes we have explored the realm of multidimensional relativistic hydrodynamics.
   - With our MUSCL code we have solved the relativistic oblique shock. In these experiments, the initial velocity of the supersonic flow (Mach 3) spans the interval $0.01 \leq v_1 \leq 0.95$ Our results accord with the analytical ones.
   - The severe Emery's step test has been overcome in its relativistic version.

The above conclusions give us confidence in the feasibility of our procedure to extend modern high-resolution shock-capturing methods to the multidimensional relativistic hydrodynamics. Some of the results shown in this lecture -in particular, those indicated in items 2 and 4- are, as far as we know, entirely new.

Acknowledgments

This work has been supported by the Spanish DGICYT (reference numbers PB90-0516-C02-02 and PB91-0648).

References

1. A.M. Anile, *Relativistic fluids and magneto-fluids*, Cambridge University Press (1989).
2. M.C. Begelman, R.D. Blandford, and M.J. Rees, Rev. Mod. Phys., **56**, 255 (1984).
3. C. Bona, J.Mª. Ibáñez, J.Mª. Martí and J. Massó, *Shock-capturing methods in 1D Numerical Relativity*, this volume (1992).
4. S. Bonazzola and J.A. Marck, in *Frontiers in Numerical Relativity*, ed. by Evans C.R. and Finn L.S., p. 239, Cambridge University Press (1989).
5. S. Bonazzola and J.A. Marck, J. Comp. Phys., **87**, 201 (1990).
6. S. Bonazzola and J.A. Marck, J. Comp. Phys., **97**, 535 (1991).
7. S. Bonazzola and J.A. Marck, in *Approaches on Numerical Relativity*, ed. by D'Inverno, Cambridge University Press (1992), *in press*.
8. H. Bondi, Proc. Royal Soc. of London, **A281**, 39 (1964).
9. G.E. Brown, H.A. Bethe, and G. Baym, Nucl. Phys., **A375**, 481 (1982).
10. J. Centrella, and J.R. Wilson, Ap. J. Suppl., **54**, 229 (1984).
11. P. Colella, and P.R. Woodward, J. Comp. Phys., **54**, 174 (1984).
12. P. Colella, and H.M. Glaz, J. Comp. Phys., **59**, 264 (1985);
13. R. Courant, E. Isaacson, and M. Rees, Comm. Pure Appl. Math., **5**, 243 (1952).
14. B. Einfeldt, SIAM J. Num. Anal., **25**, 294 (1988).
15. A.F. Emery, J. Comp. Phys., **2**, 306 (1968).
16. C. R. Evans, in *Dynamical space-times and numerical relativity* , ed. by J. Centrella (Cambridge University Press, 1986).
17. J.A. Font, J.Mª. Martí, J.Mª. Ibáñez and J.A. Miralles, Comput. Phys. Comm., submitted (1992).
18. J.A. Font, J.Mª. Ibáñez and J.Mª. Martí, *in preparation* (1992).
19. J.A. Font, J.Mª. Ibáñez A. Marquina and J.Mª. Martí, *in preparation* (1992).
20. H. Freistühler and E.B. Pitman, J. Comp. Phys., **100**, 306 (1992).
21. B.A. Fryxell, E. Müller, and W.D. Arnett, in *Numerical Methods in Astrophysics*, ed. P.R. Woodward (Academic Press) (1990).
22. R.A. Gingold and J.J. Monaghan, Mon.Not.R.Astron.Soc., **181**, 375 (1977).
23. P. Glaister, J. Comp. Phys., **74**, 382 (1988).
24. S.K. Godunov, Matematicheskii Sbornik, **47**, 271 (1959).
25. D. Gottlieb and S. Orszag, *Numerical Analysis of Spectral Methods.: Theory and Application*, Regional Conference Series Lectures in Applied Mathematics, **26** (SIAM, Philadelphia,1977).
26. E. Gourgoulhon, Astron. and Astrophys., **252**, 651 (1992).
27. A. Harten, B. Engquist, S. Osher, S. Chakravarthy, J. Comp. Phys., **71**, 231 (1987).

28.   A. Harten, J.M. Hyman, and P.D. Lax, Comm. Pure Appl. Math., **29**, 297 (1976).
29.   A. Harten and J.M. Hyman, J. Comp. Phys., **50**, 235 (1983).
30.   A. Harten, P.D. Lax and B. Van Leer, SIAM Rev., **25**, 35 (1983).
31.   A. Harten, and S. Osher, SIAM J. Num. Anal., **24**, 279 (1987).
32.   J.F. Hawley, L.L. Smarr, and J.R. Wilson, Ap. J. Suppl., **55**, 211 (1984).
33.   J.M$^{\underline{a}}$. Ibáñez, Astron. and Astrophys., **135**, 382 (1984).
34.   J.M$^{\underline{a}}$. Ibáñez, J.M$^{\underline{a}}$. Martí, J.A. Miralles and V. Romero, in *Approaches on Numerical Relativity*, ed. by D'Inverno, Cambridge University Press (1992), *in press*.
35.   A. Königl, Phys. Fluids, **23**, 1083 (1980).
36.   P.D. Lax *Regional Conference Series Lectures in Applied Math.*, **11** (SIAM, Philadelphia,1973).
37.   P.D. Lax, and B. Wendroff, Comm. Pure Appl. Math., **13**, 217 (1960).
38.   L.B. Lucy, Astron. J.,**82**, 1013 (1977).
39.   P.J. Mann, Comput. Phys. Comm., **67**, 245 (1991).
40.   A. Marquina, SIAM J. Scient. Stat. Comp., *in press* (1992).
41.   A. Marquina, J.M$^{\underline{a}}$. Martí, J.M$^{\underline{a}}$. Ibáñez, J.A. Miralles and R. Donat, Astron. and Astrophys., **258**, 566 (1992).
42.   J.M$^{\underline{a}}$. Martí, J.M$^{\underline{a}}$. Ibáñez, and J.A. Miralles, Astron. and Astrophys., **235**, 535 (1990).
43.   J.M$^{\underline{a}}$. Martí, *Ph.D. Thesis*, University of Valencia (1991).
44.   J.M$^{\underline{a}}$. Martí, J.M$^{\underline{a}}$. Ibáñez and J.A. Miralles, Phys. Rev., **D43**, 3794 (1991).
45.   F.C. Michel, Ap. Space Sci., **15**, 153 (1972).
46.   J.C. Miller and O. Pantano, Phys. Rev., **D40**, 1789 (1989).
47.   J.J. Monaghan and R.A. Gingold, J. Comp. Phys., **52**, 374 (1983).
48.   J.J. Monaghan, Ann. Rev. Astron. Astrophys., **30**, 543 (1992).
49.   T. Nakamura, Prog. Teor. Phys., **65**, 1876 (1981).
50.   T. Nakamura, K. Maeda, S. Miyama and M. Sasaki, Prog. Teor. Phys., **63**, 1229 (1980).
51.   T. Nakamura and H. Sato, Prog. Teor. Phys., **67**, 1396 (1982).
52.   W.F. Noh, J. Comp. Phys., **72**, 78 (1987).
53.   M.L. Norman, and K-H.A. Winkler, in *Astrophysical Radiation Hydrodynamics*, ed. by M.L. Norman, and K-H.A. Winkler (Reidel, 1986).
54.   O.A. Oleinik, Usp. Mat. Nauk., **12**, 3 (1957); Engl. transl., Am. Mat. Soc. Transl. Ser. 2, **26**, 95 (1963).
55.   L.I.Petrich, S.L. Shapiro and S.A.Teukolsky, Phys. Rev. Lett., **60**, 1781 (1988).
56.   T. Piran, J. Comp. Phys., **35**, 254 (1980).
57.   R. Richtmyer and K. Morton, *Difference Methods for Initial-Value Problems*, 2nd. ed., (Interscience, 1967).
58.   P.L. Roe, J. Comp. Phys., **43**, 357 (1981).
59.   S.L. Shapiro and S.A. Teukolsky, in *Black Holes, White Dwarfs and Neutron Stars* (Wiley, 1983).
60.   C.W. Shu and S.J. Osher, J. Comp. Phys., **83**, 32 (1989).
61.   R. F. Stark and T. Piran, Comp. Phys. Rept., **5**, 221 (1987).
62.   M. Steinmetz and E. Müller, preprint from Max-Planck-Institut für Astrophysik, *submitted to Astron. and Astrophys.* (1992).
63.   B. Van Leer, J. Comp. Phys., **23**, 276 (1977).
64.   B. Van Leer, J. Comp. Phys., **32**, 101 (1979).
65.   B. Van Leer, SIAM J. Sci. Stat. Comput., **5**,1 (1984).
66.   K.A. Van Riper, Ap. J., **221**, 304 (1978).

67.   J. Von Neumann, and R. Richtmyer, J. of Appl. Phys., **21**, 232 (1950).

68.   J.R. Wilson, Astrophys. J., **173**, 431 (1972).

69.   J.R. Wilson, in *Sources of gravitational radiation* , ed. L.L. Smarr (Cambridge University Press, 1979).

70.   M.J. Wilson, Mon. Not. R. Astr. Soc., **226**, 447 (1987).

71.   H.C. Yee, *VKI Lecture Notes in Computational Fluid Dynamics*, von Karman Institute for Fluid Dynamics, Belgium (1989).

Singularity-Free Spacetimes

José M. M. Senovilla

Departament de Física Fonamental, Universitat de Barcelona,
Diagonal 647, 08028 Barcelona
and
Laboratori de Física Matemàtica, Societat Catalana de Física, I.E.C.,
Barcelona, Spain.

1 Introduction

It is very well known that there are *many* singularity-free solutions of Einstein's equations. Some examples are the gravitational field of stars, galaxies, etc., which are described mathematically by means, for example, of a spherically symmetric interior solution (say of perfect fluid) matched with the vacuum Schwarzschild solution. The existence of many globally regular solutions has been studied several times, and an interesting paper on this is [1]. However, it is a widespread belief that solutions with *cosmological* properties must contain singularities if the energy and causality conditions hold. What "cosmological properties" means here is not very well defined, but we shall try to precise this later.

It should be stressed from the very begining that we do not intend to criticize the standard big-bang models of the Universe, which we believe are the best candidates to describe the large scales of the actual Universe from the nucleosynthesis time on. Our purpose is of a very different kind, namely, to raise the discussion on whether or not the big bang, *understood as the initial classical singularity*, is compulsory in standard General Relativity when quantum phenomena are not taken into consideration. By "initial classical singularity" we mean the singularity appearing in most known cosmological models which gives birth to the Universe and where the matter density, the curvature of spacetime and other physical quantities blow up. To that end, we need to know why most scientists believe that the initial singularity is unavoidable. In our opinion, this is due to two main reasons: First, because it is thought that the big-bang singularity follows from the *singularity theorems* together with the astrophysical and cosmological observations. And second, because it has been hardly ever considered the possibility of other types of singularities in cosmology (not initial big-bang-like). However, we think that these two reasons cannot stand a deeper and exhaustive scientifical analysis and in this contribution we shall try to show that they are rather naive, at least in their simplest versions.

2 The Singularity Theorems and Their Folklore Versions

To start our study, we should like to remind the reader about the beautiful and very powerful singularity theorems due to Penrose, Hawking and Geroch, among others (see Refs.[2][3][4]). As is well known, there is a large variety of those theorems but all of them have the same structure or skeleton. Thus, we can summarize them all in the following *typical singularity theorem*:

Theorem   (Typical Singularity Theorem).   *In classical General Relativity, a locally inextendible spacetime satisfying*

  i) an energy condition,
 ii) a causality condition, and
iii) a boundary or initial condition (usually the existence of a causally trapped set)

must contain, at least, one incomplete causal geodesic.

Let us explain briefly the above theorem. First of all, we want our spacetime not extendible anywhere, because a possible extension should be considered as part of the spacetime itself and could also hide the singularities. The first important assumption is the energy condition, which can be of several types (weak, dominant, strong ...). All of them assure simply, roughly speaking, that every observer measures a non-negative energy density. The implication is obviously that gravity is always attractive. The second assumption is a causality condition, which again has more than one version (examples: chronology, proper causality and stable causality conditions). More or less, they state that no-one can travel to the future and enter into a region capable of influencing his/her own past. They imply, then, that nothing can avoid the arrow of time. These two assumptions are very reasonable on classical physical grounds.

The third assumption is a little bit more complicated. A causally trapped set ζ, is a set such that $E^{\pm}(\zeta)$ is bound to be compact —by $E^{\pm}(\zeta)$ we mean the set of points in spacetime that can be reached from ζ by a future- (past-) directed null curve but cannot be so reached by a timelike curve. Some examples are closed trapped surfaces (the original Penrose hypothesis, see [2]), compact achronal sets without edge (such as space in closed cosmological models), points with reconverging lightcones, slices with bounded above (below) zero expansion, etcetera. The implication of this assumption is that there are regions in space (not spacetime) which cannot prevent having their own future (or past) completely included (or excluded) in themselves. This is something like that popular sentence "nothing can escape from a black hole".

Finally, the conclusion of the theorem is that there must exist a causal geodesic (either an observer or a photon) such that the physical quantities are not well-behaved along its worldline. The proof of the theorems vary depending on the different possibilities we have mentioned, but non-rigorously it goes like this: suppose there is a causally past trapped set, then it follows that a little time before, all the matter inside the set was in a smaller region which, being gravity always attractive, must be again trapped. The argument goes on and on like we have just shown, and as no-one can escape the arrow of time we arrive at a situation where the region is too small to have so much matter. As we can see, the energy and causality conditions are important in the theorem, but the key assumption, however, is the existence of a causally trapped set that acts like a time-bomb in the theorem.

Due to the complexity of the third assumption in the singularity theorems, there have often appeared, both in oral and written science, simplified "theorems" which are meant to be physical and to pinpoint the important clues for cosmological questions. One of these popular "theorems", which we call the sophisticated folklore cosmological singularity "theorem", reads as follows:

Conjecture (Folklore Cosmological Singularity "Theorem"). *In classical General Relativity, a locally inextendible spacetime satisfying*

i) an energy condition,
ii) a causality condition, and
iii) a boundary cosmological condition: **(now expanding) matter everywhere,**

starts in an initial big-bang singularity.

We have remarked the differences between this and the proper singularity theorem in boldface characters. There is first a difference in the boundary condition, which has been replaced for a rather obvious statement of what we are sure about the Universe. Then, there is another difference in the conclusion, which is not just one incomplete geodesic but incompleteness of every and each of the geodesics (a universal singularity). Although the above conjecture has never been proved (and in fact it is false as we shall show later), there were some good reasons for believing in it. Mainly, these reasons come from the history of the cosmological models. For example, all known *exact* cosmological models with high symmetry, say with three or more Killing vectors such as Friedman-Robertson-Walker (FRW), locally rotationally simmetric (LRS) or Bianchi models, have a big-bang singularity in the finite past (if the energy conditions hold). Thus, if one thinks that the Universe is and has always been *exactly* spatially homogeneous, then one is safe with the folklore "theorem". Actually, there is a true theorem due to Raychaudhuri [5] which assures the existence of a big-bang singularity in the finite past if the now expanding matter has neither acceleration nor rotation and the energy condition $\rho + 3p > 0$ is verified (where ρ is the energy density of matter and p its *isotropic* pressure).

One should bear in mind, however, that all these models have physical quantities depending only on time, so that the energy density and pressure of the matter contents are functions of type $\rho(t)$ and $p(t)$, where t is usually proper time along the flow-lines of matter. That means that ρ and p can become infinite only for a particular value of $t = t_o$. Therefore, which singularities could there possibly be other than big-bang? If we are to be sure of the inevitability of the initial big-bang singularity, we must perform a detailed analysis of more general models allowing for densities and pressures of type $\rho(t, \boldsymbol{x})$ and $p(t, \boldsymbol{x})$, where here $\boldsymbol{x}$ stands for generic spatial coordinates. Equivalently, we have to study models with less symmetry (spatially inhomogeneous models) and see whether or not the big-bang character of the singularity is necessary there under cosmological conditions. As noted by Ellis in Ref.[6], there is no indication that this should be the case and, in principle, one might be able to find *timelike* singularities in inhomogeneous models.

3 Diagonal G_2 Perfect-Fluid Cosmologies

The drawback in the above reasoning is that the resolution of Einstein's equations for inhomogeneous models is quite difficult. Nevertheless, some exact solutions have been found when the space of the Universe still has two symmetries (and not three, as required for homogeneous models). We restrict our analysis here to the case in which the two symmetries commute, the corresponding Killing vectors are orthogonal to each other and also hypersurface orthogonal. This is the simplest case, and the metrics are known technically as *Diagonal G_2 cosmologies*. Of course, it is very unlikely that any of these models can fit with observations of the Universe, but the crucial point is that *they serve to check, theoretically, if the folklore wisdom is right*. The first solutions of this sort were found by Wainwright and Goode in Ref.[7]. They have a realistic equation of state, satisfy the energy conditions and present a big-bang singularity in the finite past. The next solution was derived some years later in [8]. Again, the properties were similar to most known cosmological models, including the initial singularity. Therefore, the belief that this singularity was classically inevitable was reinforced.

Fortunately, this is not the end of the story. In a series of recent papers [9][10][11], a very general class of perfect-fluid Diagonal G_2 cosmologies was found and studied, and the results were quite unexpected and clarifying. For our present purposes, we consider here the following line-element [10]

$$ds^2 = T^{2m} e^{nf} e^{(1-2m)g} \left(-dt^2 + dx^2\right) + Te^g \left(T^n e^f dy^2 + T^{-n} e^{-f} dz^2\right) \qquad (1)$$

where m is a constant such that $m \geq -\frac{1}{4}$, $m \neq 1$, and $n^2 = 4m+1$. The function $T(t)$ depends only on t and is given by

$$\ddot{T} = \epsilon a^2 T, \qquad \epsilon = \pm 1, 0 \ ,$$

where a is a positive arbitrary constant, dots stand for derivatives with respect to t and ϵ characterizes the closeness or openness of the models. Equivalently, we have

$$T(t) = \begin{cases} A\cosh(at) + B\sinh(at) & \text{if} \quad \epsilon = 1 \ , \\ At + B & \text{if} \quad \epsilon = 0 \ , \\ A\cos(at) + B\sin(at) & \text{if} \quad \epsilon = -1 \ . \end{cases} \qquad (2)$$

where A and B are arbitrary constants. The line-element (1) is a solution of Einstein's equations for an energy-momentum tensor of a perfect fluid when the functions $g(x)$ and $f(x)$, which depend only on the coordinate x, satisfy the following pair of differential equations

$$f'' + f'g' = \epsilon na^2 \ ,$$
$$2(1 - m)g'' + (2m - 1)g'^2 - 2ng'f' + f'^2 = -2\epsilon(1 + m)a^2$$

where primes denote derivatives with respect to x. Then, the velocity one-form of the fluid is

$$u = -T^m e^{\frac{n}{2}f} e^{\frac{(1-2m)}{2}g} \, dt$$

so that the motion of the fluid is irrotational. The pressure takes the following form

$$p = cT^{-2m} e^{\frac{mn}{1-m}f} e^{\frac{(1-2m)m}{1-m}g} \tag{3}$$

and the equation of state of the fluid reads

$$p = \gamma\rho, \qquad \gamma = \frac{m-1}{m+1} \ . \tag{4}$$

In order to get a flavour of the behaviour of these solutions, let us define the scale factor $R(t)$ of the models, which is the analogue of the scale factor in FRW models, and is given by

$$u^\alpha \partial_\alpha R = \frac{\theta}{3} R \ , \tag{5}$$

where u is the velocity vector of the fluid and θ its expansion. This scale factor has a very simple expression:

$$R(t) = T(t)^{\frac{m+1}{3}} = T(t)^{\frac{2}{3(1-\gamma)}} \ . \tag{6}$$

As we can see from (2) and (6), the behaviour of the scale factor is similar to that in FRW models in cases $\epsilon = 0, -1$ and also for $\epsilon = 1$ with $B^2 \geq A^2$. However, there is a completely new behaviour when $\epsilon = 1$ and $A^2 > B^2$ (without loss of generality we can set $A = 1, B = 0$ in this case). For this new type of model, the energy density and pressure of the fluid are such that their time dependence goes like

$$\rho \sim \cosh^{-2\frac{1+\gamma}{1-\gamma}}(at) \ ,$$

as follows from (3) and (4). It is therefore obvious $(0 < \gamma < 1)$ that these models *do not have any spacelike (big bang) singularity*, because the physical quantities are well-behaved for all possible values of t. Using (5), (6) and (2) it follows that the time dependence of the expansion in these solutions is

$$\theta \sim a\frac{2}{1-\gamma} \frac{\sinh(at)}{\cosh^{\frac{2}{1-\gamma}}(at)}.$$

From this expression we learn that there is an expanding phase for $t > 0$ and a contracting phase for $t < 0$. At $t = 0$, where the expansion vanishes, a bounce takes place preventing the formation of a spacelike singularity by the collapsing matter and expelling that matter in an expanding bang from then on. This is possible because the matter is accelerating, a fact that, for perfect fluids, is equivalent to the existence of a spacelike gradient of pressure. In other words, there is a force which opposes the gravitational attraction.

In what follows, we summarize briefly the main *new* results for this family of solutions with no big-bang singularity. There appear several different behaviours depending on the values of the parameter γ. We present explicitly an example of solution for each of these cases.

3.1 Metrics with $\gamma > \frac{1}{3}$

If $\gamma > \frac{1}{3}$, the solutions have a *timelike* singularity (thus, not interpretable as a big bang) in both the matter quantities (density and pressure) and in the pure gravity quantities (the Weyl tensor). This is a general result in this case, and we can see how the metrics and the physical quantities behave by presenting the particular solution for $\gamma = \frac{1+\sqrt{5}}{4}$. The line-element reads [10]

$$
\begin{aligned}
ds^2 =& \cosh^{2(5+2\sqrt{5})}(at)\, e^{2(3+\sqrt{5})ax}\sinh^{2(2+\sqrt{5})}\left[(\sqrt{5}-1)\,ax\right]\left(-dt^2+dx^2\right)+ \\
&+ \cosh(at)\, e^{(2-\sqrt{5})ax} \times \\
&\sinh\left[(\sqrt{5}-1)\,ax\right]\left[\cosh^{4+\sqrt{5}}(at)\,e^{ax}\sinh^{2+\sqrt{5}}\left[(\sqrt{5}-1)\,ax\right]dy^2+ \right. \\
&\left.+\cosh^{-(4+\sqrt{5})}(at)\,e^{-ax}\sinh^{-(2+\sqrt{5})}\left[(\sqrt{5}-1)\,ax\right]dz^2\right]
\end{aligned}
$$

and the pressure is

$$
p = \text{const} \times \cosh^{-2(5+2\sqrt{5})}(at)\, e^{-(5+3\sqrt{5})ax}\sinh^{-(5+2\sqrt{5})}\left[(\sqrt{5}-1)\,ax\right] \ .
$$

As we can see, there is a singularity at $x = 0$ for the energy density and pressure, which is also a singularity in the Weyl tensor [10]. Therefore, the first thing we learn with this case is that the singularity in theoretical cosmological models do not have to be of big-bang type; rather, it can also be a timelike singularity for the matter and pure gravitational fields.

3.2 Metrics with $\gamma < \frac{1}{3}$

The case with $\gamma < \frac{1}{3}$ is a little different to the previous one. An appropriate typical example for these metrics is given by the following line-element [10]

$$
\begin{aligned}
ds^2 =& \cosh^{1+\sqrt{2}}(at)\tanh^{1-\sqrt{2}}\left(\sqrt{4+3\sqrt{2}}\,ax\right)\left(-dt^2+dx^2\right)+ \\
& \cosh(at)\sinh\left(\sqrt{4+3\sqrt{2}}\,ax\right)\cosh^{\frac{\sqrt{2}-2}{2}}\left(\sqrt{4+3\sqrt{2}}\,ax\right) \times \\
& \left[\cosh^{1+\sqrt{2}}(at)\sinh^{\sqrt{2}-1}\left(\sqrt{4+3\sqrt{2}}\,ax\right)dy^2 \right. \\
& \left.+ \cosh^{-(1+\sqrt{2})}(at)\sinh^{1-\sqrt{2}}\left(\sqrt{4+3\sqrt{2}}\,ax\right)dz^2\right] \ .
\end{aligned}
$$

From this expression it is obvious that there is a singularity at $x = 0$. However, this is not a singularity in the matter quantities, because the pressure is now

$$p = \text{const} \frac{\tanh^{1+\sqrt{2}}\left(\sqrt{4 + 3\sqrt{2}}\, ax\right)}{\cosh^{1+\sqrt{2}}(at)}$$

and the equation of state takes the form $p = \frac{4\sqrt{2}-5}{7}\rho$. Nevertheless, it can be shown [10] that $x = 0$ is a true singularity for the Weyl tensor. Thus, the next thing we learn is that the singularity, in addition to possibly being timelike, need not be a matter singularity (appearing only as a bad behaviour in the pure gravitational Weyl tensor).

3.3 Metrics with $\gamma = \frac{1}{3}$: Singularity-free Solutions

The most interesting case arises when $\gamma = \frac{1}{3}$. We present here the following line-element

$$ds^2 = \cosh^4(at)\cosh^2(3ar)\left(-dt^2 + \frac{\sinh^2(3ar)}{\cosh^2(3ar) + (K-1)\cosh^{\frac{5}{3}}(3ar) - K}dr^2\right)$$

$$+ \cosh^4(at)\frac{\cosh^2(3ar) + (K-1)\cosh^{\frac{5}{3}}(3ar) - K}{9a^2L^2\cosh^{\frac{2}{3}}(3ar)}d\phi^2$$

$$+ \cosh^{-2}(at)\cosh^{-\frac{2}{3}}(3ar)dz^2, \tag{7}$$

where $L \equiv \frac{5K+1}{6}$, K is an arbitrary constant and the range of coordinates is taken to be

$$-\infty < t < \infty, \quad 0 \leq r < \infty, \quad 0 \leq \phi \leq 2\pi$$

such that ϕ and $\phi + 2\pi$ are identified. This metric has a regular axis of symmetry $r = 0$ at all times and the so-called elementary flatness in the vecinity of the axis is satisfied. The above model is then a cylindrically symmetric cosmology.

The pressure for the above solution is

$$p = 5Ka^2\cosh^{-4}(at)\cosh^{-4}(3ar)$$

and the equation of state is $p = \frac{1}{3}\rho$. Therefore, both the energy density and the pressure are positive everywhere and with realistic equation of state if $K > 0$. Furthermore, ρ and p are *regular* over the whole spacetime, and there is no matter singularity in the solution. Let us note that the maximum value of ρ, which occurs when $t = r = 0$, is represented by the constants K and a, and thus we can choose this maximum as large as we like. It is easily seen [10] that the Weyl tensor (and in fact every physical quantity) is regular also everywhere, that is to say, the solutions do not present *any singularity at all*. The fact that the equation of state for these singularity-free models is $p = \frac{1}{3}\rho$, that is to say, realistic for radiation dominated epochs in the (so-called) very early Universe, is at least a marvelous and beautiful coincidence. Thus, the last and perhaps most interesting thing we learn is that there are cosmological solutions with no singularity whatsoever.

4 Discussion and Remarks

The reader might now be wondering what the properties of the singularity-free models are. Do they have reasonable properties? Do they satisfy causality and other requirable conditions? Well, it turns out that these models do have very good properties, an exhaustive description of which can be found in Ref.[11], where the particular case $K = 1$ of metric (7) was studied (this particular case is the first singularity-free solution found in [9]). To start with, the solutions satisfy the strongest requirement on energy conditions, namely, the density and pressure are positive everywhere and the equation of state is realistic so that $p < \rho$. These are stricter conditions than the usual so-called strong energy condition. Spacetime is also causally stable, again the strongest requirement on causality conditions, and it implies the weaker chronology and causality conditions. This means that there are no causal pathologies even if the lightcones are slightly opened up. Furthermore, spacetime is globally hyperbolic (and therefore causally simple) which is the simplest and nicest possibility. It follows the existence of global Cauchy hypersurfaces, so that there are no Cauchy horizons, the global causal structure and topology being similar to that of Minkowski spacetime. Of course, the solutions are inextendible, so we are sure we are studying the whole manifold without anything hiding somewhere. Finally, the solution is not only geodesically complete but also complete in the stricter sense that every causal curve (geodesic or not) can be extended to arbitrary values of its generalized affine parameter. This is the strongest definition of a singularity-free spacetime.

In Ref.[11] it was also proven that there is not any causally trapped set. In addition, we compared the properties of the singularity-free spacetimes with a large variety of singularity theorems and we saw that these solutions *are in accordance* with them (as it should be, being proper **theorems**). However, all solutions with no big-bang singularity (be them with timelike singularities or with no singularity at all) are counterexamples of the folklore cosmological singularity "theorem". This "theorem" fails in two respects: First, with the cosmological conditions replacing the actual boundary conditions, for the solutions here presented have matter everywhere which is expanding half the history of the model. And second with the conclusion, for the singularity may or may not be there, but there is not any theoretical reason to have an initial spacelike universal singularity. We should like to remark that any other folklore cosmological singularity "theorem" to come would fail just the same. It is obvious that *completely general* inhomogeneous models will or will not be singular, but the singularities they may have are of many different kinds, and sometimes they will not be of big-bang type. There were indications that this could be the case (see [6]), but the lack of inhomogeneous solutions had not allowed any explicit checking. The main conclusion of the above reasoning is that, on theoretical grounds, the initial big-bang singularity can certainly be avoided classically. A quite different question is whether or not the Universe we live in has originated in a big-bang singularity (possibly not singular when quantum effects are considered). In order to ascertain this question, we should try to know if the *proper* boundary conditions in the singularity theorems hold now (or did hold previously) in the Universe. Even if this could be shown beyond any doubt, it will still remain the question of the character of the singularity, and claiming it should be big-bang-like will be far from easy. Alternatively, but equivalently, one could try to construct general models both realistic (reproducing the observational large-scale data) and free of big-bang singularity. Whether this is possible or not, and despite the fact that it has been claimed the opposite hitherto, is, as yet, *an open question*.

References

1. Bonnor, W. B.: Globally regular solutions of Einstein's equations. Gen. Rel. Grav. **14** (1982) 807-821
2. Hawking, S. W., Ellis, G. F. R.: The Large Scale Structure of Spacetime, Cambridge University Press (1973) Cambridge
3. Geroch, R., Horowitz G. T.: Global structure of spacetimes, in General Relativity. An Einstein Centenary Survey, p. 212 ed. S. W. Hawking and W. Israel, Cambridge University Press (1979) Cambridge
4. Tipler, F. J. C., Clarke, J. S., Ellis, G. F. R.: Singularities and Horizons –A Review Article, in General Relativity and Gravitation. One Hundred Years After the Birth of Albert Einstein, ed. A. Held, Plenum Press (1980) New York
5. Raychaudhuri, A. K.: Relativistic cosmology, Phys. Rev. **98** (1955) 1123-1126
6. Ellis, G. F. R.: Alternatives to the big-bang, Annu. Rev. Astron. Astrophys. **22** (1984) 157-184
7. Wainwright, J., Goode, S. W.: Some exact inhomogeneous cosmologies with equation of state $p = \gamma\rho$, Phys. Rev. D **22** (1980) 1906-1909
8. Feinstein, A., Senovilla, J. M. M.: A new inhomogeneous cosmological perfect fluid solution with $p = \rho/3$. Class. Quantum Grav. **6** (1989) L89-L91
9. Senovilla, J. M. M.: New class of inhomogeneous cosmological perfect-fluid solutions without big-bang singularity. Phys. Rev. Lett. **64** (1990) 2219-2221
10. Ruiz, E., Senovilla, J. M. M.: General class of inhomogeneous perfect-fluid solutions. Phys. Rev. D **45** (1992) 1995-2005
11. Chinea, F. J., Fernández-Jambrina, L., Senovilla, J. M. M.: Singularity-free space-time. Phys. Rev. D **45** (1992) 481-486

On Radiative Solutions in General Relativity

Dietrich Kramer

Friedrich-Schiller-Universität Jena, Theoretisch-Physikalisches Institut,
O-6900 Jena, Germany

1 Radiative Vacuum Fields ($R_{mn} = 0$)

1.1 Introduction

It is a long-standing problem to construct an asymptotically flat exact solution of Einstein's field equations which describes the radiative gravitational field of an isolated source emitting gravitational waves. The nonlinearity of the field equations does not allow us to prescribe the source structure, e.g., for the restricted two-body problem (two masses moving on a line under their mutual gravitational interaction), which of course is solved in Newtonian theory, one has no guidance how to get the corresponding solution in Einstein's theory.

The main approaches to obtain exact radiative solutions are the assumption of space-time symmetries (axial or boost-rotation symmetry; spherical symmetry is excluded by Birkhoff's theorem) and/or the restriction to algebraically special gravitational fields, i.e., to fields the Weyl tensor of which has at least one repeated null eigendirection.

1.2 Penrose's Definition of Asymptotical Flatness

According to Penrose [16] a physical space-time $(\mathcal{M}, g)$ is asymptotically flat if there is an associated manifold $(\tilde{\mathcal{M}}, \tilde{g})$ with the conformally related metric

$$\tilde{g}_{mn} = \Omega^2 g_{mn} \ , \tag{1}$$

where the conformal factor Ω is smooth in $\tilde{\mathcal{M}}$ and satisfies the conditions

$$\Omega = 0 \ , \qquad \tilde{\nabla}_m \Omega \neq 0 \tag{2}$$

on the boundary $\mathcal{J}$ of $\tilde{\mathcal{M}}$. The boundary $\mathcal{J}$ is called null infinity and consists of two parts, past and future null infinity denoted by $\mathcal{J}^-$ and $\mathcal{J}^+$, respectively. The topology of $\mathcal{J}$ is assumed to be $S^2 \times R$. The geometrical concept of null infinity is crucial in gravitational radiation theory and does not depend on the choice of space-time coordinates.

The compactification of $\mathcal{M}$ which is achieved by (1) enables one to apply local geometrical techniques at infinity. Penrose's definition of asymptotical flatness does not only imply the vanishing of the curvature tensor at $\mathcal{J}$ but also its peeling-off property. The question arises whether there are at all physically reasonable radiative space-times which are asymptotically flat in the sense of Penrose. To some extent, the solutions with boost-rotation symmetry (Sec. 1.6) satisfy Penrose's criterion.

1.3 The Bondi Metric

In the famous paper by Bondi et al. [4] the space-time coordinates of an axisymmetric asymptotically flat radiative vacuum field are adapted to outward radial light rays (u, ϑ, φ are constant and r varies along the ray). The hypersurfaces $u = $ const are null. The radial coordinate r is fixed such that the area A of the 2-surfaces ($u = $ const, $r = $ const) is just $A = 4\pi r^2$. The Bondi metric takes the form

$$ds^2 = -\left(\frac{V}{r}e^{2\beta} - U^2 r^2 e^{2\gamma}\right) du^2 - 2e^{2\beta} du dr$$
$$+ r^2 \left(e^{2\gamma} d\vartheta^2 + e^{-2\gamma}\sin^2\vartheta d\varphi^2\right) - 2Ur^2 e^{2\gamma} du d\vartheta . \tag{3}$$

The four metric functions γ, β, U, and V which depend on the three variables u, r, and ϑ have to satisfy regularity conditions at the axis ($\vartheta = 0, \pi$). The expansion of the metric (3) at large distances in terms of powers of $1/r$ is possible; logarithmic terms, which proved to be a deficiency in previous investigations, are avoided in Bondi's treatment.

If one labels the coordinates as $x^m = (u, r, \vartheta, \varphi)$, $m = 0, \ldots, 3$, the vacuum field equations split into
- the four main equations:

$$R_{11} = R_{12} = R_{22} = R_{33} = 0$$

(only the standard equation $R_{33} = 0$ does contain u-derivatives),
- the trivial equation $R_{01} = 0$ which is a consequence of the main equations, and
- the supplementary conditions $R_{02} = R_{00} = 0$ which need to be satisfied only for a fixed value of r.

From the expansion

$$\gamma = \frac{c(u, \vartheta)}{r} + \ldots , \quad V = r - 2M(u, \vartheta) + \ldots \tag{4}$$

one derives the relation

$$M_0 = -c_0{}^2 + \frac{1}{2}(c_{22} + 3c_2 \cot\vartheta - 2c)_0 . \tag{5}$$

(The subscripts denote partial derivatives with respect to the corresponding coordinate.) The integration of (5) over the 2-surfaces (u = const, r = const), and the regularity conditions at the axis, lead to the important result

$$\frac{dm}{du} \leq 0, \quad m = m(u) := \frac{1}{4\pi} \iint M(u,\vartheta)\sin\vartheta \; d\vartheta d\varphi \; . \tag{6}$$

The Bondi mass $m(u)$ is a monotonically decreasing function of the retarded time u. This behaviour is due to the emission of gravitational waves. $M(u,\vartheta)$ is called the mass aspect and the function $c_0(u,\vartheta)$ in (4) which entirely controls the evolution of the radiation field is said to be the news function.

Because of the complicated form of the main equations it is unlikely to find exact solutions starting with the Bondi metric (3). Therefore, additional conditions are to be imposed. Such approaches will be briefly described in the next sections.

1.4 The Robinson-Trautman Class

The repeated null eigenvector k^m of an algebraically special curvature tensor is, for vacuum fields, necessarily geodesic and shearfree. Robinson and Trautman [19] treated the case when k^m is expanding and non-twisting. It is useful to choose coordinates which are adapted to the null congruence k^m. The Robinson-Trautman metric reads

$$ds^2 = \frac{2r^2}{p^2} \, d\zeta d\bar\zeta - 2d\sigma dr - \left[\Delta \ln P - 2r(\ln P)_{,\sigma} - \frac{2m}{r}\right] d\sigma^2$$
$$P = P(\sigma,\zeta,\bar\zeta) \, , \quad m = \text{const}, \tag{7}$$

where $K = \Delta \ln P$ is the Gaussian curvature and $\Delta = 2P^2 \partial_\zeta \partial_{\bar\zeta}$ the Laplacian of the 2-surfaces S_2 (σ = const, r = const) with the complex-conjugate coordinates ζ and $\bar\zeta$. We restrict ourselves to the physically more interesting case $m \neq 0$ (Petrov types III and N are excluded). The regular 2-surfaces S_2 are assumed to have the topology of the sphere ("spherical" Robinson-Trautman class).

The field equations reduce to the partial differential equation

$$12m \frac{P_\sigma}{P} = -\Delta K \tag{8}$$

for the real function $P = P(\sigma,\zeta,\bar\zeta)$ from which the total metric (7) can be obtained.

Using equation (8), Lukács et al. [14] were able to show that

$$\frac{d}{d\sigma} \iint_{S_2} K d\tau = 0 \, , \quad \frac{dL}{d\sigma} \leq 0 \, , \quad L := \iint_{S_2} K^2 d\tau \, , \tag{9}$$

where $d\tau$ is the surface element of S_2. The positive quantity L decreases monotonically; L is a Ljapunov functional. Moreover, $\frac{dL}{d\sigma} = 0$ implies $K = $ const, which characterizes the Schwarzschild solution. Within the Robinson-Trautman class the Schwarzschild solution is asymptotically stable. The Robinson-Traut-man solutions evolve towards that final equilibrium state. This process takes place under emission of gravitational radiation. Eventually, the gravitational field settles down to a Schwarzschild black hole. It is an open question whether the Bondi metric which is believed to cover all asymptotically flat, axisymmetric radiative vacuum fields exhibits the same behaviour.

Computer calculations [17] starting with distorted spheres at an initial value of retarded time σ as well as a linear analysis [7] also show that Robinson-Trautman solutions decay exponentially and approach the Schwarzschild solution at future null infinity. This picture is rather satisfactory. However, the solutions although well-behaved on part of $\mathcal{J}^+$ seem to be not regular at $\mathcal{J}^-$, incoming radiation might be present. To study these problems concisely, it is very desirable to find exact analytic solutions to the Robinson-Trautman equation (8).

1.5 The Axisymmetric Robinson-Trautman Class

As an additional condition which makes the integration more tractable we assume axial symmetry. The axisymmetric Robinson-Trautman class is a subclass of the Bondi metric describing axisymmetric but not only algebraically special fields. Lun and Fletcher [15] considered the complicated transformation, in series form, between the axisymmetric Robinson-Trautman metric and the algebraically special Bondi metric. (The variable r in the metrics (3) and (7) has different meaning.) The axial symmetry is associated with a hypersurface–normal Killing vector η^m with closed orbits. From a study of algebraically special gravitational fields admitting a hypersurface–normal Killing vector [12] it follows that we can confine ourselves to the case $k_m \eta^m = 0$, $k^m_{;m} = 0$. Then the metric function P in (7) has the form

$$P = P(\Theta, \sigma) , \quad \zeta = \frac{1}{\sqrt{2}}(\Theta + i\varphi) \tag{10}$$

and (8) leads to the field equation

$$12 m \dot{P} = -P^4 P'''' + P^3 P''^2 \qquad P' \equiv \frac{\partial P}{\partial \Theta}, \quad \dot{P} \equiv \frac{\partial P}{\partial \sigma} \tag{11}$$

Introducing the real coordinate ξ [18] according to

$$d\Theta = \frac{1}{u} d\xi + a\, d\sigma, \quad u := P^{-2} \tag{12}$$

the equation (11) takes the form [10]

$$u_t + u^2 u_{xxxx} = 0 \tag{13}$$

($t \equiv \sigma/12m$, and x coincides with ξ up to an arbitrary function of σ). From a solution $u = u(x, t)$ to (13) one derives the total metric:

$$ds^2 = r^2 \left[\frac{1}{u} \left(dx + \frac{1}{12m} \, uu_{xxx} d\sigma \right)^2 + u \, d\varphi^2 \right] - 2d\sigma dr$$

$$+ \left[\frac{1}{2} u_{xx} + \frac{r}{12m}(uu_{xxx})_x + \frac{2m}{r} \right] d\sigma^2 \qquad (14)$$

Special solutions to (13) are the Schwarzschild metric ($u = 1 - x^2$) and the C-metric ($u = -2x^3 + bx + c$).

The field equations for the axisymmetric Robinson–Trautman class reduce to the nonlinear evolution equation (13) in 1+1 dimensions. Do the solution–generating techniques work in this particular case?

The prolongation method [22] has been applied to (11) [8] and to (13) [10]. In both cases the result is that the arising prolongation structure leads to a finite-dimensional Lie algebra, without any closure condition, so that a spectral parameter does not occur. A linear problem which implies the nonlinear evolution equation (13), and Bäcklund transformations, do not exist.

1.6 Solutions with Boost–Rotation Symmetry

Bonnor and Swaminarayan [5] constructed an exact radiative solution by the superposition of two solutions to the potential equation which governs Weyl's class of static axisymmetric vacuum fields. One of these potentials describes flat space-time in accelerated coordinates, the other one gives the field of two Curzon particles (point-like in Weyl's coordinates). The superposition yields the gravitational field of uniformly accelerated particles. Since the particles do not remain in a finite region of space for all times, they do not provide us with an isolated source. Future null infinity is regular but not complete; there are at least two missing generators of $\mathcal{J}^+$ because the particles escape to infinity.

The Bonnor–Swaminarayan solution is a member of the more general class with the space-time metric

$$ds^2 = e^\lambda d\varrho^2 + \varrho^2 e^{-\mu} d\varphi^2$$

$$+ \frac{1}{(z^2 - t^2)} \left[e^\lambda (zdz - tdt)^2 - e^\mu (zdt - tdz)^2 \right], \qquad (15)$$

where the metric function $\mu = \mu(\varrho^2, z^2 - t^2)$ is a solution of the wave equation in flat space, and λ can be obtained by a line integral.

The metric (15) covers the solutions with boost-rotation symmetry. It is obvious from (15) that there are two commuting Killing vectors $X = \partial_\varphi$ (rotation) and $Y = z\partial_t + t\partial_z$ (boost). The orbits are spacelike ($z^2 < t^2$), timelike ($z^2 > t^2$) or null ($z^2 = t^2$). If Y is timelike the metric belongs to Weyl's class from which Bonnor and Swaminarayan started in their construction. The region ($z^2 < t^2$) in which Y is spacelike is locally equivalent to the Einstein-Rosen cylindrically symmetric gravitational waves. But the global interpretation is here completely different.

A well-known member of the class of solutions with boost–rotating symmetry is the C-metric

$$ds^2 = \frac{1}{A^2(x+y)^2}\left[-F(y)dt^2 + \frac{dy^2}{F(y)} + G(x)dz^2 + \frac{dx^2}{G(x)}\right]$$
$$G(x) = 1 - x^2 - 2mAx^3 \tag{16}$$
$$F(y) = -G(-y), \quad A, m = \text{const},$$

which represents the field of uniformly accelerated black holes moving in different directions along the axis of symmetry, with nodal singularities between the black holes or from the black holes to infinity. For a discussion of the C-metric, see Ashtekar and Dray [1].

Other examples of solutions with boost-rotation symmetry were studied by Bicak et al. [3].

2 Pure Radiation Fields

2.1 The Vaidya Metric

The energy–momentum tensor of a pure radiation field is $T_{mn} = nk_m k_n$; the radiation propagates along the null vector k^m. The unique spherically symmetric pure radiation solution is due to Vaidya [21],

$$ds^2 = -\left(1 - \frac{2m(u)}{R}\right)dR^2 - 2dudR + R^2(d\vartheta^2 + \sin^2\vartheta d\varphi^2). \tag{17}$$

For non-negative energy density ($n \geq 0$), one finds that $m = m(u)$ must be a monotonically decreasing function of the retarded time u, $dm/du \leq 0$. For constant m, the Vaidya metric (17) goes over to the Schwarzschild solution. The Robinson-Trautman class (Sec. 1.4) can be generalized to include pure radiation fields. Solutions of that extended class approach the Vaidya solution in the remote future [2].

2.2 Rotating Generalization of the Vaidya Solution

The Kerr solution with the mass and angular momentum parameters m and a belong to the class of algebraically special vacuum solutions with an expanding and **twisting** repeated null eigendirection of the Weyl tensor. In general, allowing the parameters m and a to become any functions of retarded time one does not arrive at a rotating pure radiation field; the energy–momentum tensor turns out to have a different algebraic structure.

The problem to find a pure radiation solution which is a rotating generalization of the Vaidya solution in the stationary limit was treated by Kramer [9]. The underlying metric for algebraically special pure radiation fields is given, e.g., in Kramer et al. [13], eq. (26.14) (with $\phi_1^0 = 0$) and the particular solution, in these coordinates, is determined by

$$P = 1 + \frac{\zeta\bar{\zeta}}{2}, \quad L = -ia\bar{\zeta}P^{-2}, \quad M = 0$$
$$a = a_0(1 + cu)^{1/2}, \quad m = m_0(1 + cu)^{-3/2} \tag{18}$$
$$m_0, a_0, c \quad \text{constants}.$$

This pure radiation solution is a radiating generalization of the Kerr solution ($c = 0$) and a rotating generalization of the Vaidya solution ($a_0 = 0$). The solution is axisymmetric and asymptotically flat and it has Kerr–Schild form

$$g_{mn} = \eta_{mn} - 2Hk_mk_n , \tag{19}$$

η_{mn} being the Minkowski metric. The quantities m and a have a restricted dependence on u.

2.3 Interior Vaidya Solution

Recently we discussed a (spherically symmetric) regular model of an interior Vaidya solution which consists of a shearfree fluid with radial heat flow [11]. The junction conditions [20] give rise to a temporal evolution of the model.

Starting with any static spherically symmetric perfect fluid solution and admitting the parameters to become time-dependent, one obtains a solution with non-zero heat flow.

We applied that generation trick to the interior Schwarzschild solution which describes our model in the remote past. Afterwards the radiating sphere collapses; the area of its boundary surface decreases monotonically. Finally, the model runs into a physical singularity.

In terms of comoving isotropic coordinates the solution obtained is analytically given by

$$ds^2 = -A^2 dt^2 + B^2(dr^2 + r^2 d\Omega^2),$$
$$d\Omega^2 = d\vartheta^2 + \sin^2 \vartheta \, d\varphi^2 \tag{20}$$
$$A = \frac{1 + 2yr^2 - 2y - r^2y^2}{(1+y)(1+yr^2)} , \quad B = \frac{(1+y)^3}{1+yr^2} , \quad 0 \le r \le 1$$

where $y = y(t)$ is determined by the first-order differential equation

$$y' = -\frac{1-y}{(1+y)^3} \ln\left(\frac{1-y}{1-y_0}\right), \quad y_0 = \text{const.} \tag{21}$$

Fayos et al. [6] have studied the matching of a general spherically symmetric metric to the Vaidya solution.

References

1.   Ashtekar, A. and Dray, T., *Comm. Math. Phys.* **79**, (1981) 581
2.   Bičák, J. and Perjés, Z., *Class. Quantum Grav.* **4**, (1987) 595
3.   Bičák, J., Hoenselaers, C. and Schmidt, B.G., *Proc. Roy. Soc. Lond.* A **390**, (1983) 411
4.   Bondi, H., van der Burg, M.G.J. and Metzner, A.W.K., *Proc. Roy. Soc. Lond.* A **269**, (1962) 21
5.   Bonnor, W.B. and Swaminarayan, N.S., *Z. Phys.* **186**, (1965) 222

6.   Fayos, F., Jaén, X., Llanta, E., and Senovilla, J.M.M., *Phys. Rev.* D **45** (1992) 2732

7.   Foster, J. and Newman, E.T., *J. Math. Phys.* **8**, (1967) 189

8.   Glass, E.N. and Robinson, D.C., *J. Math. Phys.* **25**, (1984) 3382

9.   Kramer, D., *Third Soviet Grav. Conf. Erevan*, Tezisy, (1972) 321

10.   Kramer, D., The nonlinear evolution equation for Robinson-Trautman space-times, *Relativity today*, ed. Z. Perjés, World Scientific, Singapore (1988)

11.   Kramer, D., *J. Math. Phys.* **33**, (1992) 1458

12.   Kramer, D. and Neugebauer, G., *Comm. Math. Phys.* **7**, (1968) 173

13.   Kramer, D., Stephani, H., MacCallum, M.A.H., and Herlt, E., *Exact Solutions of Einstein's Field Equations*, Cambridge University Press, Cambridge (1980)

14.   Lukács, B., Perjés, Z., Porter, J. and Sebestyén, Á., *Gen. Rel. Grav.* **16**, (1984) 691

15.   Lun, A.W.C. and Fletcher, S.J., *13th Int. Conf. on General Relativity and Gravitation (GR13)*, Abstracts of Contributed Papers, (1992) 53

16.   Penrose, R., *Phys. Rev. Lett.* **10**, (1963) 66

17.   Perjés, Z., Dynamics of Robinson–Trautman space-times, *Non-perturbative methods in field theory*, ed. Z. Horváth et al., World Scientific, Singapore (1988)

18.   Robinson, I. and Robinson, J.R., Equations of motion in the linear approximation, *Studies in Relativity* (Papers in honour of J.L. Synge), ed. L.O'Raifeartaigh, Clarendon Press, Oxford, (1972) 151

19.   Robinson, I. and Trautman, A., *Proc. Roy. Soc. Lond.* A **265**, (1962) 463

20.   Santos, N.O., *Mon. Not. R. Astron. Soc.* **216**, (1985) 403

21.   Vaidya, P.C., *Proc. Indian Acad. Sci.* A **33**, (1951) 264

22.   Wahlquist, H.D. and Estabrook, F.B., *J. Math. Phys.* **16**, (1975) 1

Application of Wahlquist-Estabrook Method to Relativity Vacuum Equations with One Killing Vector

B. Kent Harrison

Department of Physics and Astronomy, Brigham Young University,
Provo UT 84602, USA

1 Introduction

Several ways of solving the vacuum Einstein equations with two Killing vectors, usually written as the Ernst plus other equations, were found around the year 1978. Most of these accomplished the task by solution generation methods, in which a new solution was found by a technique starting with a known solution. One of these techniques employed a Bäcklund transformation [2,3,5]. It would be very useful if one could find a similar Bäcklund transformation (BT) for the equations with one Killing vector. We describe here a search for such a BT; we note, however, that the existence of such a transformation may be judged unlikely because most such BT's apply only to equations with two independent variables.

An established method for search for a BT is that of Wahlquist and Estabrook (WE) [6,7]. To see how that is used, we review it here as applied to the Ernst equation. However, we will consider here only the first half of the method, which involves searching for a "prolongation structure" (PS).

2 The Ernst Equation in Differential Forms

To set up the Ernst equation, we choose a metric in standard form:

$$ds^2 = -f(dt + \omega d\varphi)^2 + \rho^2 f^{-1} d\varphi^2 + e^{2\gamma} f^{-1}(d\rho^2 + dz^2) \ . \tag{1}$$

We define a potential ϕ by, where subscripts mean derivation:

$$\phi_\rho = \rho^{-1} f^2 \omega_z, \quad \phi_z = -\rho^{-1} f^2 \omega_\rho \ . \tag{2}$$

Then the Ernst potential is $E = f + i\phi$ and the Ernst equation is:

$$E_{\rho\rho} + \rho^{-1}E_\rho + E_{zz} = f^{-1}(E_\rho^2 + E_z^2) \ . \tag{3}$$

It is convenient to write it in terms of differential forms. We define a linear Hodge star operator $*$ by $*d\rho = dz$ and $*dz = -d\rho$; then the Ernst equation may be written as two 2-form equations:

$$d(*df) + \rho^{-1}d\rho \wedge *df = f^{-1}(df \wedge *df - d\phi \wedge *d\phi) \tag{4}$$

$$d(*d\phi) + \rho^{-1}d\phi \wedge *d\phi = 2f^{-1}df \wedge *d\phi \ . \tag{5}$$

These can be satisfied formally by taking ω as a potential and using a new potential η. We write (2) as

$$*d\phi = \rho^{-1}f^2 d\omega \tag{6}$$

and also write

$$*df = \rho^{-1}f(d\eta + \omega d\phi) \ . \tag{7}$$

η and ω are not arbitrary, since they must satisfy their own equations. Using the various variables, one can now introduce a set of six 1-forms ξ_k. The exterior derivatives of these 1-forms, and the Ernst equation itself, can be written in terms of products of the ξ_k with constant coefficients. The equations for the ξ_k are denoted as an ideal I. We do not write them out here because they are given in the references [2,3,4].

3 Prolongation Procedure for the Ernst Equation

Application of the WE method now proceeds as follows. We write a column vector of 1-forms,

$$\Omega = -dq + (B^k\xi_k)q \ , \tag{8}$$

where q is a column of 0-forms (functions), the six B_k are square matrices, and the ξ_k are the 1-forms introduced above. There is a sum on k. The dimension of the q space is not specified at this point; it is determined by the size of the representation discussed below. One now requires that the exterior derivative of Ω is included in the augmented ideal made up of I and Ω itself:

$$d\Omega \subset \{I, \Omega\} \ . \tag{9}$$

Substitution of Ω into this equation gives a set of equations involving the ξ_k and B^k. The requirement that these be in the ideal I can most easily be satisfied by substituting for the $d\xi_k$ and by replacing certain products of the ξ_k by their equivalent from the equations in I. Alternatively, one requires that $d\Omega$ be written as linear combinations of Ω and the equations in I. The resultant equations for the matrices B^k can be solved in terms of an incomplete Lie algebra. (Note: for this problem we must take the matrices to be functions of a variable ζ, which is invariant under the invariance group of the Ernst equation.) If one constructs a representation of the incomplete Lie algebra, then setting Ω equal to zero in (8) provides what is called a PS for the Ernst equation.

4 Equations for the One-Killing Vector Case

Before attempting the same procedure for this case, we again write the equations as differential forms. This will avoid several complications associated with specifying a metric in detail. We begin with the metric [1] :

$$ds^2 = \lambda f(dx^k + g_A dx^A)^2 + \lambda f^{-1}\gamma_{AB}dx^A dx^B , \qquad (10)$$

where $\lambda = \pm 1$ and $k = 0$ or 1. A and B then take values 2, 3, and m (which equals that one of 0 or 1 which is not equal to k.) The metric quantities are functions of x^A.

The field equations are then

$$\Delta_2(f) = f^{-1}[\Delta_1(f) - \Delta_1(\phi)] \qquad (11)$$

$$\Delta_2(\phi) = 2f^{-1}\Delta_1(f,\phi) \qquad (12)$$

$$P_{AB} = (1/2)f^{-2}(f_{,A} f_{,B} + \phi_{,A} \phi_{,B}) , \qquad (13)$$

where P_{AB} is the Ricci tensor in the three-dimensional space with metric γ_{AB}, commas mean partial differentiation, and Δ_1 and Δ_2 are defined by, for any G and H,

$$\Delta_1(G) = \gamma^{AB}G_{,A} G_{,B} \qquad (14)$$

$$\Delta_1(G, H) = \gamma^{AB}G_{,A} H_{,B} \qquad (15)$$

$$\Delta_2(G) = \gamma^{AB}G_{;AB} . \qquad (16)$$

In these equations, ϕ is a potential defined in terms of the g_A [1] . One sees that (11) and (12) are similar to the Ernst equation.

To write these equations as differential forms, we define a set of basis 1-forms θ_A by [4]

$$\gamma_{AB}dx^A dx^B = \Gamma_{AB}\theta^A \otimes \theta^B \qquad (17)$$

where

$$\Gamma_{AB} = \text{diag}(-1, \lambda, \lambda) . \qquad (18)$$

It was convenient in the derivation to use a Hodge star operator defined in this case by, where ϵ_{ABC} is the Levi-Civita symbol and $\epsilon_{m23} = 1$,

$$*\theta^A = -(1/2)\Gamma^{AB}\epsilon_{ABC} \theta^C \wedge \theta^D , \qquad (19)$$

although it is not used here. We raise and lower with Γ_{AB}. Note that

$$\epsilon^{ABC} = -\Gamma^{AD} \Gamma^{BE} \Gamma^{CF}\epsilon_{DEF} . \qquad (20)$$

We define first derivatives of f and ϕ, projected onto the 1-form basis, where there is a sum on A:

$$df = f_A\theta^A, \quad d\phi = \phi_A\theta^A . \qquad (21)$$

We define connection forms ω_{AB} by

$$d\theta_A = -\omega_{AB} \wedge \theta^B, \quad \omega_{BA} = -\omega_{AB} \ . \tag{22}$$

Integrability of df and $d\phi$ gives

$$(df_A - f_B\omega^B{}_A) \wedge \theta^A = 0 \tag{23}$$

$$(d\phi_A - \phi_B\omega^B{}_A) \wedge \theta^A = 0 \ . \tag{24}$$

We write the curvature forms as, since the Riemann tensor in three dimensions may be expressed in terms of the Ricci tensor:

$$d\omega_{AB} + \omega_{AC} \wedge \omega^C{}_B = \theta_A \wedge \psi_B + \psi_A \wedge \theta_B \tag{25}$$

where

$$\psi_A \equiv (1/2)f^{-2} \left[(f_A f_B + \phi_A \phi_B)\theta^B - (1/4)(f^B f_B + \phi^B \phi_B)\theta_A \right] \ . \tag{26}$$

This is equivalent to (13) above. Finally, (11) and (12) become

$$\epsilon_{ABC} \left[(1/2)d\phi^A \wedge \theta^B \wedge \theta^C - \phi^A \omega^B{}_D \wedge \theta^D \wedge \theta^C \right] - 2f^{-1}f_A\phi^A\sigma = 0 \ , \tag{27}$$

$$\epsilon_{ABC} \left[(1/2)df^A \wedge \theta^B \wedge \theta^C - f^A \omega^B{}_D \wedge \theta^D \wedge \theta^C \right] - f^{-1}(f_A f^A - \phi_A\phi^A)\sigma = 0 \ , \tag{28}$$

where

$$\sigma = \theta^m \wedge \theta^2 \wedge \theta^3 \ . \tag{29}$$

5 Equations with Constant Coefficients

The author has attempted to find a PS for this system, but the matrices used (somewhat like the B^k above) must be considered to be functions of the eight variables f, ϕ, f_A, and ϕ_A. Solving the resultant equations for these matrices, therefore, has been difficult. One asks: can a reformulation of the differential form equations be found such that all coefficients are constant, so that one might at least guess that the matrices in the PS also are constant?

The answer is yes. We define 1-forms $\sigma^C, \alpha, \beta, \xi_A$, and η_A by the following equations:

$$\alpha = f^{-1}df, \quad \beta = f^{-1}d\phi \tag{30}$$

$$\omega_{AB} = \epsilon_{ABC}\,\sigma^C \tag{31}$$

$$\xi_A = (1/2)f^{-1}\epsilon_{ABC}\,f^B\theta^C, \quad \eta_A = (1/2)f^{-1}\epsilon_{ABC}\,\phi^B\theta^C \ . \tag{32}$$

Then it is trivial to see that

$$d\alpha = 0, \quad d\beta = -\alpha \wedge \beta \ . \tag{33}$$

Several relations result from the definitions (30) and (32):

$$\alpha \wedge \theta_A = 2\epsilon_{ABC}\,\xi^B \wedge \theta^C \tag{34}$$

$$\beta \wedge \theta_A = 2\epsilon_{ABC}\,\eta^B \wedge \theta^C \ , \tag{35}$$

which are 2-forms, and 3-forms

$$\xi_A \wedge \xi_B \wedge \theta^B = 0 \tag{36}$$

$$\eta_A \wedge \eta_B \wedge \theta^B = 0 \tag{37}$$

$$\eta_A \wedge \xi_B \wedge \theta^B + \xi_A \wedge \eta_B \wedge \theta^B = 0 \tag{38}$$

$$2\theta^A \wedge \theta^B \wedge \xi_C + (\delta_C^A \theta^B - \delta_C^B \theta^A) \wedge \theta^D \wedge \xi_D = 0 \tag{39}$$

$$2\theta^A \wedge \theta^B \wedge \eta_C + (\delta_C^A \theta^B - \delta_C^B \theta^A) \wedge \theta^D \wedge \eta_D = 0 \tag{40}$$

The exterior derivatives of θ_A and σ_C now become

$$d\theta_A = \epsilon_{ABC} \, \sigma^B \wedge \theta^C \tag{41}$$

and

$$d\sigma_C = (1/2) \left[\epsilon_{ABC} \, \sigma^A \wedge \sigma^B + \xi_C \wedge \alpha + \eta_C \wedge \beta \right) \\ + \Gamma_{CD} \, \epsilon^{ABD} (\xi_A \wedge \xi_B + \eta_A \wedge \eta_B] \quad . \tag{42}$$

The Ernst-like equations (27) and (28) become

$$d\xi_A \wedge \theta^A - \epsilon_{ABC}(2\eta^A \wedge \eta^B + \sigma^A \wedge \xi^B) \wedge \theta^C = 0 \tag{43}$$

$$d\eta_A \wedge \theta^A + \epsilon_{ABC}(2\xi^A - \sigma^A) \wedge \eta^B \wedge \theta^C = 0 \quad . \tag{44}$$

The closure of (41), $dd\theta_A = 0$, is automatically satisfied by virtue of (34) through (37), (41), and (42). The closure of (42), $dd\sigma_C = 0$, yields the equation

$$2(\xi_B \wedge \xi_C + \eta_B \wedge \eta_C + \xi_B \wedge \theta_C - \xi_C \wedge \theta_B + \eta_B \wedge \theta_C - \eta_C \wedge \theta_B) \wedge \sigma^B \tag{45}$$
$$+ 2\Gamma_{CD} \, \epsilon^{ABD}(d\xi_A \wedge \xi_B + d\eta_A \wedge \eta_B) + \alpha \wedge \beta \wedge \eta_C + d\xi_C \wedge \alpha + d\eta_C \wedge \beta = 0 \quad .$$

Finally, taking the exterior derivative of the 2-forms (34) and (35) gives these equations:

$$\epsilon_{ABC} \, d\xi^B \wedge \theta^C + (\sigma_B \wedge \xi_A + \xi_B \wedge \sigma_A) \wedge \theta^B = 0 \tag{46}$$

$$\epsilon_{ABC} \, d\eta^B \wedge \theta^C + (\sigma_B \wedge \eta_A + \eta_B \wedge \sigma_A) \wedge \theta^B + 2\eta^B \wedge (\xi_A \wedge \theta_B - \xi_B \wedge \theta_A) \quad . \tag{47}$$

Equations (33) to (47) constitute the ideal I.

We note an interesting fact. The 2-forms $\tau \equiv \xi_A \wedge \theta^A$ and $\rho \equiv \eta_A \wedge \theta^A$ may be written, by using (30) and (36) - (38), as

$$\rho = f d\nu, \quad \tau = d\mu - \phi d\nu \quad , \tag{48}$$

where ν and μ are 1-form potentials (reminiscent of the potentials defined for the 2-variable case. However, it has not proved possible to employ these in any useful way.

6 General Prolongation Procedure for Three-Variable Case

The prolongation procedure for cases in which there are two independent variables, such as for the Ernst equation, using (8) and (9), is very straightforward. However, if there are more that two variables, the procedure is less well defined. The following outlines a procedure which was motivated by a paper by H. C. Morris [8] and which is equivalent to the calculations done in that paper.

We write a column 1-form, similar to (8) of the previous treatment (in which $\Omega = 0$), by

$$d\zeta = \psi \mu + \gamma \zeta \ , \tag{49}$$

where ζ and μ are columns of 0-forms and ψ and γ are matrices of 1-forms. Closure gives

$$\begin{aligned}
0 = dd\zeta &= d\psi\,\mu - \psi \wedge d\mu + d\gamma\,\zeta - \gamma \wedge d\zeta \\
&= d\psi\,\mu - \psi \wedge d\mu + d\gamma\,\zeta - \gamma \wedge (\psi\,\mu + \gamma\,\zeta)
\end{aligned}$$

where we have substituted for $d\zeta$. (This is equivalent to requiring Ω to be part of the ideal in the previous case.) We now multiply this equation on the left by ψ and require

$$\psi \wedge \psi = 0 \ . \tag{50}$$

(This is not identically equal to zero since ψ is a matrix.) We now require that the coefficients of μ and ζ be in the ideal I:

$$\psi \wedge (d\psi - \gamma \wedge \psi) \subset I \tag{51}$$

$$\psi \wedge (d\gamma - \gamma \wedge \gamma) \subset I \ . \tag{52}$$

7 Prolongation for the Einstein Equation One-Killing Vector Case

To specialize to the problem at hand, we must guess how to construct ψ and γ for use in (50) - (52). The best choice seems to be

$$\psi = M_A \theta^A \tag{53}$$

and

$$\gamma = P_A \theta^A + Q_A \sigma^A + R_A \xi^A + S_A \eta^A + U\alpha + V\beta \ , \tag{54}$$

where the matrices $M_A, P_A, Q_A, R_A, S_A, U$ and V are assumed constant.

Equation (50), with (53), yields, for all A and B–since antisymmetrizing the matrix coefficient of $\theta^A \wedge \theta^B$ produces a commutator:

$$[M_A, M_B] = 0 \ . \tag{55}$$

To expand (51), one constructs the expression $\psi \wedge (d\psi - \gamma \wedge \psi)$ by substituting for ψ and γ from (53) and (54). One must satisfy the requirement that it be in the ideal I by equating it to appropriate terms with arbitrary multipliers from the forms in I. In this case one needs to use only (39) and (40); however, one must substitute for occurrences of terms $\alpha \wedge \theta_A$ and $\beta \wedge \theta_A$, on the left hand side of the equation, from (34) and (35), as well as for $d\theta^A$ from (41). The resulting equation is

$$M_C\theta^C \wedge \left[\epsilon_{ABD} M^A \sigma^B \wedge \theta^D - (P_B\theta^B + Q_B\sigma^B + R_B\xi^B + S_B\eta^B) \wedge M_A\theta^A\right]$$
$$- 2M_C\theta^C \wedge UM^A \epsilon_{ABD}\, \xi^B \wedge \theta^D - 2M_C\theta^C \wedge VM^A \epsilon_{ABC}\, \eta^B \wedge \theta^D$$
$$= \lambda_{AB}{}^C \left[2\theta^A \wedge \theta^B \wedge \xi_C + (\delta_C^A\theta^B - \delta_C^B\theta^A) \wedge \theta^D \wedge \xi_D\right]$$
$$+ \mu_{AB}{}^C \left[2\theta^A \wedge \theta^B \wedge \eta_C + (\delta_C^A\theta^B - \delta_C^B\theta^A) \wedge \theta^D \wedge \eta_D\right] . \tag{56}$$

One sees that there are four types of terms, those with three θ^A and those with two θ^A and one of the 1-forms σ^B, ξ^B, or η^B. Setting the coefficients for these four groups of terms to zero yields, after elimination of the multipliers $\lambda_{AB}{}^C$ and $\mu_{AB}{}^C$,

$$\epsilon^{ABC} M_A P_B M_C = 0 \tag{57}$$
$$M_A Q_B M_C - M_C Q_B M_A = (\epsilon_{DBC} M_A - \epsilon_{DBA} M_C)M^D \tag{58}$$
$$M_A R^A M_C - M_C R^A M_A = 2\epsilon_{BDC} M^B U M^D \tag{59}$$
$$M_A S^A M_C - M_C S^A M_A = 2\epsilon_{BDC} M^B V M^D . \tag{60}$$

The equations obtained from (52) are long and complicated. Equations (36) - (40), (43), (44), (46), and (47) of the ideal I must be represented on the right hand side, and (30), (41), and (42) must be used in expansion of $d\gamma$ on the left hand side. The equation becomes

$$M_D\theta^D \wedge \left[\epsilon_{ABC}P^A \sigma^B \wedge \theta^C + (1/2)\epsilon_{ABC}\, Q^C(\sigma^A \wedge \sigma^B - \xi^A \wedge \xi^B - \eta^A \wedge \eta^B)\right.$$
$$+ (1/2)Q^C(\xi_C \wedge \alpha + \eta_C \wedge \beta) + R_A d\xi^A + S_A d\eta^A - V\alpha \wedge \beta$$
$$- (P_A\theta^A + Q_A\sigma^A + R_A\xi^A + S_A\eta^A + U\alpha + V\beta)$$
$$\wedge (P_B\theta^B + Q_B\sigma^B + R_B\xi^B + S_B\eta^B + U\alpha + V\beta)\big]$$
$$= K_{AB}{}^C \left[2\theta^A \wedge \theta^B \wedge \xi_C + (\delta_C^A\theta^B - \delta_C^B\theta^A) \wedge \theta^D \wedge \xi_D\right]$$
$$+ L_{AB}{}^C \left[2\theta^A \wedge \theta^B \wedge \eta_C + (\delta_C^A\theta^B - \delta_C^B\theta^A) \wedge \theta^D \wedge \eta_D\right]$$
$$+ Y^A \xi_A \wedge \xi_B \wedge \theta^B + W^A \eta_A \wedge \eta_B \wedge \theta^B + U^A(\eta_A \wedge \xi_B + \xi_A \wedge \eta_B) \wedge \theta^B$$
$$+ \lambda \left[d\xi_A \wedge \theta^A - \epsilon_{ABC}(2\eta^A \wedge \eta^B + \sigma^A \wedge \xi^B) \wedge \theta^C\right]$$
$$+ \mu \left[d\eta_A \wedge \theta^A + \epsilon_{ABC}(2\xi^A - \sigma^A) \wedge \eta^B \wedge \theta^C\right]$$
$$+ \zeta^A \left[\epsilon_{ABC}\, d\eta^B \wedge \theta^C + (\sigma_B \wedge \eta_A + \eta_B \wedge \sigma_A) \wedge \theta^B + 2\eta^B \wedge (\xi_A \wedge \theta_B - \xi_B \wedge \theta_A)\right]$$
$$+ \psi^A \left[\epsilon_{ABC}\, d\xi^B \wedge \theta^C + (\sigma_B \wedge \xi_A + \xi_B \wedge \sigma_A) \wedge \theta^B\right] . \tag{61}$$

The multipliers are $K_{AB}{}^C, L_{AB}{}^C, Y^A, W^A, U^A, \lambda, \mu, \psi^A$, and ζ^A. Substitution for $\alpha \wedge \theta_A$ and $\beta \wedge \theta_A$ from (34) and (35) must be made on the left hand side as before; because of length we do not write the result out. Equation (45) does not appear on the right hand side because terms similar to its terms do not appear on the left hand side.

When we collect terms and set the coefficients of various terms equal to zero, twelve equations result, as follows. One sees the frequent appearance of matrix commutators, much as in (55) above. From the $\theta\theta\theta, \theta\theta\sigma$, and $\theta\sigma\sigma$ equations we get

$$\epsilon^{ABC} M_C[P_A, P_B] = 0 \tag{62}$$

$$M_D[P_C, Q_B] - M_C[P_D, Q_B] = (\epsilon_{ABD} M_C - \epsilon_{ABC} M_D) P^A \tag{63}$$

$$M_D([Q_A, Q_B] - \epsilon_{ABC} Q^C) = 0 . \tag{64}$$

The $\theta d\xi$ and $\theta d\eta$ equations give

$$M_B R_A = \lambda \Gamma_{AB} + \epsilon_{ABC} \psi^C \tag{65}$$

$$M_B S_A = \mu \Gamma_{AB} + \epsilon_{ABC} \zeta^C . \tag{66}$$

From the $\theta\sigma\xi$ and $\theta\sigma\eta$ equations we get

$$M_C[Q_A, R_B] - 2\epsilon_{BCD} M^D[U, Q_A] = \lambda \epsilon_{ABC} + \Gamma_{BC} \psi_A - \Gamma_{AC} \psi_B \tag{67}$$

$$M_C[Q_A, S_B] - 2\epsilon_{BCD} M^D[V, Q_A] = \mu \epsilon_{ABC} + \Gamma_{BC} \zeta_A - \Gamma_{AC} \zeta_B . \tag{68}$$

The $\theta\theta\xi$ and $\theta\theta\eta$ equations yield, after elimination of the multipliers $K_{AB}{}^C$ and $L_{AB}{}^C$,

$$M_B[P_A, R^A] - M_A[P_B, R^A] = 2\epsilon_{ACB} M^A[U, P^C] \tag{69}$$

$$M_B[P_A, S^A] - M_A[P_B, S^A] = 2\epsilon_{ACB} M^A[V, P^C] . \tag{70}$$

The $\theta\xi\xi$ equation results in

$$M_C(\epsilon^{ABD} Q_D - [R^A, R^B]) + \delta_C^A Y^B - \delta_C^B Y^A$$
$$+ \epsilon_{DEC} M^D [\Gamma^{BE}(Q^A + 2[U, R^A]) - \Gamma^{AE}(Q^B + 2[U, R^B])] = 0 . \tag{71}$$

From the $\theta\eta\eta$ equation we get

$$M_C(\epsilon^{ABD} Q_D - [S^A, S^B]) - \delta_C^B W^A + \delta_C^A W^B - 4\lambda\Gamma_{CD} \epsilon^{ABD}$$
$$+ \epsilon_{DEC} M^D [\Gamma^{BE}(Q^A + 2[V, S^A]) - \Gamma^{AE}(Q^B + 2[V, S^B])] = 0 . \tag{72}$$

Finally, the $\theta\xi\eta$ equation yields

$$M_C[R_A, S_B] + 4(\Gamma_{AB} M_C - \Gamma_{BC} M_A)(V + [U, V]) + \Gamma_{BC} U_A - \Gamma_{AC} U_B$$
$$+ 2\epsilon_{DAC} M^D[U, S_B] - 2\epsilon_{DBC} M^D[V, R_A] + 2\epsilon_{ABC}\mu + 2\Gamma_{AB} \zeta_C - 2\Gamma_{BC} \zeta_A = 0 . \tag{73}$$

The equations to be satisfied by the various matrices are now (55), (57)–(60), and (62) – (73). In the above, the multipliers $Y^A, W^A, U^A, \lambda, \mu, \psi^A$, and ζ^A have been left in the equations for simplicity. It is sometimes helpful to see how they occur in the equations, especially since we often want them to be nonzero. For example, we expect that λ and μ will be nonzero since they multiply the Ernst-like equations. If they were forced to equal zero, then one would expect that one's final PS would be trivial, not involving the content of those equations. It may be that multipliers ψ^A and ζ^A are less important and could equal zero, since they multiply the forms from (46) and (47), which result from closure; but this question is not decided.

Elimination of the remaining multipliers can be done as follows. If one has some expression like (73) represented symbolically as

$$F_{ABC} = \Gamma_{AC}U_B - \Gamma_{BC}U_A \, , \tag{74}$$

then one sees by contraction with Γ^{AC} that $U_A = (1/2)F_{BA}{}^B$, and this can be substituted into the previous equation, thus eliminating the U_A. Similarly, one can obtain explicit expressions for λ and ψ^C from (65), e.g., by contracting with Γ^{AB} and ϵ^{ABD} respectively.

8 Remarks on the Solution of the Matrix Equations

The M_A satisfy (55)–they commute with each other. The trace of the left hand side of (58) equals zero, because cyclic permutation of matrices in a trace does not change the value and because the M_A commute. Thus the trace of the right hand side of (58) also equals zero. By considering different combinations of indices one sees quickly that

$$Tr(M_A M_B) = 0 \tag{75}$$

for all A and B, including the case when they are equal. This forbids the choice of any one M_A to be the identity matrix (and incidentally forbids the removal of M_A from an equation like (64).) We also wish the M_A to be real and to be linearly independent in order that all basis forms be represented linearly in (53).

One can now explore the dimension of possible representations for the M_A. In this exploration, it is helpful to note that a similarity transformation of the M_A leaves (55) and (75) invariant. Thus we can assume a similarity transformation in order to simplify the form of the M_A.

To consider 2×2 representations, we use the theorem that a set of matrices may be triangularized by a similarity transformation. The eigenvalues appear on the diagonal. Equation (75) for $A = B$ then requires that the sum of the squares of the eigenvalues equals zero, so that the eigenvalues must either be all zero or must be complex. Further calculation shows that the matrices cannot satisfy all three conditions (55), (75), and independence.

For 3×3 representations, it is easily proved that if all M_A can simultaneously be put into their Jordan forms, that such a representation is impossible (inconsistent with the three conditions). The proof is more difficult if the matrices are merely triangular and has not been completed. We will assume that a 3×3 representation is impossible.

One can easily construct 4×4 matrices that satisfy the conditions on the M_A–e.g., by taking all 2×2 corners of these matrices to vanish except for one judiciously chosen in each matrix. It is not yet known whether the many other equations to be satisfied by the various matrices allow any nontrivial solution.

If no solution is obtained, one may assume some relation among the various differential forms–an Ansatz–which may allow a reformulation of the PS search. One interesting possible version of this comes from the relations, which one easily sees from the definitions (32):

$$f^A \xi_A = 0, \quad \phi^A \eta_A = 0 . \tag{76}$$

If one allows the variables f^A and ϕ^A to be reintroduced, then one could write

$$\xi_m = -(f^2/f^m)\xi_2 - (f^3/f^m)\xi_3 \tag{77}$$

and a similar equation for η_m. Then one could rewrite the forms by using these equations and could define the matrices in the PS as functions of the variables f^2/f^m, etc. It is clear, however, that this would be complicated. It is also clear that this would bring us full circle, in a sense, to the differential form equations given in Section 4!

References

1. Harrison, B. K.: New solutions of the Einstein-Maxwell equations from old. J. Math. Phys. **9** (1968) 1744–1752
2. Harrison, B. K.: Bäcklund transformation for the Ernst equation of general relativity. Phys. Rev. Lett. **41** (1978) 1197–1200
3. Harrison, B. K.: Study of the Ernst equation using a Bäcklund transformation. Proc. Second Marcel Grossmann Meeting on General Relativity, R. Ruffini, ed. (North-Holland, 1982) 341–352
4. Harrison, B. K.: Prolongation structures and differential forms. Solutions of Einstein's equations: Techniques and Results, C. Hoenselaers and W. Dietz, eds. (Springer, 1983) 26–54
5. Neugebauer, G.: Bäcklund transformations of axially symmetric stationary gravitational fields. J. Phys. A: Math. Gen. **12** (1979) L67–L70
6. Wahlquist, H. D., Estabrook, F. B.: Bäcklund transformation for solution of the Korteweg-de Vries equation. Phys. Rev. Lett. **31** (1973) 1386–1390
7. Wahlquist, H. D., Estabrook, F. B.: Prolongation structures of nonlinear evolution equations. J. Math. Phys. **16** (1975) 1–7
8. Morris, H. C.: Prolongation structure and nonlinear evolution equations in two spatial dimensions. J. Math. Phys. **17** (1976) 1870–1872

On the Regularity of Spherically Symmetric Static Spacetimes

Thomas W. Baumgarte

Max-Planck-Institut für Astrophysik, Karl-Schwarzschild-Str. 1, W-8046
Garching bei München, Fed. Rep. of Germany

Abstract: It is shown that the regularity of spherically symmetric static spacetimes is independent of the monotonic behaviour of the equation of state and the isotropy of the pressure.

1. Introduction

Although the equations governing spherically symmetric static stellar models are very well known [1], existence of global solutions has only been shown for isotropic pressure and a monotonic equation of state [2]. The aim of this article, which is based on a work done by A. D. Rendall and myself [3], is to show that these assumptions are not necessary, so that regular solutions do exist independently of the isotropy of the pressure and the monotonic behaviour of the equation of state.

After presenting the equations and briefly reviewing some earlier results, we shall treat the case of isotropic pressure in section 2. In section 3 the generalization to non-isotropic pressure will be sketched, which may be regarded as the most general spherically symmetric static case.

2. Isotropic Pressure

In this section we will consider a perfect fluid. In this case, the stress-energy tensor has only two distinct eigenvalues, namely the energy density ρ and the pressure p. Then the usual approach to construct a stellar model is the following (see e.g. [1] or [5] for more details):

1. The pressure distribution is governed by the Oppenheimer-Volkoff equation[1]

---

[1] We use geometrized units throughout this paper.

$$\frac{dp}{dr} = -\frac{(\rho + p)}{r^2}\frac{(m(r) + 4\pi p r^3)}{(1 - 2m(r)/r)}, \tag{1}$$

where $m(r)$ is the "mass-energy up to r":

$$m(r) = 4\pi \int_0^r \rho(p(s))s^2 ds. \tag{2}$$

2. Choose an equation of state

$$\rho = \rho(p) \tag{3}$$

and, as starting value for (1), a central pressure $p_0 > 0$.
3. Integrate equations (1) – (3) out from $r = 0$ up to the boundary R of the star defined as the radius of vanishing pressure: $p(R) = 0$.
4. If $R < \infty$, join on an exterior Schwarzschild solution.

A solution to the coupled set of equations (1) – (3) determines the whole geometry of spacetime. However, very few exact solutions are known (see [4] and references therein), and it is far from obvious that the above "recipe" always works. As has been shown in [2], regular solutions exist in a neighbourhood of $r = 0$, which is a singular point of (1), but it might still happen that the right hand side of (1) diverges before p itself vanishes. This would be the case if $m(r)/r$ tends to 1/2. Also, if $m(R)/R \geq 1/2$, it is impossible to join on a static exterior Schwarzschild solution. Thus, a global solution only exists, if $m(r)/r < 1/2$ for all $r \leq R$.

Treating this problem most authors assumed the equation of state (3) to be monotonic ($d\rho/dp > 0$ for $p > 0$), which, in view of the Oppenheimer-Volkoff equation (1), implies

$$\frac{d\rho}{dr} \leq 0. \tag{4}$$

In fact, this is a very reasonable assumption from a physical point of view, as "it is difficult to imagine that a fluid sphere with a larger density near the surface than near the center would be stable" [5]. Assuming the existence of a global solution Buchdahl [6] showed that the compactness $m(R)/R$ of such stars is always bounded by 4/9. The existence and uniqueness of a global solution to every central density and every monotonic equation of state was established by Rendall and Schmidt [2]. Generalizing Buchdahl's result, Bondi [7] showed that for any positive density distribution, not necessarily obeying (4), the compactness is bounded by $6\sqrt{2} - 8 \simeq 0.485$. However, just like Buchdahl, he assumed that global solutions do exist. This is the gap we want to close.

The aim of this section is to show that the existence of global solutions to (1) – (3) is indeed independent of assumption (4). To prove this, we will assume that ρ is given as a function of r instead of p (i.e. we will assume to be given a density profile instead of an equation of state). This may not be very sensible from a physical point of view, but does not reduce generality and simplifies the problem considerably, as equation (3) can now be integrated immediately, so that (1) becomes an ordinary differential equation for p. This enables us to prove the following theorem:

Theorem 1. *Let $\rho : \mathbb{R}_0^+ \to \mathbb{R}_0^+$, $\rho \in C^0$, be a non-negative, continuous[2] function and $p(r)$ a solution to equation (1) with starting value $p(0) = p_0$, $0 < p_0 < \infty$. Then[3]*

$$\frac{m(r)}{r} < \frac{1}{2}$$

for all $r \leq R$. This implies that dp/dr remains finite for all non-negative p.

Proof. An expansion in terms of r can be used to show regularity of both $m(r)/r$ and the right hand side of (1) at $r = 0$. Therefore we only have to show that $1 - 2m(r)/r$ does not vanish for $r \leq R$. Let r_0 be defined as

$$r_0 = \min\left\{ r \in \mathbb{R} \mid \frac{m(r)}{r} = \frac{1}{2} \right\}$$

(if no such point exists we are already done). We will show in two steps that both assumptions $r_0 < R$ and $r_0 = R$ lead to a contradiction.

Assume $r_0 < R$. This implies that $p(r_0) > 0$, so that the numerator of the right hand side of (1) can be estimated to be greater than a positive constant at $r = r_0$. As $\rho \in C^0$, m must be C^1, therefore it can be written as $m(r) = m(r_0) + (r - r_0)S(r)$, where $S(r) \in C^0$ is a continuous remainder term. Hence the denominator of the right hand side of (1) becomes

$$r^2 \left(1 - 2m(r)/r\right) = (r - r_0)r \left(1 - 2S(r)\right). \tag{5}$$

Here $r_0 \left(1 - 2S(r_0)\right)$ can be estimated to be smaller than a positive constant, so that the right hand side of (1) must have a first or higher order singularity at $r = r_0$. This means that the integral from $r = 0$ up to $r = r_0$ diverges, implying that p approaches negative infinity (with p_0 finite). Therefore p must have vanished before, so that $R < r_0$, contradicting our assumption.

Now assume $r_0 = R$. If $\rho(R) > 0$ the numerator in (1) is still positive and the same argument as before can be repeated. Therefore assume $\rho(R) = 0$. As $\rho \in C^0$, an r_* can be found for every $\epsilon > 0$ with $R - \epsilon < r_* < R$, such that $\rho_* = \rho(r_*) \geq \rho(s)$ for all $s \in [r_*, R[$. Using the estimate $m(R) \leq m(s) + 4\pi R^2 \rho_* \delta s$, where $\delta s = R - s$, and $m(s)/s < 1/2$, we can expand $m(r)/r$ around s and find the following relation for $m(R)/R$:

$$\frac{1}{2} = \frac{m(R)}{R} \leq \frac{m(s) + 4\pi R^2 \rho_* \delta s}{s + \delta s} < \frac{1}{2} + \left(4\pi R^2 \rho_* - \frac{m(s)}{s}\right)\frac{\delta s}{s} + O(\delta s^2). \tag{6}$$

With K being a suitably chosen constant this implies

$$\rho_* + \frac{K}{4\pi R^2}\delta s > \frac{m(s)}{4\pi R^2 s}, \tag{7}$$

---

[2] Actually, it would have been enough to assume ρ to be a non-negative, piecewise continuous function.

[3] Obviously, in (2) the argument of ρ must now be s instead of $p(s)$.

which has to hold for all $s \in [r_*, R[$. This, however, is a contradiction again, because ρ_* and δs can be chosen arbitrarily small, whereas the right hand side is always greater than $m(r_*)/(4\pi R^3)$. Therefore the assumption $r_0 = R$ must be wrong, which completes the proof. $\square$

This theorem guarantees that the Oppenheimer-Volkoff equation can be integrated out until p vanishes. If this is the case for a finite radius R, Bondi's result can be applied, the compactness $m(R)/R$ of the fluid sphere is bounded by $6\sqrt{2} - 8$ and a regular exterior Schwarzschild solution can be joined on.

A simple example shows that the above theorem cannot be generalized to negative pressures. Consider the equations of state $p = -\rho$ or $p = -1/3\rho$. For both, $p = const$ (and therewith $\rho = const$) is a solution to equations (1) – (3) for any starting value p_0, which describes a homogeneous and isotropic, globally extended fluid. The compactness is then given by $m(r)/r = 4\pi\rho r^2/3$, which obviously is not bounded.

3. Non-isotropic pressure

In more general matter models, e.g. the collisionless gas or the elastic solid, we might have to relax the isotropy of the pressure. In a spherically symmetric spacetime we then have to distinguish between the radial pressure p_r and the tangential pressure p_t. In this case the Oppenheimer-Volkoff equation generalizes to

$$\frac{dp_r}{dr} = -\frac{(\rho + p_r)}{r^2}\frac{(m(r) + 4\pi p_r r^3)}{(1 - 2m(r)/r)} - \frac{2}{r}(p_r - p_t). \tag{8}$$

The boundary of the "star" is now defined by $p_r = 0$. Note that for $r = 0$ full isotropy is recovered, so that $p_r(0) = p_t(0)$. Also, for $p_r = p_t$, equation (8) reduces to (1). The difference to the previous situation is only the additional term in (8) and, having one more variable, the need for another given function. Therefore one can prove a very similar theorem:

Theorem 2. Let $\rho : \mathbb{R}_0^+ \to \mathbb{R}_0^+$, $\rho \in C^0$ be a non-negative, continuous function, $p_t : \mathbb{R}_0^+ \to \mathbb{R}$, $p_t \in C^1$ a continuously differentiable function and $p_r(r)$ a solution to equation (8) with starting value $p_r(0) = p_t(0)$. Then

$$\frac{m(r)}{r} < \frac{1}{2}$$

for all $r \leq R$.

The proof in the last section needs only few extensions to hold for this situation and can be found in [3]. This especially includes a discussion of the regularity at $r = 0$, as well as some remarks on the assumptions and the existence of global solutions, which might depend on the choice of the matter model.

4. Summary

Having assumed ρ to be given as a function of r we have shown that all earlier assumptions about the equation of state are not necessary for the regularity of the solutions. This method can be generalized to non-isotropic pressure, which was not possible until now. In addition, the above results can be used to establish existence and uniqueness of solutions p and p_r, so that this approach is both a simplification as well as a generalization of earlier arguments.

Acknowledgements

I would like to thank J. Ehlers, A. D. Rendall and B. G. Schmidt for their contribution of good ideas.

References

1. J. R. Oppenheimer, G. Volkoff, Phys. Rev. **55** (1939) 374
2. A. D. Rendall, B. G. Schmidt, Class. Quantum Grav. **8** (1991) 985
3. T. W. Baumgarte, A. D. Rendall, MPA-preprint 674, to appear in Class. Quantum Grav.
4. D. Kramer, H. Stephani, M. MacCallum, E. Herlt, *Exact Solutions of Einstein's Field Equations*, Cambridge University Press (1980)
5. S. Weinberg, *Gravitation and Cosmology*, Wiley, New York (1972), p. 332
6. H. A. Buchdahl, Phys. Rev. **116** (1959) 1027
7. H. Bondi, Proc. Roy. Soc. **282** (1964) 303

Shock capturing methods
in 1D Numerical Relativity

C. Bona [1], J.M. Ibáñez [2], J.M. Martí [2] and J. Massó [1] .

[1]Departament de Física, Universitat de les Illes Balears,
E-07071 Palma de Mallorca, Spain
[2]Departament de Física Teórica, Universitat de Valencia,
46100 Burjassot (Valencia), Spain

Abstract: A numerical code is presented which uses modern shock capturing methods to evolve spherically symmetric perfect fluid space-times. Harmonic slicing is used to ensure singularity avoidance, which is crucial in strong field situations. Some tests are presented, including an application to the stellar collapse problem.

1 Introduction

In this work, we describe a numerical code which has been developed to apply modern shock capturing methods in the evolution of spherically symmetric space-times. Spherical symmetry implies that the space dependence of the dynamical quantities is given by the radial coordinate r (1D problem). Both the metric and hydrodynamic equations are put as a hyperbolic system of balance laws, allowing us to apply the same numerical methods to both systems. The resulting code is named NOSTRUM, which is an acronym for Numerically Observing Space-Times with Relativistic Upwind Methods and it is also an homage to the *mare nostrum*, our beloved Mediterranean sea.

We will start with the generic diagonal form of a spherically symmetric line element

$$ds^2 = -\alpha^2(t,r)dt^2 + X^2(t,r)dr^2 + r^2Z^2(t,r)d\Omega^2, \qquad (1)$$

where we have chosen to deal with the function Z instead of the standard choice $Y = rZ$ to display explicitly the inverse powers of r near the coordinate singularity at the origin and to cancel out singular terms appearing in the derivation of the equations.

The matter content of the space-time will be described as a perfect fluid with the following energy-momentum tensor

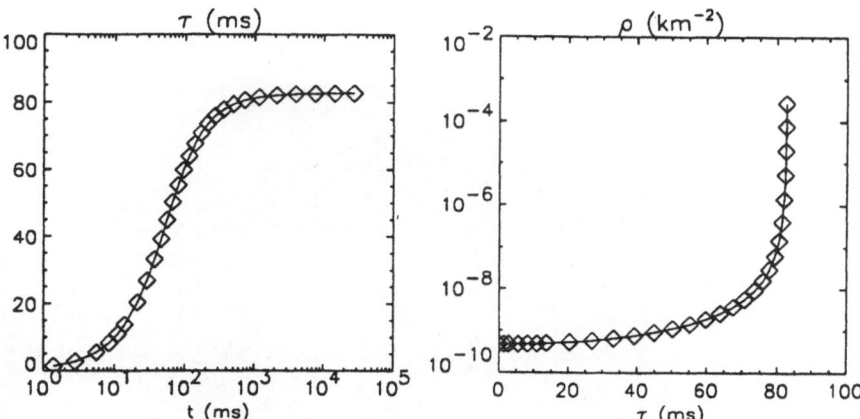

Fig. 1. *Collapse of a homogeneous ball of dust* (closed Friedmann-Robertson-Walker model) starting at the instant of time symmetry. In the left hand side the proper time coordinate τ is plotted versus the corresponding harmonic time coordinate t to demonstrate the singularity avoidance properties of our gauge. We have taken parameters corresponding to a white dwarf: $M = 1.38 M_\odot$ and $R = 1000$ km, so that the final singularity is reached at $\tau = 82$ ms. As it seen in the right hand side plot, the code is able to track the collapse while the star density ρ changes six orders of magnitude. Solid lines correspond to the analytic solution. The simulation uses the standard MacCormack method in an equally spaced grid with only twenty zones. The algorithm stops due to stability problems at the external boundary point, which, in this test, is simply extrapolated from the interior.

$$T_{ab} = (\mu + p)u_a u_b + p g_{ab}, \qquad (2)$$

which is moving with respect to our coordinate frame with velocity

$$v = \frac{X u^r}{\alpha u^t}, \qquad (3)$$

so that the Lorentz factor can be written $W = \alpha u^t$.

The well known code of May and White [1] is written in a comoving coordinate system ($v = 0$). Other codes [2] are based in the Schwarzschild choice $Z = 1$. As it is well known, all these choices are prone to develop coordinate singularities when applied to strong field situations. We are interested in building a code which could be used to study black hole formation. We have then chosen a harmonic time coordinate, that is

z^{-1}

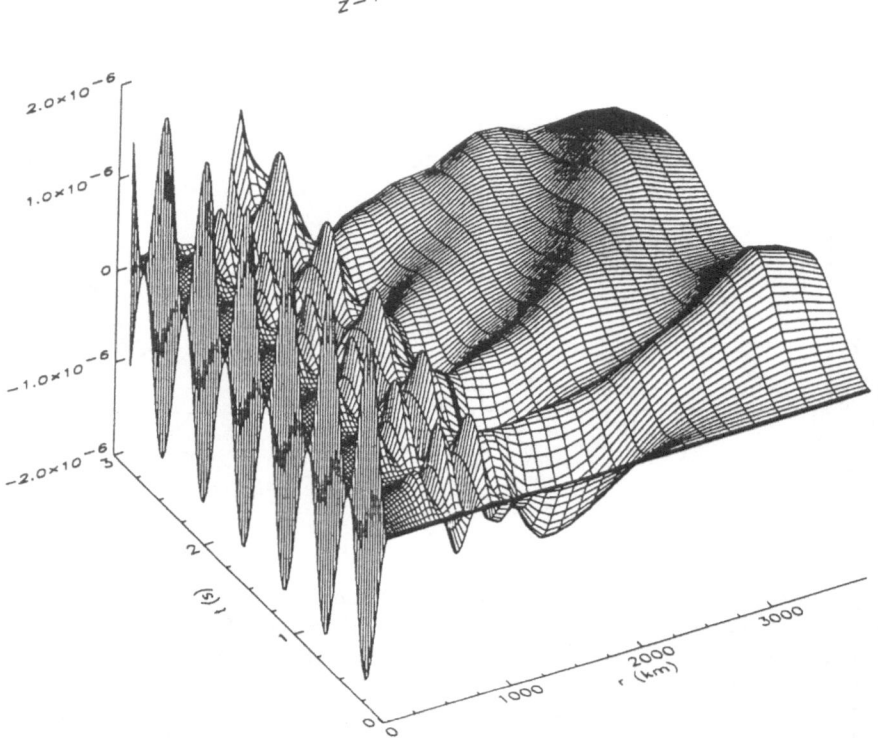

Fig. 2. *Oscillations around a white dwarf equlibrium model* which was generated using an eigth order Runge-Kutta method to integrate the Oppenheimer-Volkov equation in a grid of 7000 poits. The equation of state is taken from Salpeter and Zapolski [5] and we have chosen parameters $M = 1.23M_\odot$ and $R = 3500km$, inside the stability zone. A static Schwarzschild metric is matched at the star surface. This initial model is put into a geometrical grid consisting of 50 interior plus 10 exterior points (only two shown in the figure). Truncation errors initiate an oscillatory motion around the equilibrium values. The simulation uses a modified MacCormack method including shock capturing corrections [4] and is carried over 10^5 iterations (2.9 s). The figure displays the relative errors in Z (which is equal to one everywhere in the equilibrium model). Only the geometric quantities are evolved in this test, but similar results are obtained if the hydrodynamic quantities are evolved instead [6].

$$\alpha = b(r)XZ^2 \tag{4}$$

because of the singularity avoidance properties of this algebraic condition [3]. There is a price to pay for this, as one can not simplify the Einstein field equations to the extent that the other choices do; this will be made clear in the following section.

2 Writing down the equations

Allowing for (4), there are only two independent metric coefficients and we have chosen α and Z as the basic quantities to be evolved. we want to express the second order Einstein field equations as a first order system, Therefore, we introduce the following additional quantities to represent the time and space derivatives of the metric functions, respectively:

$$Q^z = \frac{\dot{Z}}{cZ}, \quad Q^\alpha = \frac{\dot{\alpha}}{c\alpha} \tag{5}$$

$$\Gamma^z = \frac{Z'}{Z}, \quad \Gamma^\alpha = \frac{\alpha'}{\alpha} \tag{6}$$

where $c = \frac{\alpha}{X}$ is the local light speed in our reference frame.

The Einstein field equations allow us to compute the time derivatives of Q^z, Q^α and Γ^z, namely

$$\partial_t Q^z = c[\Gamma^\alpha \Gamma^z + Q^z(Q^\alpha - 3Q^z) + \frac{\Gamma^\alpha}{r}$$
$$- 4\pi X^2(W^2 - 1)\mu + 8\pi W^2 p + \frac{2m}{r^3 Z^3}], \tag{7}$$

$$\partial_t Q^\alpha - \partial_r(c\Gamma^\alpha) = c[2\Gamma^\alpha \Gamma^z + 2Q^z(Q^\alpha - 3Q^z) + 2\frac{\Gamma^\alpha}{r}$$
$$+ 4\pi X^2(\mu - p - 2W^2(\mu + p))], \tag{8}$$

$$\partial_t \Gamma^z = c[\Gamma^z(Q^\alpha - 3Q^z) + \Gamma^\alpha Q^z + \frac{Q^\alpha - 3Q^z}{r}$$
$$+ 4\pi X^2(\mu + p)W^2 \frac{v}{c} \tag{9}$$

where we have denoted by m the *Bondi mass function*,

$$\frac{2m}{rZ} = 1 + \left(\frac{Z}{X}\right)^2 [(rQ^z)^2 - (1 + r\Gamma^z)^2]. \tag{10}$$

Note that the appearance of the quantity $\Psi = \frac{2m}{r^3 Z^3}$ in the right hand side of (7) is a source of numerical problems near the origin. To remedy this, one can replace α by Ψ as a basic quantity whose time derivative is given by

$$\partial_t \Psi = c\left[-Q^Z(8\pi((W^2 - 1)\mu + W^2 p) + 3\Psi) - 8\pi(\mu + p)W^2 \frac{v}{c}(\frac{1}{r} + \Gamma^z)\right], \tag{11}$$

where we have used (7, 9). The evolution system is completed by noting that

$$\partial_t \Gamma^\alpha - \partial_r(cQ^\alpha) = 0, \tag{12}$$

which follows easily from (5, 6).

Equations (5,11,7,8,9, and 12) give us an evolution system for the complete set of metric quantities

$$U = (Z, \Psi, Q^\alpha, Q^z, \Gamma^z, \Gamma^\alpha), \tag{13}$$

respectively. It is easy to see that the system is hyperbolic and has the balance law structure:

$$\partial_t U + \partial_r F(U) = S(U). \tag{14}$$

This is the same structure as the hydrodynamic equations [2] which follow from the conservation of the energy-momentum tensor (2) and that of the mass current

$$J^a = \rho u^a, \tag{15}$$

where ρ is the proper mass density of the fluid.

One can then apply to the combined system the current numerical methods of computational fluid dynamics, as it is actually done in the next section.

3 Numerical Implementation

We have used a second order operator splitting approach (Strang splitting) in order to deal with the transport (flux) and source terms separately. When the flux terms are switched out, the system becomes a set of ordinary differential equations, where metric and hydrodynamic quantities are strongly coupled. We have used a standard second order Runge-Kutta method to evolve the resulting ODE system.

In the transport step, when the source terms are switched out, the set of dynamical quantities reduces to Q^α, Γ^α plus the subset of hydrodynamic variables. Moreover, the hydrodynamic and metric systems of equatioons are decoupled and one can even apply different numerical methods to both parts. The hydrodynamic sector is treated with a MUSCL method [2] which is very robust and has good shock capturing properties. The metric sector is a linear system and we know the exact charateristic speed (the light speed c). We have chosen then a simpler method proposed by Yee [4], consisting in a standard MacCormack predictor-corrector plus a TVD (total variation diminishing) correction which can be applied as a postprocessor to avoid oscillations caused by the hydrodynamical shocks.

The numerical problems near the coordinate origin are solved by arranging the points of the numerical grid so that $r = 0$ be the first cell left interface. A Taylor development gives the parity of the metric quantities under reflection at the origin (only Γ^z and Γ^α are odd) and, therefore, we can deduce all the

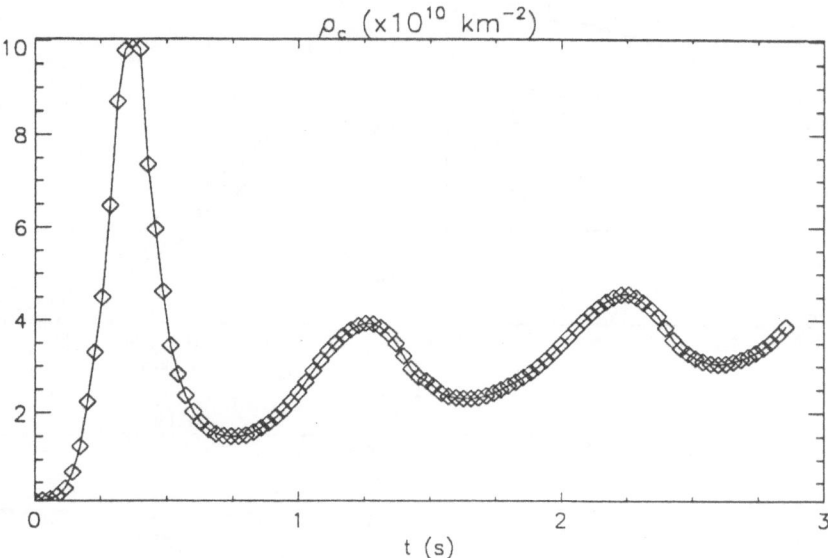

Fig. 3. *Dynamical evolution towards equilibrium* of a initial uniform density model with the same mass, radius and equation of state as in the previous test (see Figure 2). This initial non-equilibrium model has no pressure gradient to balance gravity so that it starts to collapse. The figure shows the evolution of the central density, which increases quickly in the initial infall phase becoming much greater than the equilibrium value. The star begind then an expansion phase until it reaches a value of the central density too low and collapses again, and so on, oscillating around the equilibrium value. The simulation is carried over 10^5 iterations using the same grid and numerical method than Figure 2. There is a tendency to reduce the star radius due to the lack of resolution at the surface and the central density rises accordingly.

values at the "mirror images" (negative r) of the first point and use them as inner boundary condition. A first test of algorithm is shown in Figure 1.

The outer boundary can be treated in a much simpler way because there is no singularity there. In most astrophysical applications, the outer mesh zones do not contain matter and the metric there is static (Birkhof theorem) so that the same initial values at the outer boundary point can be used every time. This is the case of the tests described in Figures 2 and 3, where either we start from or we approach to an equilibrium situation. One has to face numerical accuracy problems in these cases, where the dynamics is governed by the post-Newtonian terms. In an equilibrium model, our variable Γ^α has a non vanishing Newtonian limit:

$$V \equiv \Gamma^\alpha - r X^2 \frac{\psi}{2} = \text{post-Newtonian terms} \ldots \tag{16}$$

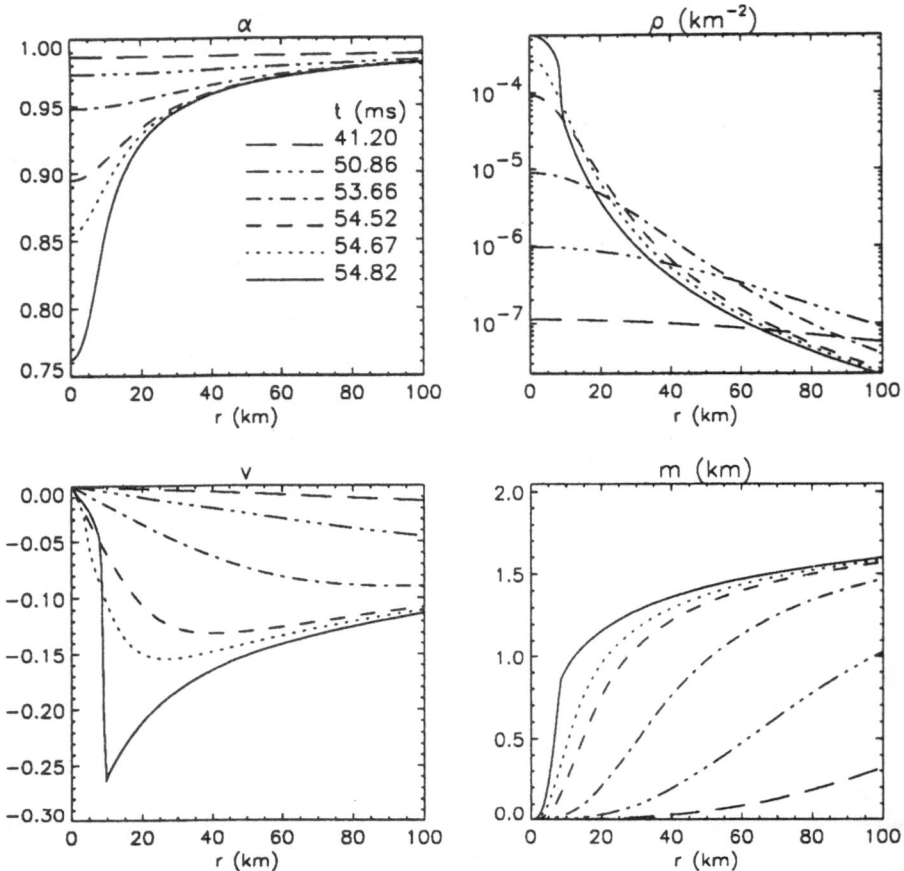

Fig. 4. *Stellar collapse (infall phase)* The initial white dwarf equilibrium model of the second test (see Figure 2) is put into a geometrical grid of 200 points. The collapse is initiated by substituting the original equation of state by that of a polytrope with adiabatic index $\gamma_0 = 1.3$, below the critical one. The upper left plots display the evolution of the lapse function, showing that the singularity avoidance properties of the gauge are not crucial here. This is due to the fact that the density (upper right) increases very quickly up to the nuclear matter value $\rho_b = 2.015 \times 10^{-4}$ km $^{-2}$. At the points where this happens, the adiabatic index is taken to be $\gamma = \gamma_0 + log(\rho/\rho_b)$ to simulate the stiffening that occurs due to the nuclear forces [7]. This stops the matter infall in the inner region and causes a shock discontinuity in the velocity profiles (lower left). An inner core is then formed which, for our choice of parameters, contains $0.8 M_\odot$ within a radius of 10 km (lower right mass plots). In our grid, this core contains only 15 mesh zones and we are unable to follow the shock wave propagation due to this lack of resolution which causes numerical problems at the core surface.

which masks the really dynamical terms. We have chosen to evolve the post-Newtonian variable V instead of Γ^α, by making the appropiate modifications in the evolution equation (12) (the structure of the system is not affected by these changes).

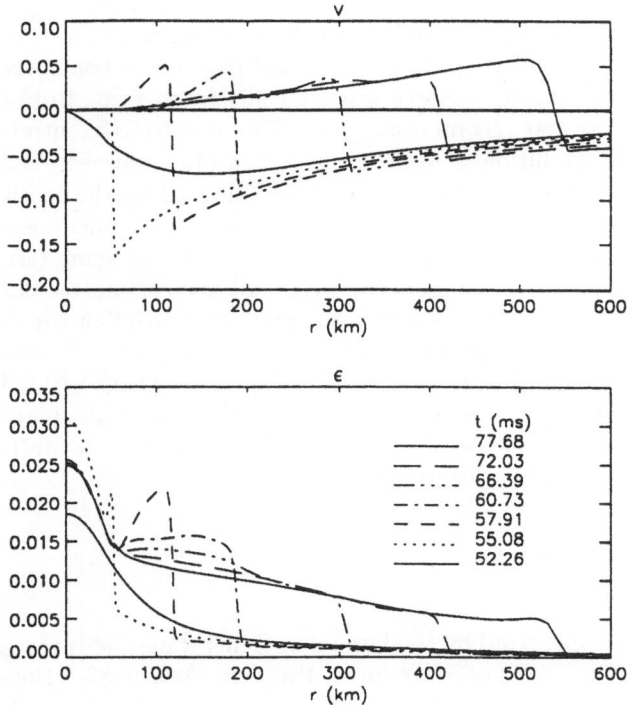

Fig. 5. *Stellar collapse (shock wave propagation).* The same test as in Figure 4 but with a lower (unrealistic) bounce density $\rho_b = 3 \times 10^{-6}$ km $^{-2}$ which causes the core to form within a radius of 50 km giving us more resolution. the upper plots show the evolution of the velocity profile starting at the begining of the shock formation. We can see the change in the velocity sign (the bounce) and the subsequent shock wave propagation. Note that the inner core remains near an equilibrium state. The specific internal energy ϵ, defined by $\mu = \rho(1 + \epsilon)$, is displayed in the lower plots, showing a clear increasing just behind the shock due to the heating of the fluid which crosses the shock wave.

4 Application to spherical collapse

The spherical collapse problem is very demanding from the numerical point of view, because one needs a numerical grid which fully covers the initial 10^3 km white dwarf model while keeping enough resolution in the 10 km central region where the remnant neutron star core is likely to form. This is not to be achieved with an evenly spaced fixed numerical mesh, because of the large number of points needed and also because the stability of our numerical methods is restricted by the causality condition

$$\Delta t < \frac{\Delta r}{c} \qquad (17)$$

so that higher resolution implies smaller timesteps.

As a way out of this situation, we actually use a nonuniform grid in which the size of every cell increases geometrically when going from the center to the exterior of the star (geometrical grid). This way we concentrate our grid points where more resolution is needed. Condition (17), however, can not be avoided that way because the timestep is actually limited by the smaller central zones. The results obtained starting with a realistic initial model are shown in figure 4.

We present also in Figure 5 a second run of the same test, where we have considered an unrealistic (too low) value of nuclear matter density to obtain a larger core. This way our grid resolution allows us to follow the shock propagation while the inner neutron star core keeps oscillating around equilibrium.

It is clear to us that the use of a fixed grid is the major NOSTRUM drawback and that the use of an adaptive numerical mesh is mandatory if one wants to study highly dynamical problems like black hole formation. This will be considered in the next NOSTRUM version.

5 Acknowledgment

This work is supported by the Dirección General para la Investigación Científica y Técnica (DGICYT) of Spain under Projects PB87-0583, PB90-0516 and PB91-0335.

References

1.   M.M. May, R.H. White: Phys. Rev. **141**, 1232 (1966)
2.   J.M. Ibáñez: "Numerical Relativistic Hydrodynamics" (this volume)
3.   C. Bona, J. Massó: Phys. Rev. D **38**, 2419 (1988)
4.   H.C. Yee:"A Class of High Resolution Explicit an Implicit Shock Capturing Methods", in "Computational Fluid Dynamics", Lectures given at the von Karman Institute for Fluid Dynamics, Belgium (1989)
5.   E.E. Salpeter, H.S. Zapolsky: Phys. Rev. **158**, 876 (1967)
6.   J.M. Martí: Ph.D. thesis, Universitat de Valencia (1991)
7.   K.A. Van Riper: Ap. J. **221**, 304 (1978)

Invariance Transformations of the Class $y'' = F(x)\, y^N$ of Differential Equations Arising in General Relativity

E. Herlt and H. Stephani

Friedrich-Schiller-Universität Jena, Theoretisch-Physikalisches Institut,
Max-Wien-Platz 1, O-6900 Jena, Germany

1 The Problem

In the context of relativistic spherically symmetric perfect fluids, differential equations of the form

$$\frac{d^2y}{dx^2} = F(x)\, y^N \quad N \neq 0 \tag{1}$$

arise on two occasions.

The first (with $N = 2$) is that of a fluid in shearfree motion (see e.g. Kramer et al. [1]) the second with $N = -1/3$ for a certain class of shearing perfect fluids.

In both cases the differential equation (1) is the key for obtaining exact solutions: if it is solved, pressure, mass density and metric can easily be calculated.

The function $F(x)$ is arbitrary: to fix it means essentially to fix the equation of state. But it is crucial that for a given $F(x)$ solutions can be found which depend on at least one arbitrary parameter (constant of integration), since in the derivation of equation (1) it was assumed that y does not only depend on the radial coordinate x, but also on time, and the only possibility for this time dependence is via the constants of integration.

Also Emden's equation for gravitating gaseous spheres is contained in (1) for $F = x^{1-N}$. The same structure have other important differential equations in physics and especially in general relativity, for instance the Thomas-Fermi equation in quantum mechanics, the background equation for algebraically special type III vacuum solutions, the differential equation for a Friedman-like cosmological model, algebraically special vacuum fields with time-dependent mass aspect.

Overview on Emden-Fowler-Equation $y'' = F(x) y^N$ in Physics and General Relativity

$y'' = F(x) y^2$	Shearfree spherically symmetric solutions in general relativity
$y'' = F(x) y^{-1/3}$	Spherically symmetric solutions with shear in general relativity
$y'' = y^{-1/3} x^{-2}$	Friedman-like cosmological models with volume viscosity
$y'' = y^{-1/2} x^{-1}$	*Background equation for type III vacuum fields in general relativity*
$y'' = y^{-3} x^{-2}$	Algebraically special vacuum fields with time dependent mass aspect
$y'' = y^N x^{1-N}$	Emden's differential equation for gravitating gaseous spheres
$y'' = y^{3/2} x^{-1/2}$	Differential equation of Thomas and Fermi in quantum mechanics
$y'' = y^M x^N$	equivalent to the linear damped harmonic oscillator in a nonlinear potential in point mechanics

2 The Goal

Solutions of the differential equation $y'' = F(x) y^N$ are rare; they have been found only for rather restricted classes of functions $F(x)$. The "obvious" way of solving Eq. (1) by prescribing $y(x)$ and calculating $F(x)$ from it does not work; in general, the (necessary!) constants occuring in $y(x)$ will also appear in $F(x)$ – which is forbidden.

It is the aim of this paper to ask wether the knowledge of the general solution $y(x)$ for a special function $F(x)$ can be used for constructing the general solution for a wider class of functions $\tilde{F}(\tilde{x})$. The answer will be positive.

3 Invariance Transformations of Emden - Fowler Equations $y'' = F(x) y^N$ and Their Generators

Given a differential equation

$$\frac{d^2 y}{dx^2} = F(x) y^N \quad N \neq 0 ,$$

(2)

is there a mapping $\{x, y, F\} \rightarrow \left\{\tilde{x}, \tilde{y}, \tilde{F}\right\}$ which maps (2) into

$$\frac{d^2 \tilde{y}}{d\tilde{x}^2} = \tilde{F}(\tilde{x})\, \tilde{y}^N \ , \tag{3}$$

such that every solution $y(x)$ of (2) is mapped into a solution $\tilde{y}(\tilde{x})$ of (3) and that $\tilde{F}$ does not depend on the constants of integration contained in $y(x)$? Or, in other words, will there exist an invariance transformation? In general, the two functions $\tilde{F}(\tilde{x})$ and $F(\tilde{x})$ will be different. If they coincide, then the mapping under consideration is a symmetry of the differential equation and we know how to search for such mappings: we assume that the mapping is an element of a Lie point group of transformations and construct the generator by well established methods.

If F and $\tilde{F}$ are different, we can use a similar approach by assuming that there exists a whole group of transformations

$$\tilde{x} = \tilde{x}(x, y, F, \lambda), \quad \tilde{y} = \tilde{y}(x, y, F, \lambda) \ , \quad \tilde{F} = \tilde{F}(x, y, F, \lambda) \tag{4}$$

λ being the group parameter, all of them leaving $y'' = F(x)\, y^N$ invariant. This group is fully characterized by its generator

$$X = \xi(x, y, F)\frac{\partial}{\partial x} + \eta(x, y, F)\frac{\partial}{\partial y} + \phi(x, F)\frac{\partial}{\partial F} \ , \tag{5}$$

whrere the functions ξ, η and ϕ are defined by

$$\xi = \frac{d\tilde{x}}{d\lambda}\bigg|_{\lambda=0} \ , \quad \eta = \frac{d\tilde{y}}{d\lambda}\bigg|_{\lambda=0} \ , \quad \phi = \frac{dF}{d\lambda}\bigg|_{\lambda=0} \ . \tag{6}$$

This generator will lead to an invariance transformation if it leaves the differential equation (2) invariant

$$X\left(y'' - F(x)\, y^N\right) \equiv 0 \qquad (\text{mod } y'' = F(x)\, y^N) \ . \tag{7}$$

There are two ways of dealing with this invariance condition, which we shall discuss now in detail. The first idea is to take F as an extra variable, i.e. to consider X as a generator of a Lie point transformation in the space $\{x, y, F\}$ and, therefore, to demand that Eq. (7) holds identically in all variables which cannot be eliminated by $y'' = F(x)\, y^N$, i.e. identically in $\{x, y, F, y', F', F''\}$. Eq. (7) then reads in full

$$\begin{aligned}
&\eta_{xx} + (2\eta_{xy} - \xi_{xx})y' + (\eta_{yy} - 2\xi_{xy})y'^2 - \xi_{yy}y'^3 \\
&+ (\eta_y - 2\xi_x - 3\xi_y y')\, F y^N \\
&+ 2\left[\eta_{xF} + (\eta_{yF} - \xi_{xF})y' - \xi_{yF}y'^2 - \xi_F F\, y^N\right] F' \\
&+ (\eta_{FF} - y'\xi_{FF})F'^2 + (\eta_F - y'\xi_F)F'' - \phi\, y^N - \eta y^{N-1}\eta F \equiv 0 \ .
\end{aligned} \tag{8}$$

Identically in F'' this can be true only for

$$\xi_F = 0 = \eta_F \tag{9}$$

and therefore

$$\eta_{xx} + (2\eta_{xy} - \xi_{xx})y' + (\eta_{yy} - 2\xi_{xy})y'^2 - \xi_{yy}y'^3$$
$$+ (\eta_y - 2\xi_x - 3\xi_y y')\,Fy^N - \phi y^N - \eta y^{n-1}F \equiv 0. \tag{10}$$

Since ξ and η do not depend on y', the coefficients of the different powers of y' must vanish separately. This leads to

$$\xi = y\alpha(x) + \beta(x)\,; \quad \eta = y^2\alpha' + y\left(\frac{\beta'}{2} + c\right) + \delta(x) \tag{11}$$

and

$$\alpha'' = \alpha F\, y^{N-1}$$
$$y^2\alpha''' + \frac{y}{2}\beta''' + \delta'' - y^{N+1}\eta F\alpha' + y^N F\left[(1-N)C - \frac{3+N}{2}\beta'\right]$$
$$- y^N\phi - y^{N-1}\eta\delta F \equiv 0 . \tag{12}$$

As the functions α,β, and ϕ do not depend on y, these equations can hold identically in y only if $\alpha = 0$ and

$$N = 1 \qquad \delta'' = \delta F\,, \qquad\qquad \phi = \frac{1}{2}\beta''' - 2F\beta' \tag{13a}$$

$$N = 2 \qquad \delta'' = 0\,, \beta''' = 4\delta F\,, \quad \phi = -F\left(\frac{5}{2}\beta' + c\right) \tag{13b}$$

$$N \neq 0,1,2 \quad \delta = 0\,, \qquad\qquad \phi = F\,[(1-N)c - (3-N)\beta'\eta] \tag{13c}$$

In the last step, we demand that Eqs. (13a-c) hold identically in F. Since only ϕ may depend on F, we arrive at

$$N = 1 \qquad \alpha = 0,\ \delta = 0, \qquad\qquad \phi = \frac{1}{2}\beta''' - 2F\beta'$$
$$N \neq 0,1 \quad \alpha = 0,\ \delta = 0,\ \beta''' = 0 \quad \phi = F\left[(1-N)C - (3+N)\frac{\beta'}{2}\right] \tag{14}$$

To summarize: the generator (5) gives rise to an invariance transformation if its components satisfy (11) and (14).

The second idea is to consider F as a function of x. Then the extra dependence of $\tilde{x}$ and $\tilde{y}$ on F does not make sense, cp. Eqs. (4)–(5), and we start already with (9), i.e. $\xi_F = 0 = \eta_F$. $\tilde{F}$ (and ϕ) should depend on F, of course, since the possible change of F is exactly what we are interested in.

The calculations run as in the foregoing case, and we again come from the invariance condition (10) to (13a–c) and $\alpha = 0$. But now Eqs. (13a–c) are no longer identities in F and x, but constraints on the functions $F(x)$, $\delta(x)$, $\beta(x)$, $\phi(x)$, which have to be satisfied along the group orbit. Obviously the second approach is the more general one, and we will use it from now on. The results obtained so far can be summarized as follows:

Theorem 1. *If the components of the generator*

$$X = \beta(x)\frac{\partial}{\partial x} + \left[y\left(\frac{\beta'(x)}{2} + c\right) + \delta(x)\right]\frac{\partial}{\partial y} + \phi\frac{\partial}{\partial F}$$

satisfy the condition

$N = 1$	$\delta'' = \delta F$,	$\phi = \dfrac{1}{2}\beta''' - 2F\beta'$
$N = 2$	$\delta = 0$, $\beta''' = 4\delta F$,	$\phi = -F\left(\dfrac{5}{2}\beta' + c\right)$
$N \neq 0,1,2$	$\delta = 0$, $\beta''' = 0$,	$\phi = F\left[(1 - N)c - (3 + N)\dfrac{\beta'}{2}\right]$

then all finite group transformations generated by X leave the class $y'' = F(x)\, y^N$ of differential equations invariant.

To give the explicit form of the solutions $\tilde{y}(\tilde{x})$ for a function $F(x)$, we need the finite transformations. In principle, they can be constructed straightforwardly; one has to integrate

$$\frac{d\tilde{x}}{d\lambda} = \tilde{\beta}(\tilde{x}) , \quad \frac{d\tilde{y}}{d\lambda} = \frac{1}{2}\tilde{y}\tilde{\beta}'(x) + \tilde{\delta}(\tilde{x}) , \quad \frac{d\tilde{F}}{d\lambda} = \phi(\tilde{x}, \tilde{F}) \qquad (15)$$

with initial values $\tilde{x}_0 = x$, $\tilde{y}_0 = y$. We will not do this in some detail, but we will only give the results.

4 Results – The Finite Invariance Transformations

If we introduce the abreviations.

$$\psi(x) \equiv \beta^{1/2}(x)\,\beta^{-1/2}(\tilde{x}) , \quad \tilde{x} = \tilde{x}(x)$$

$$H(x) \equiv -\beta^{1/2}(x)\int_{x}^{\tilde{x}} \frac{\tilde{\delta}(\tilde{x})}{\beta^{3/2}(\tilde{x})}\, d\tilde{x} , \qquad (16)$$

we then arrive at the following theorems.

Theorem 2. *If $y(x)$ is a solution of $y'' = F(x)\, y^2$ for a given function $F(x)$, then $\tilde{y}(\tilde{x}) = \dfrac{1}{\psi(x)}\left[y(x) - H(x)\right]$ is a solution of $\tilde{y}''(\tilde{x}) = \tilde{F}(\tilde{x})\,\tilde{y}^2$ provided*

$$\frac{d\tilde{x}}{dx} = \frac{1}{\psi^2(x)} \quad \text{and}$$

$$\tilde{F}(\tilde{x}) = \psi^5(x)F(x)$$

$$\psi'' = 2H(x)F(x)\psi(x)$$

$$H''(x) = F(x)H^2(x) .$$

This law for constructing the new solution exhibits a rather strange feature: it has the form of a superposition principle where one has to substract (and then to manipulate, of course) two solutions (y and H) of the *same* differential equation!

More specifically, in order to obtain the general solution $\tilde{y}(\tilde{x})$, one has to take a general solution $y(x)$ and an arbitrary, special solution $H(x)$ to start with.

Because of its importance, a vast literature exist for the case $N = 2$, i.e. for the differential equation $y'' = F(x)\,y^2$, where solutions and their physical properties are discussed, see e.g. [1], Srivastava [2], Wyman [3] for further references.

At first glance it looks rather promising to take all the different functions $F(x)$ for which $y'' = F(x)\,y^2$ has been solved, and to apply our invariance transformations to each of them, thus obtaining a plethora of new solvable cases $\tilde{F}(\tilde{x})$. But a closer look at the known solvable cases is rather disappointing (from our point of view): it turns out that nearly all of them can be obtained from $F = \text{const}$ by means of our theorem. New seed solutions $F \neq \text{const}$, -5, $-5/2$, $-15/7$, $-20/7$ are needed to get new exact solutions of $y'' = F(x)\,y^2$.

What about the diffential equation $y'' = F(x)\,y^N$ $N \neq 0, 2, 1$? This question is answered by the following theorem.

Theorem 3. *If $y(x)$ is a solution of $y'' = F(x)\,y^N$, then $\tilde{y}(\tilde{x}) = \dfrac{1}{\psi(x)}\,y(x)$ is a solution of $\tilde{y}'' = \tilde{F}(\tilde{x})y^N$ with a function $\tilde{F}$, provided $\psi(x) = b_1 x + b_2$,*
$$x = x(\tilde{x}) = \frac{d - b\tilde{x}}{a\tilde{x} - c} \ , \quad F(\tilde{x}) = \psi^{(3+N)}(x)\,F(x) \ .$$

Theorem 4. *If $y(x)$ is a solution of the linear diffential equation $y'' = F(x)y$, then $\tilde{y}(\tilde{x}) = \dfrac{1}{\psi(x)}y(x)$ is a solution of $\tilde{y}'' = \tilde{F}(\tilde{x})y$ with a new function $\tilde{F}$ provided $\dfrac{d\tilde{x}}{dx} = \psi^{-2}(x)$, $\tilde{F}(\tilde{x}) = \psi^4(x)F(x) - \psi^3(x)\psi''(x)$.*

5 Applications

5.1 Example

To illustrate the power of the transformation method, we shall give now a simple example. The starting point is the differential equation

$$y'' = 6y^2 \tag{17}$$

whose general solution can be given in terms of elliptic functions. To apply Theorem 2 to this case $F(x) = \text{const.} = 6$ we solve $H'' = 6H^2$ by $H = x^{-2}$ and

$$\psi'' = 12x^{-2}\psi \quad \text{by}$$
$$\psi = C_1 x^4 + C_2 \frac{x^{-3}}{7} \ . \tag{18}$$

Then we get ($C_2 \neq 0$)

$$C_2 \tilde{x} = C_0 + \left(C_1 + \frac{C_2 x^{-7}}{7} \right)^{-1}$$

and the transformed F becomes

$$\tilde{F}(\tilde{x}) = 6\psi^5(x)$$

$$= 6 \left[C_1 \left(\frac{C_2(C_2\tilde{x} - C_0)}{7 - 7C_1(C_2\tilde{x} - C_0)} \right)^{4/7} + \frac{C_2}{7} \left(\frac{C_2(C_2\tilde{x} - C_0)}{7 - 7C_1(C_2\tilde{x} - C_0)} \right)^{-3/7} \right]^5 \quad (19)$$

It is certainly not obvious that the solution of $\tilde{y}'' = \tilde{F} y^2$ with the above $\tilde{F}(\tilde{x})$ can be given in terms of elliptic functions!

5.2 A Remark on Classes of Abelian Differential Equations

If the differential equation $y'' = F y^N$ admits a Lie-point symmetry, it can be reduced to a first order differential equation (and a quadrature). If $\tilde{y}'' = \tilde{F}\tilde{y}^N$ inherits this symmetry Theorems 2 and 3 will also relate the corresponding first order differential equations and their solutions.

As an example, we take the two equations

$$\frac{d^2y}{dx^2} = x^n y^2 \quad \text{and} \quad (20)$$

$$\frac{d^2\tilde{y}}{d\tilde{x}^2} = \tilde{x}^{\tilde{n}}\tilde{y}^2 \quad \text{with} \quad (21)$$

$$\tilde{n} = -\frac{5}{2} \mp \frac{1}{2}; \, [2 - (2N+5)^{-2}]^{-1/2} \, |(2N+5)|\sqrt{2} < 1 \quad (22)$$

which are related by Theorem 2, choosing

$$H = x^{-(2+n)}, \quad \tilde{x} = x^\beta, \quad 2\beta = 1 \pm [8n^2 + 40n + 49]^{1/2} . \quad (23)$$

Substituting $\zeta = \ln x$, $y = \eta(\zeta) e^{-(n+2)\zeta}$, $\frac{d\eta}{d\zeta} = P(\eta)$ into (20) leads to the Abelian differential equation

$$P\frac{dP}{d\eta} - (2N+5)P + (N+2)(N+3)\eta - \eta^2 = 0 \quad (24)$$

If a solution of this equation was known for a certain n, a solution can also be constructed for $\tilde{n}$ given by (22).

6 Summary

We have found all transformations $\{x, y, F\} \rightarrow \{\tilde{x}, \tilde{y}, \tilde{F}\}$ which leave the class $y'' = F(x)\, y^N$ of differential equations form-invariant (and are elements of a Lie group) by constructing their generators (Theorem 1). The corresponding finite transformations are written down.

For $N = 1$ (i.e. for the linear case $y'' = F(x)\, y$), these transformations are not of much interest and value. Besides scalings and the linear superposition of solutions, they consist of transformations which transform any two functions $F(x)$ and $\tilde{F}(\tilde{x})$ into each other – but to map, say $y'' = F(x)\, y$ into $\tilde{y}'' = 0$, one has to solve exactly the original differential equation.

For $N \neq 0, 1, 2$ the general transformation can be explicitly given (Theorem 3).

The most interesting case is $N = 2$. Here the transformation law includes a kind of superposition principle; again, the finite transformations could be explicitly given. It turns out that in the vast literature on this case $N = 2$, the transformation method has in fact been used in a disguised form, without reference to Lie groups.

References

1.   Kramer, D., Stephani, H., MacCallum, M.A.H. and Herlt, E.: *Exact Solutions of Einstein's Field Equations*, Cambridge University Press, Cambridge (1980)

2.   Srivastava, D.C.: *Exact Solutions for shearfree motion of spherically symmetric perfect fluid distributions in general relativity*, Class. Quantum Grav. 4 (1987) 1083-1117

3.   Wyman, M.: *Jeffery-Williams Lecture. Nonstatic radially symmetric distributions of mathes*, Math. Bull. 19 (1976) 343-357

Relativistic Kinetic Theory and Cosmology

D. R. Matravers

School of Mathematical Studies, University of Portsmouth,
Portsmouth PO1 2EG, England

1 Introduction

The sources of my current interest in Kinetic Theory were two apparently unrelated sets of papers and ideas they generated. The first set concerned Penrose's initial conditions for the universe [1], [2], [3] and [4] and Tod's conjecture [5] drawn from them. Penrose's Weyl Curvature Hypothesis is that the Weyl tensor tends to zero at the initial singularity. Calculations by Goode and Wainwright [6] to elucidate the idea enabled Tod to conjecture that the universe was initially Friedman-Robertson-Walker (FRW). Newman [7] has since proved that: *a $\gamma = 4/3$ perfect fluid space time which evolves from a spacelike conformal singularity subject to the Weyl Curvature Hypothesis is necessarily FRW near the singularity.* The Hartle and Hawking [8] quantum cosmology programme leads to similar conclusions and so we are led to consider the possibility that the universe began with a geometry that was exactly FRW. These pictures of the early universe do not preclude inflation although it is not necessary because the special nature of the universe is set by the big bang itself.

The second set of papers centred round the question of how much freedom there is for a spacetime that begins with FRW geometry to change during evolution. Arguments which point out that the Bianchi type cannot change and that the symmetry in the Cauchy data severely limits any other freedom were set out very strongly in a paper by Ellis [9]. Even a discontinuity in a deterministic equation of state can only change the time evolution of the universe and not the symmetry of its space sections so isometry and homogeneity cannot be lost [10].

The inflation scenario gets round this apparent impasse of a universe which is intially isotropic and homogeneous but now has large scale structure by invoking quantum fluctuations to generate perturbations [11] which then evolve to give the structure that is observed. These arguments although plausible are not totally closed and some inputs from those using the model are required.

As the Penrose argument is essentially classical, in the sense of not involving quantum mechanics, it is interesting to look for a classical solution. To construct such a model we assume that the evolution of universe is described by a solution to the Einstein-Boltzmann system of equations, ie, the matter is described by a kinetic theory model and the relation between matter and the geometry is described by Einstein's field equations. The idea then is to find models which evolve from homogeneity and isotropy, ie having FRW geometry, to inhomogeneity and anisotropy. This is achieved by having a particle distribution function which:

1. is initially inhomogeneous and anisotropic but is compatible with FRW geometry, and
2. the distribution is initially collision free or collision dominated with effectively mass zero particles, then either
3. as the temperature drops collisions occur which have an appreciable effect on the distribution function and change its structure through the collision terms in the Boltzmann equation or
4. the particles which were initially effectively mass zero cool as the universe expands and acquire mass. In this case constraints on the distribution function are violated and interactions have to occur.

In both scenarios inhomogeneity and anisotropy of the distribution are communicated to the geometry via anisotropic pressures etc that are generated. Suitable models to achieve this are described in [12], [13] and [14].

These proposals for the evolution of the universe have an advantage in that they completely bypass the conditions contained in the following theorems which were the subject of some intense debate in the 1970's. The theorems preclude most of the obvious models for the evolution of the universe if one takes a strict mathematical view. The way out of course is to argue that the models are only an approximation to reality and so the theorems should not be applied in a precise way [15]. While this attitude is reasonable it is not in keeping with the intention of this programme.

Firstly we have a theorem which states that: If a spacetime contains a mixture of gases in collision dominated equilibrium then the ratios of the chemical potentials to the temperature have no time dependence measured in the rest frame of the gas mixture. The gas motion is always shear free and if one of the components of the gas has non-zero rest mass then only non-expanding non-shearing motion is possible and the spacetime is stationary [16].

In the picture presented here the particles have zero rest mass in the collision dominated era if there is one, drop out of equilibrium and subsequently move to become collisionless so the theorem does not apply at any time in the evolution.

The second result is the theorem due to Ehlers, Geren and Sachs [17] in which they show that if the distribution is isotropic and either collision free or detailed balancing occurs and $m = 0$ then the spacetime is either stationary or Robertson-Walker. Since the distribution we employ is not isotropic the theorem poses no restrictions.

If the scenario developed here is applied to Bianchi models then the assumption that matter is described by a perfect fluid for all time is probably not reasonable. The heat flow and the anisotropic pressure vanish only if conditions on the components F_{ab} and F_a are satisfied and to maintain these for all time if there is shear and once collisions begin would be very difficult if possible at all. One could consider the possibility that the critical moments F_a, F_{ab}, F_{abc} are all zero however a theorem proved in [18] shows that the space-time must then be FRW. This result provides a motivation for investigating Bianchi models with heat flow and anisotropic pressures; not to find new exact solutions as is quite common but to establish the asymptotic behaviour. Of course to be useful in this context one has to be careful to ensure the fluids conform to the Kinetic Theory model which gives the equation of state etc. Such a programme would be a valuable addition to the work done by Ellis and Collins [19] among others.

Having put the above case it is unlikely that the exact solutions and the model on which the arguments are based are realised in the actual universe but in looking for models for the origin and evolution of the universe there are few criteria for selection. So we should be prepared to consider the phase space of possibilities which will include this picture.

Relativistic Kinetic Theory is not a topic that is included in most introductory courses in General Relativity and or Cosmology and very few, if any, of the standard texts mention the subject. There are excellent references, eg the articles by Ehlers [20], [21] but to make this paper relatively self contained I shall begin with a brief outline to establish the notation.

To get a physical picture imagine a spacetime filled with a system of many particles. The particles may be elementary, atoms, ..., galaxies or even clusters of galaxies and the interactions between them are separated into *long range* and *short range*. The mean free path between short range interactions is much larger than their range. The long range interaction is modelled by an external mean field in which each particle moves as a test particle. The short range interactions are modelled as point collisions governed by special relativistic theory. The typical average behaviour of systems which satisfy these conditions is the subject of kinetic theory.

Ehlers [21] points out that the average properties could provide a microscopic description of a gas whose particles obey quantum laws. The appropriateness and limitations of the approximation are discussed in the non- relativistic case in [22]. So the solutions are not totally unrealistic from a physical point of view.

2 Notation and the Homogeneous Case

We will be concerned with a region of spacetime which can be modelled as a coordinate neighbourhood of a differentiable manifold with coordinates on the cotangent bundle given by: $\{(x^i, p_i),$ where $i, j, ... = 1, .., 4\}$. We assume that the spacetime admits a nonsingular metric (g_{ij}) which is adequately differentiable and a symmetric Christoffel connection (Γ^i_{jk}). Partial derivatives will be denoted by a comma and covariant ones by a semi-colon.

The geometry is taken to be determined by the matter distribution through the Einstein Field Equations:

$$R^{ij} - \frac{1}{2}g^{ij}R = T^{ij} \tag{1}$$

where R^{ij} is the Ricci tensor etc.

The state of a particle at any instant is given by its position and momentum so the space of concern here is the seven dimensional phase space:

$$B_+(x) = \{p^i \in T_x : p_0 > 0, p^i p_i \leq 0\} \tag{2}$$

or if we restrict attention to particles of a single mass the phase space is the mass shell:

$$B_m(x) = \{p^i \in T_x : p_0 > 0, p^i p_i = -m^2\} \ . \tag{3}$$

Without loss of generality we will consider only single mass particles to make the arguments easier to follow.

The one particle distribution function $f(x^i, p_i)$ is defined on phase space and is required to be positive, differentiable and to satisfy convergence conditions which will become apparent later. Details of the derivation and interpretation of f are given in Ehlers [15]. In the case of no collisions or detailed balancing the function f must be constant on the phase space trajectories, ie it must satisfy the Liouville equation:

$$0 = Lf := \left(p^i \frac{\partial}{\partial x^i} - \Gamma^i_{jk} p^j p^k \frac{\partial}{\partial p^i} \right) f \ . \tag{4}$$

If collisions are taking place then the behaviour of f is governed by the Boltzmann equation:

$$Lf = C(f) \tag{5}$$

where C is a functional which models the collisions. The general Boltzmann problem gives rise to an integro-differential equation.

Once one has solved for f the properties of the matter distribution can be obtained from the moments. In particular, the energy momentum tensor:

$$T^{ij}(x) = \int f(x, p) p^i p^j \pi_m \tag{6}$$

where π_m is the volume element on the relevant fibre restricted to the mass shell. The Einstein-Boltzmann problem is to solve the system of equations (1) and (6).

From f we can obtain a particle 4-current density:

$$N^i = \int f p^i \pi_m \tag{7}$$

and hence a mean flow 4-velocity v^i (kinematical mean velocity) defined by:

$$N^i = n v^i, \qquad v^i v_i = -1 \tag{8}$$

or one can decompose the energy momentum tensor T^{ij} to get a unique (dynamical mean) velocity u^i using

$$T^{ij} = \rho u^i u^j + w^{ij}, \tag{9}$$

where $u_i u^i = -1$, u^i is future directed and $w_{ij} u^j = 0$. In general the velocities u and v will not coincide. Some implications of this difference are discussed in [25]. Here we will use u to model the Hubble flow.

Using u we can make a $(3+1)$ decomposition of the space-time and introduce orthonormal tetrad coordinates. These coordinates will be denoted by indices from the beginning of the alphabet. The momentum can be described in terms of an orthonormal dyad (u, e) where $u^a e_a = 0$ and $e^a e_a = 1$;

$$p^a = Eu^a + \lambda e^a \tag{10}$$

where e^a are the direction cosines in the tetrad frame, $E = -p^a p_a > 0$ is the particle energy relative to u and λ is the magnitude of the momentum relative to u. Clearly,

$$\lambda^2 = E^2 - m^2 . \tag{11}$$

The distribution function $f(x^i, p^a)$ for a gas of identical particles can be written: $f(x^i, p^a) = g(x^i, m, E, e^a)$ using the decomposition (10) and the function g can be expanded in covariant harmonics,

$$g = F + F_a e^a + F_{ab} e^a e^b + \dots \tag{12}$$

where the harmonics $F_{ab\dots r}(x^i, m, E)$ are symmetric and trace free tensors orthogonal to u^a [23], [24].

In a $k = 0$ FRW spacetime with length scale factor $S(t)$ the Liouville equation (4) for the harmonics (12) takes the form:

$$SE \left(\frac{\partial F_{ab\dots r}}{\partial t} \right) - \dot{S}(E^2 - m^2) \left(\frac{\partial F_{ab\dots r}}{\partial E} \right) = 0 \tag{13}$$

and it is easy to see that a solution is given by:

$$G_{ab\dots r}(X) = F_{ab\dots r}(x^i, m, E) \tag{14}$$

where

$$X = (E^2 - m^2)^{1/2} S(t) . \tag{15}$$

In [25] a solution to the Einstein-Liouville system of equations for a $k = 0$ FRW spacetime is constructed using the harmonics $F_{ab\dots r}$ subject to the conditions: if $m \neq 0$ (case (a))

$$\int_0^\infty G_a(X) \, dX = 0 \tag{16}$$

$$G_{ab} = 0 \tag{17}$$

or
if $m = 0$ (case (b))

$$\int_0^\infty G_a(X) \, dX = 0 . \tag{18}$$

$$\int X^3 G_{ab}(X)\,dX = 0 \tag{19}$$

These equations have to be satisfied to ensure that the heat flow and anisotropic pressure are zero as required in an FRW spacetime. Notice that (16), or for $m = 0$ (18), does not require that G_a be zero and hence [22] that the matter flow follow the geometrically preferred velocity u. For comments on the significance of this for cosmological modelling see [24]. In FRW models these peculiar velocities decay rapidly as the universe expands but they could have an effect on the very early evolution and on initial conditions.

The condition (19) permits non-zero solutions for G_{ab}. The zeroth harmonic, which determines the density and pressure, and all harmonics higher than the second are essentially arbitrary subject to suitable convergence of the expansion and $f(x^i, p^a)$ being positive.

In [12],[13] the spacetime is assumed to start off with FRW geometry and a distribution function which has Bianchi I symmetry; it is therefore spatially homogeneous but anisotropic. As the universe cools, in scenario (3) the particles whose distribution initially satisfied (16,17) start to collide and so generate non-zero moments in the collision functional. Consequently (13) determining F_{ab} becomes;

$$E\frac{\partial F_{ab}}{\partial t} - \frac{\dot{S}}{S}\lambda^2\frac{\partial F_{ab}}{\partial E} = b_{ab} \tag{20}$$

which will no longer be satisfied by the function (14) in the case of $m = 0$ or by zero for $m \neq 0$; it must contain at least an additive part to allow for the b_{ab}. This means that in general (19) and/or (17) will no longer be satisfied, ie, the anisotropic pressure

$$\pi_{ab} = \frac{8\pi}{15R^4}\int_0^\infty \lambda^3 F_{ab}\,dE \neq 0 \ . \tag{21}$$

The initial conditions for the next phase of the evolution come from the FRW era and are that ω, σ_{ab}, u^a and the conformal curvature are zero and so equation (4.16) of Ellis [26] shows that the non-zero π_{ab} will generate shear in the geometry as the system evolves. The Boltzmann equations for the first two moments of the distribution function then become;

$$\frac{2}{15}\lambda^{-1}\left(\frac{\partial(\lambda^3\sigma^{ab}F_{ab})}{\partial E}\right) - E\dot{F} + \frac{1}{3}\lambda^2\theta\frac{\partial F}{\partial E} = -b \tag{22}$$

$$\frac{6}{35}\lambda^{-2}\frac{\partial(\lambda^4\sigma^{bc}F_{abc})}{\partial E} + Eh_a^d\dot{F}_d - \frac{1}{3}\lambda^2\theta\left(\frac{\partial F_a}{\partial E}\right) - \frac{2}{5}\lambda^{\frac{1}{2}}\left(\frac{\partial(\lambda^{\frac{3}{2}}F_d\sigma_a^d)}{\partial E}\right) = b_a \ . \tag{23}$$

Clearly any anisotropy in the F_{abc} which would not have been "seen" by the metric previously gets passed to the lower moments and to the density, pressure etc and hence to the geometry. And similarly anisotropies in the higher moments cascade down.

Alternatively, in scenario (4), particles whose distribution initially satisfied (18,19) pick up mass; consequently the required conditions change to (16, 17), and the latter will not be satisfied in general. This evolution is demonstrated explicitly in [13]. The final effect is the same as above with shear generated by the change in F_{ab} and the rest following from there. In a realistic situation one would expect both processes (a) and (b) to occur.

In either case, failure to satisfy (17) causes an anisotropic pressure which pushes the geometry away from the FRW form, the anisotropy generating shear. This then feeds backs through the Boltzmann equations, to increase the anisotropy in the distribution function; but on the other hand the expansion of the universe tends to decrease it. Depending on the details of the situation, the result may or may not be to isotropise the higher moments. It would be interesting to have some specific results to demonstrate this point and the timescale over which it occurs.

3 The Anisotropic and Inhomogeneous Case

In [27] Maharaj and Maartens obtained solutions to the Einstein-Liouville equations with an FRW geometry and a distribution function which is inhomogeneous and anisotropic. The solutions depend on constants of the motion K_a constructed from the Killing vectors, and C, generated by conformal Killing vectors if $m = 0$. In [27] a general solution and a more specific one that is easier to analyse because the distribution is spherically symmetric about the origin of coordinates, are given.

The analysis carried out in the previous case goes through for the specific solution, except for the part which we will now consider. In place of the solution (14) to the Liouville equation the specific solution has the form

$$F_{ab....r}(x^i, m, E) = H_{ab....r}(t, r, m, X) \tag{24}$$

where X is defined as above, $r^2 = x^2 + y^2 + z^2$ and the r and t enter through the constant of motion $C = C(X, r, t)$. The conditions that gave rise to (16-19) become:

$$\int_0^\infty X^3 H_a(t, r, X, m) \, dX = 0 \tag{25}$$

and

$$\int_0^\infty X^4 [X^2 + m^2 S^2(t)]^{-1/2} H_{ab}(t, r, X, m) \, dX = 0 . \tag{26}$$

These conditions are weaker than those in the previous discussion; because H is explicitly dependent on t, (26) does not require that H_{ab} be zero everywhere even when $m \neq 0$. Clearly the conditions on $H_{ab....r}$ are different depending on whether m is zero or not and thus in general the same function will not satisfy (26) in both cases. Hence the argument that the universe will be driven away from an FRW geometry as it expands holds. This can happen either through collisions or through mass becoming non-negligible.

The argument developed for the specific form of the distribution function which is spherically symmetric about the origin of coordinates can be extended to a more general solution as long as the solution is effectively collision-free. One can combine such solutions linearly with different origins to get new, more complex solutions of the Einstein-Liouville equations in the FRW regime. This will generate a distribution of spherically symmetric inhomogeneities, arbitrarily placed and in principle capable of giving any desired mass distribution. It is possible to do this because the Liouville equation is linear in the distribution function or its moments, as are (25, 26). Consequently we can add distribution functions with constant coefficients, still obtaining solutions of the Liouville equation. There is resultant non-linearity in the field equations but it will cause no problem in practice.

Even more general solutions can be obtained by using the generic solution of Maharaj and Maartens, rather than the restricted one employed here. These will allow tilted matter flows (cf. [28]) that include the possibility of rotation, which cannot occur with the above solution.

In [12] the continuation of the solution was achieved by assuming Bianchi I symmetry in the collisions and matching a Bianchi I model to an FRW model across a spacelike boundary at the onset time using the Darmois conditions. The present situation is much more complicated because the continuation is to be inhomogeneous and anisotropic and so we do not have much chance of constructing an exact solution.

The existence of such a solution, ie one that continues after the onset of the anisotropic pressure and the introduction of both inhomogeneity and anisotropy into the geometry, is assured by the work of Bichteler, Bancel and Choquet-Bruhat ([29] and references cited there) if a suitable collision functional is used. The required theorem is: There exists (subject to mild conditions) a domain Ω in $\mathbb{R}^4$, a metric g^{ab} and a function f on $\Omega \times \mathbb{R}^3$ such that,

1. g^{ab} and f are solutions to the Einstein-Boltzmann equations
2. g^{ab} and f take the given Cauchy data.

These solutions are unique in Ω and depend continuously on the data. (Ω may be chosen to be globally hyperbolic for the metric g^{ab} and to admit a Cauchy surface).

The mild conditions refer to the Sobolev class in which the solutions fall and the model for the collision functional is for binary, elastic collisions and probably can be used here. Also the consistency of the Cauchy data on a spacelike surface at the onset is assured by the previous FRW history.

The inhomogeneous initial conditions produced by the mechanisms discussed above will produce an inhomogeneous space-time with seeds that could in principle lead to a plausible distribution of galaxies.

References

1. Penrose, R.: in *Proc First Marcel Grossman Meeting*, ed Ruffini, R.: North Holland, Amsterdam (1977) 173-181

2. Penrose, R.: Singularities and spacetime asymmetry in *General Relativity, An Einstein Centenary Volume* eds Hawking, S. W. and Israel, W. , Camb. Univ. Press, Cambridge (1979) 581-638

3. Penrose, R.: in *Quantum Gravity II* eds Isham, C. J., Penrose, R. and Sciama, D. W.: OUP, Oxford (1981)

4. Penrose, R.: in *Quantum Concepts in Space and Time* ed Penrose, R. and Isham C. J.: OUP, Oxford (1986)

5. Tod, K. P.: *Class. Quantum Gravity* **4** (1987) 1457- 1468

6. Goode and Wainwright: *Class. Quantum Grav.* **2** (1985) 99-115

7. Newman, R. P. A. C.: *On the structure of conformal singularities in classical general relativity II: Evolution equations and a conjecture of K. P. Tod*. Preprint (1992) - Mathematical Institute, Oxford.

8. Hartle, J. B. and Hawking, S. W.: Wave Function of the Universe. *Phys Rev* **D 28** (1983) 2960

9. Ellis, G. F. R.: Note on "Symmetry Changes in Friedmann Cosmologies" by S. S. Bayin. *Ap. J.* **314** (1987) 1-2

10. Ellis, G. F. R. and King, A.R.: Was the big-bang a whimper. *Comm. Math. Phys.* **38** (1974) 119-156

11. Bardeen, J.M., Steinhart, P.J., and Turner, M.S.: *Phys Rev* **D 28** (1983) 679

12. Matravers, D. R. and Ellis, G. F. R.: Evolution of anisotropies in Friedmann cosmologies. *Class. Quantum Grav.* **6** (1989) 369-381

13. Ellis, G. F. R. and Matravers, D. R.: A note on the evolution of anisotropy in a Robertson-Walker cosmology. *Class. Quantum Grav.* **7** (1990) 1869-1873

14. Ellis, G. F. R. and Matravers, D. R.: Inhomogeneity and anisotropy generation in FRW cosmologies. Preprint (1992) Cape Town.

15. Schucking, E. L. and Spiegel, E. A.: *Comments on Astrophysics* **2** (1970) 121 - 125

16. Stewart, J. M.: *Non-equilibrium relativistic kinetic theory.* Lecture Notes in Physics Vol 10 (1971) Springer-Verlag Berlin

17. Ehlers, J., Geren, P. and Sachs, R. K.: *J. Math. Phys.* **9** (1968) 1344

18. Ellis, G. F. R., Treciokas, R. and Matravers, D. R.: Anisotropic solutions of the Einstein-Boltzmann field equations: II. Some exact properties of the equations. *Ann Phys* NY **150** (1983) 487-503

19. Collins, C. B. and Ellis, G. F. R.: Singularities in Bianchi Cosmologies. *Phys Reports* **56**, 2 (1979) 65 - 105

20. Ehlers, J.: General relativity and kinetic theory, in *General Relativity and Cosmology (XLVII Enrico Fermi Summer School Proc.)* ed R. K. Sachs: Academic Press, NY (1971) 1-70.

21. Ehlers, J.: Survey of General Relativity Theory, in *Relativity, Astrophysics and Cosmology* ed Israel, W.: Reidel, Dordrecht (1972) 1-125

22. Kadanoff, L. B. and Baym, G.: *Quantum Statistical Mechanics*, W. A. Benjamin, New York (1962)

23. Tauber, G. E. and Weinberg, J. W.: *Phys Rev* **122** (1961) 1342-1365

24.   Ellis, G. F. R., Matravers, D. R. and Treciokas, R.: Anisotropic solutions of the Einstein-Boltzmann field equations: I. General formalism. *Ann Phys* NY **150** (1983) 455-486.

25.   Ellis, G. F. R., Matravers, D. R. and Treciokas, R.: An exact solution of the Einstein-Liouville equations. *Gen. Rel. Grav.* **15** (1983) 931-944.

26.   Ellis, G. F. R.: Relativistic Cosmology, in *General Relativity and Cosmology (XLVII Enrico Fermi Summer School)* ed R. K. Sachs, Academic Press, NY. (1971) 104-182

27.   Maharaj, S. D. and Maartens, R.: Exact inhomogeneous Einstein- Liouville solutions in Robertson-Walker spacetimes. *Gen. Rel. Grav.* **15** (1987) 931-944.

28.   King, A. R. and Ellis, G. F. R.: Tilted Homogeneous Cosmological Models, *Comm. Math.Phys.* **31** (1973) 209-242

29.   Bancel, D. and Choquet-Bruhat, Y.: Existence and Uniqueness and Local Stability for the Einstein-Maxwell-Boltzmann System. *Comm. Math. Phys.* **33** (1973) 83-96

Colliding Gravitational Waves with Variable Polarization

[1] N. Breton

Dpto. Física Teórica, Universidad del País Vasco, Bilbao, Spain

Abstract: A family of solutions to Einstein's field equations representing collisions between plane gravitational waves with variable polarization is presented. The boundary conditions on the null hypersurfaces and the behaviour of the solutions near the focussing surface are analysed.

1 Introduction

Spacetimes representing the phenomena of gravitational plane waves interaction are of interest due to the nonlinear effects that these processes exhibit. The colliding plane wave spacetimes are usually represented in a bidimensional space in (u, v) null coordinates, where four regions are separated by null hypersurfaces. Throughout all spacetime it is assumed the existence of a pair of commuting spacelike Killing vectors ∂_x and ∂_y. Being spacetimes possessing two Killing vectors, all the Ernst potential formalism (see Hoenselaers' lecture) can be adapted to them. Consequently the related numerous generating techniques are available. For example, Ehlers and Harrison transformations have been used to generate new colliding wave solutions.

The Gutsunaev-Manko algorithm[1] based on Kramer-Neugebauer transformations has been used to generate stationary axisymmetric solutions possessing gravitational as well as angular multipole moments. In this report we use the Gutsunaev-Manko algorithm combined with the Ernst potential approach, adapted for the purposes of the colliding wave problem, to construct new noncollinear solutions. Starting with a seed solution with constant polarization we obtain a family of solutions describing collision between gravitational waves with variable polarization such that their main properties can be anticipated. We work "backwards" in the sense that we first determine the general solution in the interaction region and then apply the proper boundary conditions along the null surfaces $u = 0$ and $v = 0$.

---

[1] On leave from Depto. de Física, CINVESTAV del IPN, Mexico

2 The Solution Generating Algorithm

The line element describing the interaction region produced in the collision of two plane gravitational waves can be written as

$$ds^2 = 2e^{-M}\,dudv - e^{-U}(e^V \cosh W dx^2 - 2\sinh W dxdy + e^{-V}\cosh W dy^2), \quad (2.1)$$

where the metric functions (U, M, V, W) depend only on null coordinates (u, v). The vacuum Einstein equations restrict the function e^{-U} to satisfy the one dimensional wave equation. Consequently the transitivity surface area has to be of the form $e^{-U} = f(u) + g(v)$. In order to derive the standard form of the Ernst equation we transform to the timelike coordinate t and the spacelike coordinate z related to (u, v) by

$$t = \sqrt{\frac{1}{2} - f}\sqrt{\frac{1}{2} + g} + \sqrt{\frac{1}{2} - g}\sqrt{\frac{1}{2} + f}$$

$$z = \sqrt{\frac{1}{2} - f}\sqrt{\frac{1}{2} + g} - \sqrt{\frac{1}{2} - g}\sqrt{\frac{1}{2} + f}. \quad (2.2)$$

With this coordinate transformation one readily derives the Ernst equation

$$(Z + \overline{Z})\{[(1 - t^2)Z_{,t}]_{,t} - [(1 - z^2)Z_{,z}]_{,z}\} = 2\{(1 - t^2)Z_{,t}^2 - (1 - z^2)Z_{,z}^2\}, \quad (2.3)$$

where

$$Z \equiv \chi + i\omega, \quad \chi \equiv e^{-V}\operatorname{sech}W,$$

and

$$\omega \equiv e^{-V}\tanh W, \quad \sinh W = ImZ/ReZ. \quad (2.4)$$

When the metric function W globally vanishes the Ernst potential Z becomes real $(Z = \chi)$ and (2.3) reduces to

$$[(1 - t^2)\chi_{,t}]_{,t} - [(1 - z^2)\chi_{,z}]_{,z} = 0, \quad (2.5)$$

which is the linear equation of the collinear case. The main idea of the Gutsunaev-Manko algorithm is that it relates in a simple way to every solution of the collinear case a particular solution of the non-linear equation (2.3). Thus, if a given function $\psi(t, z)$ is a solution of (2.5), then the function

$$Z = e^{2\psi}\frac{t(1 - ab) + iz(a + b) - (1 + ia)(1 - ib)}{t(1 - ab) + iz(a + b) + (1 + ia)(1 - ib)}, \quad (2.6)$$

is the solution of the complex Ernst equation (2.3), with the functions $a(t, z)$ and $b(t, z)$ of (2.6) satisfying the linear equations

$$(\ln a)_{,t} = \frac{2}{t - z}[(tz - 1)\psi_{,t} + (1 - z^2)\psi_{,z}]$$

$$(\ln a)_{,z} = \frac{2}{t - z}[(1 - t^2)\psi_{,t} + (tz - 1)\psi_{,z}]$$

$$(\ln b)_{,t} = \frac{-2}{t+z}[(tz+1)\psi_{,t} + (1-z^2)\psi_{,z}]$$

$$(\ln b)_{,z} = \frac{-2}{t+z}[(1-t^2)\psi_{,t} + (tz+1)\psi_{,z}]. \tag{2.7}$$

Once the seed function $\psi(t,z)$ is given, then the functions $a(t,z)$ and $b(t,z)$ which one needs to construct the Ernst potential in (2.3) can be found by simple integration. We now consider a general class of solutions of (2.5) obtained by separation of variables. The function $\psi(t,z)$ given by

$$\psi(t,z) = c\ln\left[(1-t^2)(1-z^2)\right]$$
$$+ \sum_{n=0}^{\infty}\{a_n P_n(z)P_n(t) + q_n P_n(z)Q_n(t) + p_n P_n(t)Q_n(z)\} \tag{2.8}$$

where c, a_n, p_n, and q_n are arbitrary constants to be restricted by the boundary conditions. After tedious but straightforward integration one obtains for the functions $a(t,z)$ and $b(t,z)$ of the Ernst potential

$$\left.\begin{array}{c}a(t,z)\\b(t,z)\end{array}\right\} = \exp\left\{2\sum_{n=1}^{\infty}\left(a_n\{-(\pm 1)^n P_1(t) + P_n(z)P_{n-1}(t)\right.\right.$$

$$-\sum_{k=1}^{n-1}(\pm 1)^k P_{n-k}(z)[P_{n-k+1}(t) - P_{n-k-1}(t)]\}$$

$$-q_n\{(\pm 1)^n\frac{1}{2}\ln\frac{(t\mp z)^2}{1-t^2} - (\pm 1)^n Q_1(t) + P_n(z)Q_{n-1}(t)$$

$$-\sum_{k=1}^{n-1}(\pm 1)^k P_{n-k}(z)[Q_{n-k+1}(t) - Q_{n-k-1}(t)]\}$$

$$\pm p_n\{(\pm 1)^n\frac{1}{2}\ln\frac{(t\mp z)^2}{1-z^2} - (\pm 1)^n Q_1(z) + P_n(t)Q_{n-1}(z)$$

$$\left.-\sum_{k=1}^{n-1}(\pm 1)^k P_{n-k}(t)[Q_{n-k+1}(z) - Q_{n-k-1}(z)]\}\right)$$

$$\left.-q_o\ln\frac{(t\mp z)^2}{1-t^2} \pm p_o\ln\frac{(t\mp z)^2}{1-z^2}\right\}\left(\frac{(1+t)(1\mp z)}{(1-t)(1\pm z)}\right)^{2c}. \tag{2.9}$$

Once the Ernst potential is known one is able to determine all metric functions (in principle). In [2] the detailed derivation of this solution as well as the expressions for the metric components are given.

3 The Boundary Conditions on the Null Surfaces

We now turn to the boundary conditions that must be satisfied in order that the solution found represents the interaction region of two colliding plane waves. The appropriate boundary conditions for the colliding wave problem, in terms of the Ernst potentials, were formulated by Ernst and Hauser and by Griffiths [3]. For our solutions these conditions amount to

$$2k_1 = \lim_{\substack{t \to 0 \\ z \to 0}} [1 - \psi_{,t} - \psi_{,z}]^2, \quad 2k_2 = \lim_{\substack{t \to 0 \\ z \to 0}} [1 + \psi_{,z} - \psi_{,t}]^2, \qquad (3.1)$$

where the constants k_1 and k_2 must satisfy the inequalities

$$\frac{1}{2} \le k_1 < 1, \qquad \frac{1}{2} \le k_2 < 1, \qquad (3.2)$$

imposed by the vacuum Einstein's field equations. Using (2.8) and the properties of the Legendre polynomials one gets

$$\left\{ 1 - \sum_{n=0}^{\infty} (p_n + q_n)[P_n(0)Q_n'(0) + Q_n(0)P_n'(0)] \right\}^2 = 2k_1,$$

$$\left\{ 1 + \sum_{n=0}^{\infty} (p_n - q_n)[P_n(0)Q_n'(0) - Q_n(0)P_n'(0)] \right\}^2 = 2k_2, \qquad (3.3)$$

Since clearly k_1 can be different from k_2 one can therefore obtain approaching plane waves with different wavefronts [5].

4 Behaviour of the Gravitational Field Near the Focussing Surface

It is well known that when two plane gravitational waves collide, they focus each other and either a curvatute singularity or a Cauchy horizon is produced in the course of the collision. The focussing surface in the coordinates we use is given by the equation $t = 1$. Here the invariants of the gravitational field either blow up or a Cauchy horizon is formed which prevents to determine future development of the solution uniquely. From the arguments of Belinski, Khalatnikov and Lifshitz repeated recently by Yurtsever [4] in the colliding wave context, these solutions either tend to a degenerate Kasner type when the spatial dependence is suppressed or to a standard Kasner solution. The degenerate Kasner solution is flat spacetime and thus when the focussing surface is approached no curvature singularity is formed. On the other hand if the solution in the interaction region tends to a general Kasner-like regime near the focussing surface, then a curvature singularity is developed at $t \to 1$. To determine which is the case, it is sufficient to analyse the behaviour of the longitudinal expansion function $e^{\gamma(t,z)}$ at $t \to 1$.

After analysis one arrives to the conclusion that the condition of nonsingular behaviour is:

$$c - \frac{q_0}{2} = -\frac{1}{2}, \qquad q_i = 0, \qquad i \neq 0, \tag{4.1}$$

for arbitrary a_n and p_n of (2.8). One may now make the following observation. Whereas for the collinearly colliding waves the condition to avoid curvature singularity [5] reduces to $c - \frac{q_0}{2} = \pm\frac{1}{2}$; $\quad q_i = 0$, $\quad i \neq 0$, the conditions given by (4.1) are more restrictive in the sense that the nonsingularity of the seed solution is necessary for the nondiagonal solution to be regular. However this is not sufficient since all nonsingular diagonal seeds complying with the equation $c - \frac{q_0}{2} = \frac{1}{2}$; $\quad q_i = 0$, $\quad i \neq 0$, will give rise to a curvature singularity at $t \to 1$.

5 Conclusions

The general solution presented contains as particular cases the generalization to variable polarization of solutions as Khan-Penrose, Szekeres, Ferrari-Ibáñez, Griffiths, Feinstein-Ibáñez and others. The corresponding generalizations can be obtained by tuning adequately the coefficients c, a_n, p_n and q_n in the seed solution (2.8).

Like in the collinear polarized case we see that there exists an infinite dimensional family of solutions which do not result in a curvature singularity. However *for every nonsingular solution there exists an infinite dimensional family of solutions which do develope strong singularity on the focussing surface.* In [2] these solutions are related to the unpolarized (nondiagonal) Gowdy cosmologies and some issues on the strong cosmic censorship conjecture are discussed.

Acknowledgements

N.B. gratefully acknowledges financial support from Dirección General de Investigación Científica y Técnica (DGICYT), Spain.

References

1.   T. S. Gutsunaev and V. Manko: Gen. Rel. Grav. **20** 327 (1988)
2.   N. Bretón, A. Feinstein and J. Ibáñez: Class. and Quantum Grav. (1992) in press.
3.   F. J. Ernst, A. García and I. Hauser: J. Math. Phys. **28** 681 (1988);
     J. B. Griffiths: "Colliding Plane Waves in General Relativity" (Oxford, Oxford University Press, 1991).
4.   U. Yurtsever: Phys. Rev. D **40** 329 (1989)
5.   A. Feinstein and J. Ibáñez: Phys. Rev. D **39** 470 (1989)

The Monopole – Quadrupole solution
of Einstein's Equations

J. Martín, J.L.H. Pastora

Grupo de Física Teórica, Universidad de Salamanca,
37008 Salamanca, Spain

Abstract: We present a method for generating exact solutions of the static axisymmetric vacuum Einstein equations, in the form of uniformly convergent series. The relevance of this method lies in the possibility of determining *a priori* the desired Geroch–Hansen multipole moments of the solution.

1 Introduction

Static axisymmetric exact solutions of the vacuum Einstein equations are described by Weyl's metric [1]:

$$ds^2 = -e^{2\Psi} dt^2 + e^{-2\Psi} \left[e^{2\gamma}(d\rho^2 + dz^2) + \rho^2 d\varphi^2 \right] \quad , \tag{1}$$

where Ψ is any function depending on the coordinates (ρ, z) which satisfies the 3–dimensional Laplace's equation

$$\Delta\Psi \equiv \Psi_{\rho\rho} + \frac{1}{\rho}\Psi_\rho + \Psi_{zz} = 0 \quad , \tag{2}$$

the general asymptotically flat solution of which is

$$\Psi = \sum_{n=0}^{\infty} a_n \frac{P_n(\omega)}{r^{n+1}} \quad , \tag{3}$$

where $r^2 = \rho^2 + z^2$, $\omega \equiv \cos\theta = z/r$ and a_n are arbitrary parameters. We will refer to these parameters as "Weyl's moments". The other metric function γ, also depending on coordinates (ρ, z), is determined from Ψ by quadratures.

It is well known that the Schwarzschild solution describes the exterior field of a spherically symmetric static mass. However, this solution does not correspond to the a_0 Weyl moment, but is contained in (3) as a combination of some particular values of all the moments a_n. The resulting function Ψ of the Schwarzschild solution is well known, and discussed already by Darmois: it would represent the gravitational potential generated by an uniform rod located on the symmetry axis, with its center of mass at the origin of coordinates, of length $2GM/c^2$. This picture looks like far away from spherical symmetry.

On the other hand the Weyl solution corresponding to the "monopole" a_0-term possesses all higher-order Geroch–Hansen [2] multipole moments. For all these reasons, although the function Ψ defines any static vacuum solution, it is physically meaningless itself.

With the aid of the known multipole moment analysis, one can obtain the Schwarzschild metric by looking for the Weyl solution possessing only the monopole relativistic moment. Our purpose is to generalize this idea, i.e., classify Weyl solutions by means of its multipole moments, which is both coordinate–invariant and physically relevant.

We shall be looking for a static axisymmetric vacuum solution possessing the monopole moment M and mass quadrupole moment Q. Such solution will describe the exterior gravitational field generated by a static object with a small deformation from the spherical symmetry.

We shall obtain, so far, a metric with two independent parameters, M and Q, like in the case of the Erez-Rosen metric [3], but possessing only the first two pure gravitational moments.

2 Methodology

A procedure to obtain the required result consists in imposing the condition over the Geroch–Hansen multipole moments, computed from the general static axisymmetric Weyl solution, that they vanish except for the monopole and quadrupole ones.

Such limitation implies some conditions on the coefficients a_n, which can be expressed in terms of the exempted multipole moments. This relation, a_n versus moments, thus fully defines two metric functions Ψ, γ, and hence, the metric. The convergency of the series is always fulfilled provided that the absolute value of the constants a_n are limited [4].

The steps to be followed are

1. Looking for the harmonic coordinates.
2. Write down the initial coordinates (Weyl or prolate) and the metric function Ψ in the harmonic coordinates.
3. Compute the Geroch–Hansen multipole moments in terms of the coefficients a_n.
4. Invert expressions to obtain the coefficients a_n in terms of moments.

2.1 The Harmonic Coordinates

Starting from the Weyl spherical coordinates $\{r, \theta\}$ or cylindrical ones $\{\rho, z\}$, we look for the spherical harmonic coordinates $\{\hat{r}, \hat{\theta}\}$ or cylindrical $\{\hat{\rho}, \hat{z}\}$, such that, the corresponding Cartesian coordinates verify the equation

$$\Box \, \hat{x}^{(\lambda)} \equiv g^{\mu\nu} \nabla_\mu \nabla_\nu \hat{x}^{(\lambda)} = 0 \quad . \tag{4}$$

A reasonable solution (with a good asymptotic behaviour) for $\hat{z}$ is

$$\hat{z}(r, \theta) = r \cos \theta \equiv z \quad . \tag{5}$$

And the solution for $\hat{\rho}$ may be expressed in the form

$$\hat{\rho}(r, \theta) = r \sin \theta \sum_{l=0}^{\infty} \frac{1}{r^l} H_l(\cos \theta) \quad , \tag{6}$$

where H_l are functions of $\cos \theta \equiv \omega$, which satisfy the following differential equation

$$(1 - \omega^2)\{l(l - 3)H_l(\omega) - 4\omega H_l'(\omega) + (1 - \omega^2)H_l''(\omega)\} =$$

$$= \begin{cases} 0 \quad , \quad l \leq 1 \\ \displaystyle\sum_{k+n=l-2} B_k(\omega)H_n(\omega) \quad , \quad l \geq 2 \end{cases} \tag{7}$$

where

$$B_l(\omega) = \sum_{k=0}^{(l-\delta_l)/2} \frac{2^{k+1}}{(k+1)!} B_{l-2k}^{(k+1)}(\omega) \quad \begin{cases} \delta_l = 0 : l \text{ even} \\ \delta_l = 1 : l \text{ odd} \end{cases} \tag{8}$$

$$B_l^{(n)}(\omega) = \sum_{i_1 + \ldots + i_n = l} B_{i_1}^{(1)} \ldots B_{i_n}^{(1)} \tag{9}$$

$$B_l^{(1)}(\omega) = \sum_{k+n=l} A_{kn}(\omega) \tag{10}$$

$$\gamma(r, \omega) = \sum_{k,n=0}^{\infty} \frac{(k+1)(n+1)}{k+n+2} a_k a_n \frac{1}{r^{k+n+2}} \times$$

$$[P_{k+1}(\omega)P_{n+1}(\omega) - P_k(\omega)P_n(\omega)]$$

$$\equiv \sum_{k,n=0}^{\infty} A_{kn}(\omega) \frac{1}{r^{k+n+2}} \tag{11}$$

2.2 The Coordinate Inversion

After solving the differential equation for each order, we determine the harmonic coordinates

$$\hat{z}(r,\theta) \equiv \hat{r}\cos\hat{\theta} = r\cos\theta \tag{12}$$

$$\hat{\rho}(r,\theta) \equiv \hat{r}\sin\hat{\theta} = r\sin\theta \sum_{l=0}^{\text{order}} H_l(\omega)\frac{1}{r^l} \tag{13}$$

and try to obtain the expressions $r = r(\hat{r},\hat{\theta})$ and $\theta = \theta(\hat{r},\hat{\theta})$.

The complicated but straightforward calculations have been carried out by computer by using the MATHEMATICA^{TM} software implementation.

2.3 The Multipole Moments

The definition we are using is similar to that of Thorne [5], and equivalent to the Geroch–Hansen definition, leading to the results which agree with those obtained earlier by Hoenselaers [6] [7].

We expand the metric component $g_{00} = -exp\{2\Psi\}$ in terms of the coefficient $GM/c^2\hat{r}$, where $\hat{r}$ is the radial harmonic coordinate. The mass multipole moment of order l corresponding to the solution Ψ is the factor related to the term $2P_l(\cos\hat{\theta})/\hat{r}^{l+1}$ of the expansion, where P_l are the Legendre polynomials.

2.4 The Inversion of the Moments

The last step consists in expressing the "Weyl coefficients" a_n in terms of multipole moments. From the study of the successive orders, we have got an analytical series for its general term, the coeficient a_n.

3 The Results

We searched for the structure of the series a_n such that the corresponding solution had only the first two multipole moments. We assumed the solution to have the symmetry with respect to the equatorial plane in order to make easier the calculation, without lack of generality.

The required structure of the solution was found to be

$$a_{2n} = -\frac{M^{2n+1}}{2n+1} - \sum_{\alpha=1}^{(2n-q)/3} M^{2n+1-3\alpha}Q^\alpha F(\alpha,n) \quad , \quad q = \begin{cases} -1 & : n = 3j+1 \\ 0 & : n = 3j+3 \\ 1 & : n = 3j+2 \end{cases}$$

$$a_{2n+1} = 0 \tag{14}$$

$$\Psi = -\sum_{n=0}^{\infty} \frac{M^{2n+1}}{2n+1} \frac{P_{2n}(\omega)}{r^{2n+1}} - \sum_{k=0}^{\infty} Q^{2k+1} F(2k+1, 6k+2) \frac{P_{6k+2}(\omega)}{r^{6k+3}} +$$

$$- \sum_{v=0}^{2} \sum_{j=0}^{\infty} \sum_{\alpha=1}^{2j+v} Q^{\alpha} M^{6j-3\alpha+3+2v} F(\alpha, 6j+2+2v) \frac{P_{6j+2+2v}(\omega)}{r^{6j+3+2v}} \quad,$$

$$\tag{15}$$

where M is the total mass, Q is the quadrupole moment, and $F(\alpha, n)$ is a function depending on the power of the quadrupole parameter and the order of the Weyl coefficient. We are able to reach the order 13 in $1/\hat{r}$ (the moment of order $l = 12$), and to adjust the coefficients it is necessary to expand $F(\alpha, n)$ up to that order. As the result we have

$$F(\alpha, n) = \frac{(2n)!}{(2n+2\alpha+1)!!(2n-2\alpha)!!} P^{(\alpha)}[n] \tag{16}$$

$$P^{(\alpha)}[n] = \begin{cases} \frac{5}{2}(n+2) & : \quad \alpha = 1 \\ \frac{5}{8}(n-2)(5n+21) & : \quad \alpha = 2 \\ \frac{25}{48}(n-3)(5n^2+18n-98) & : \quad \alpha = 3 \end{cases} \tag{17}$$

and we hope that the calculations carried out for higher orders will enable us to establish the exact form of these polynomials for arbitrary α.

The first sum in (15) represents a pure monopole solution, which is obviously the Schwarzschild solution

$$\Psi = -\sum_{n=0}^{\infty} \frac{M^{2n+1}}{2n+1} \frac{P_{2n}(\omega)}{r^{2n+1}} = \frac{1}{2} \log\left(\frac{\hat{r}-M}{\hat{r}+M}\right) \quad . \tag{18}$$

The third term represents the interacting Monopole–Quadrupole solution, and the second sum is the pure Quadrupole Solution. In this latter case, where $F(\alpha = 2k+1, n = 6k+2)$, the function Ψ may be expressed as

$$\Psi = -\sum_{k=0}^{\infty} Q^{2k+1} \left[\frac{(6k+2)!15}{(10k+5)!!(2k)!!} \frac{(4k+1)!!}{(2k+2)!!} \right] \frac{P_{6k+2}(\omega)}{r^{6k+3}} \quad . \tag{19}$$

References

1.   H. Weyl: Ann. Phys. (Leipzig) **54** 117 (1917)
2.   R. Geroch: J. Math. Phys. **11** 2580 (1970) ;
      R. O. Hansen: J. Math. Phys. **15** 46 (1974)
3.   G. Erez, N. Rosen: Bull. Res. Counc. **8F** 47 (1959)
4.   H. Quevedo: Fortschritte der Physik **38** 10, 733–840 (1980);
      H. Quevedo: Phys. Rev. D **39** 10 (1989)
5.   K. S. Thorne: Rev. Mod. Phys. **52** 299 (1980)
6.   C. Hoenselaers, G. Fodor, Z. Perjés: J. Math. Phys. **30** 10 (1989)
7.   Y.Gürsel: Gen. Rel. Grav. **12** 1003 (1983)

Effective Action Methods in Cosmology: The Back-reaction Problem

Antonio Campos and Enric Verdaguer

Grup de Física Teòrica, Universitat Autònoma de Barcelona,
08193 Bellaterra (Barcelona), Spain

1 Introduction

Gravitational quantum effects in the Early Universe have been used in the last years to improve our understanding of the present status of the Universe. Particle production, topological defects formed at phase transitions, or various inflation mechanisms have been claimed to explain, respectively, the generation of entropy, the formation of structure in the universe, or the observed isotropy and homogeneity of non-causally connected regions. More recently, general quantum arguments have been employed by Hawking [1] to give a chronology protection conjecture.

Quantum Field theory in curved space [2] is the semiclassical theory of gravity where most of this work has been carried on. In this incomplete theory, the gravitational background field is considered purely classical whereas the matter fields are treated quantum mechanically. It may be viewed as a good test theory of the, still unknown, full Quantum Theory of Gravity interacting with matter fields.

As a first step, one can work in the test field approximation, where it is assumed that the gravitational field remains unmodified by the action of the matter fields. In this approximation, the semiclassical theory of gravity allow us to study the particle production and vacuum polarization due to the dynamical evolution of a quantum field over the fixed classical gravitational background in several cases. Physical situations in which this approximation is relevant are those involving background geometries with event horizons, like the Rindler space (Unruh effect [3]), the black hole geometry (Hawking radiation [4]) where the creation of particles (radiated with a thermal spectrum) is predicted, or those involving expanding geometries where the rapid time variation of the gravitational field can produce particles (Cosmological particle creation).

In the last sixties, Parker [5] studied the particle production in a Friedmann-Robertson-Walker (FRW) space-time using the method of the Bogoliubov's coefficients. He proved that only particles not conformally coupled can be produced. Later work tried to analize the production of particles when the conformal invariance is broken by weak perturbations in the early universe. Particle production due to anisotropies [6] was studied with an approximation method based in Parker's work, but the inhomogeneous case [7] needs a very different method which is based in the perturbative evaluation of the scattering S-matrix. This is because the mode separation used to find the Bogoliubov's coefficients is not appropiate in this more general case.

A more difficult question, however, is to analize and discuss the dynamical evolution of the gravitational field as a consequence of its interaction with quantum matter. This is known as the back-reaction problem. The study of the back-reaction of Hawking radiation over the background geometry gives us a lot of information about the thermodynamical properties of the black holes (black hole evaporation). In Cosmology, the gravitational entropy generation and the damping of weak primordial perturbations in a dissipative process can be modeled and analyzed in this context. Other applications to the study of back-reaction in the early universe are those which consider the avoidance or the modification of the structure of the initial singularity and those which investigate the dynamical evolution of topological defects.

In the framework of a semiclassical theory of gravity the action of the quantum field over the space-time geometry can be considered by inserting the vacuum expectation value of the stress-energy tensor in a given quantum state into the classical Einstein's equations [8],

$$G_{\mu\nu} = < \xi |T_{\mu\nu}| \xi > , \tag{1}$$

where $G_{\mu\nu}$ is the Einstein's tensor. The solution of these semiclassical equations involves the specification of a space-time geometry $g_{\mu\nu}(x)$ together with a quantum state $|\xi >$. As it is well known in Quantum Field theory in flat space, the energy-momentum tensor is a product of distributions and some renormalization technique is needed to obtain finite answers in the theory. These methods become more involved when they are extended to curved space.

Early work considering the evolution of the metric, to study the back-reaction of particles created by anisotropies in a cosmological context, was done by Lukash and Novikov [9]; but they assumed very special conditions near the Planck time. In 1978, Hu and Parker [10] considered a Bianchi Type-I model conformally coupled to a scalar field. They calculated the expectation value of the stress tensor in the low frequency approximation and computed the resulting modified Einstein's equations numerically. The results of such work indicate that the dynamical mechanism of particle production is sufficient to achieve a rapid damping of anisotropy if the calculations are extrapolated to the Planck era.

Since the exactly solvable models are usually too simple to have physical significance, one often must resort to perturbative techniques in order to derive concrete expressions for the vacuum expectation value of a stress-energy tensor. For instance, Horowitz [11], by applying the axiomatic arguments outlined by Wald [8], gave a general expression to first order for the stress tensor in the case of a smooth perturbation around the Minkowski background and around

the spatially flat FRW backgrounds. Also, Davies and Unruh [12] developed
an approximation method using an iteratively evaluated mode decomposition.
Finally, a rather different approach based in the functional formalism of the
effective action was developed by Hartle and Hu [13] to compute the back-
reaction equations due to anisotropy in a Bianchi Type-I model.

In the last method one can obtain, in principle, the rate of particle production, the spectrum of the created particles, information about the role of the
trace anomaly in the structure of the cosmological singularity, the vacuum energy density and the damping of the anisotropy near the Planck time. However,
it is very difficult to give a physical interpretation to some of the results because
the expression of the effective equations of motion is non-local, non-causal and
complex. This is a consequence of the fact that the usual effective action formalism (also called in-out formalism) evaluates transition amplitudes between
two different quantum states instead of expectation values. Thus the semiclassical Einstein's equations derived with this formalism would be an (incorrect)
equation like,

$$G_{\mu\nu} = \frac{< out|T_{\mu\nu}|in >}{< out|in >} \ , \tag{2}$$

rather than an equation like (1).

There are two possible ways to solve this problem. On the one hand, one can
find the Bogoliubov transformations between unequivalent vacua states [2] and
modify the right hand side of equation (2) to write a true expectation value by
inserting the corresponding Bogoliubov coefficients, but this may not be very
practical in many applications. On the other hand, one can look directly for
a functional method well suited to obtain expectation values of an operator in
Quantum Field theory. Such a formalism was first proposed by Schwinger [15],
the relativistic theoretical framework and some statistical mechanical results
were developed by Chou *et al.* [16]. Jordan [17] showed that the effective
equations of motions are real and causal to second loop order in curved space-
times. This method has been used in cosmology to evaluate the back-reaction due
to particle creation in a spatially flat FRW space-time with small anisotropies
by Jordan [18] and by Calzetta and Hu [19], independently.

Here we discuss the back-reaction problem of particle production in inhomogeneous cosmologies. We use Schwinger's formalism, or in-in formalism, to obtain
a real and causal effective action for the classical gravitational field interacting
with a quantum scalar field. In this formalism one can derive the semiclassical
Einstein's equations (1) in a natural way. The next section is devoted to explain
briefly the in-in formalism to obtain expectation values instead of transition
amplitudes. In the third section the method is applied to spatially flat FRW
space-time perturbed with a smooth inhomogeneity and coupled with a conformal massless scalar field. Finally, some conclusions and remarks are given in the
last section.

2 In-In Functional Formalism of the Effective Action

Quantum corrections to a classical field theory can be studied with the help of the effective action. For simplicity, we consider the quantization of a scalar field $\phi(x)$. The usual in-out effective action is based in the generating functional which is defined as the vacuum persistence amplitude in the presence of some classical source $J(x)$. This generating functional, $W[J]$, carries all the quantum information of the connected graphs of the theory and can be expressed as a path-integral with certain vacuum boundary conditions ($\phi \longrightarrow e^{\mp i\omega t}$, when the time $t \rightarrow \pm\infty$, and $\omega > 0$) as,

$$e^{iW[J]} \equiv < 0, out|0, in >_J \equiv \int \mathcal{D}[\phi] e^{i(S[\phi]+J\phi)} , \qquad (3)$$

where $S[\phi]$ is the classical action of the field theory, and we have used the common shorthand notation $J\phi$ for the complete expression $\int d^n x J(x)\phi(x)$. By differentiating with respect to the source one generates matrix elements from $W[J]$,

$$\frac{\delta W[J]}{\delta J(x)} = \frac{< 0, out|\phi(x)|0, in >_J}{< 0, out|0, in >_J} \equiv \bar{\phi}[J] . \qquad (4)$$

If we assume that the above expression can be reversed, the effective action can be defined as the Legendre transformation of the generating functional,

$$\Gamma[\bar{\phi}] = W[J] - J\bar{\phi} . \qquad (5)$$

This new functional is the generator of the one-particle-irreducible graphs (graphs that remain connected when any internal line is cut) and it accumulates all the quantum corrections to the classical action. Finally, the dynamical equations for the effective mean field $\bar{\phi}[0]$, *i.e.* the matrix element of the field ϕ in the absence of the source $J(x)$, are deduced from

$$\frac{\delta \Gamma[\bar{\phi}]}{\delta \bar{\phi}} \Big|_{\bar{\phi}=\bar{\phi}[0]} = 0 . \qquad (6)$$

These effective equations express the quantum corrections to the classical equations as a variational problem of the effective action.

In order to work with expectation values rather than matrix elements one defines a new generating functional where the dynamics is determined by two different external classical sources, J_+ and J_-,

$$e^{iW[J_+, J_-]} = \sum_\alpha < 0, in|\alpha, t >_{J_-} < \alpha, t|0, in >_{J_+} . \qquad (7)$$

Here, $\{|\alpha, t >\}$ is a complete basis of eigen states of the field operator $\phi(x)$ at time t. This in-in generating functional has an integral representation, with special boundary conditions in order that the new sources do not increase the number of degrees of freedom,

$$e^{iW[J_+,J_-]} = \int \mathcal{D}[\phi_+]\mathcal{D}[\phi_-]e^{i\{S[\phi_+]+J_+\phi_+ -S[\phi_-]-J_-\phi_-\}} \, , \tag{8}$$

where the sum is over all fields ϕ_+, ϕ_- with negative and positive frequency modes, respectively, in the remote past but which coincide at time t. This integral can be thought of as the path sum of two different fields evolving in two different time branches [20], one going forward in time in the presence of J_+ from the "in" vacuum to a time t, and the other backward in time in the presence of J_- from time t to the "in" vacuum with the constraint $\phi_+ = \phi_-|_t$. Because of such a path integral representation, this formalism is also called close-time-path formalism.

Now, the functional $W[J_+, J_-]$ generates expectation values of the field instead of matrix elements like in equation (4),

$$\frac{\delta W[J_+, J_-]}{\delta J_+}\Big|_{J_\pm = J} = <0, in|\phi(x)|0, in>_J \, . \tag{9}$$

It generates not only the desired expectation values of time ordered $(T^{(t)})$ field operators but also the anti-time-ordered $(T^{(a)})$ ones in the same footing

$$\frac{\delta e^{iW[J_+,J_-]}}{i\delta J_+(x_1)\cdots(-i)\delta J_-(y_1)\cdots}\Big|_{J_\pm = J} = <0, in|T^{(a)}(\phi(y_1)\cdots)T^{(t)}(\phi(x_1)\cdots)|0, in> . \tag{10}$$

In analogy with the in-out formalism the in-in effective action is defined as the Legendre transform of the new generating functional as,

$$\Gamma[\bar\phi_+, \bar\phi_-] = W[J_+, J_-] - J_+\bar\phi_+ + J_-\bar\phi_- \, , \tag{11}$$

where the sources are functions of the fields $\bar\phi_+$ and $\bar\phi_-$, through the definitions

$$\pm\frac{\delta W[J_+, J_-]}{\delta J_\pm} \equiv \bar\phi_\pm[J_+, J_-] \, . \tag{12}$$

If we take $J_\pm = 0$ in (12) we recover the expectation value of the mean field that satisfies the in-in effective equations of motion,

$$\frac{\delta\Gamma[\bar\phi_+, \bar\phi_-]}{\delta\bar\phi_+}\Big|_{\bar\phi_\pm = \bar\phi_\pm[0,0]} = 0 \, , \tag{13}$$

where $\bar\phi_\pm[0, 0] = <0, in|\phi(x)|0, in>$.

Jordan [17] has demostrated that these equations are real and causal up to two loop order in the perturbative expansion for quantum scalar fields in curved space-time and he has also checked the unitarity of the formalism restricted to vacuum in states.

As an example, one can look for a concrete expression of the in-in generating functional in the simple case of a free scalar field in flat space-time. Using the Schwinger-Dyson equations determined by (8) and the adequate boundary conditions outlined before it is not difficult to show that the in-in generating functional in flat space, $W_o[J_+, J_-]$, can be written as [19] ,

$$W_o[J_+, J_-] = -\frac{1}{2} \mathcal{J}^T \mathcal{G} \mathcal{J} \ , \tag{14}$$

where $\mathcal{J}^T \equiv (J_+, J_-)$, and $\mathcal{G}$ is the matrix constructed with the Feynman (Δ_F), Dyson (Δ_D) and Wightman ($\Delta^{\pm}$) propagators in the following way,

$$\mathcal{G} \equiv \begin{pmatrix} \Delta_F & \Delta^+ \\ -\Delta^- & -\Delta_D \end{pmatrix} \ . \tag{15}$$

If we restrict ourselves to the first component in (14) we recover the usual in-out generating functional,

$$W_o[J, J] = -\frac{1}{2} J^T \Delta_F J \ . \tag{16}$$

Henceforth, if we make $J_+ = J_- = J$ in the same equation we see that $W_o[J, J] = 0$. This is a general property of the in-in generating functional which can be deduced from its definition (7) and provides a rough prove of unitarity [17].

Obviously, this formalism treats all the propagators in the same way and, as a consequence, this leads to an increase in the number of Feynman graphs. However, this does not involve more technical complications in the renormalization procedure because the new infinities are all of the same kind as those found in the in-out formalism, *i.e.* if the in-out theory is renormalizable then the in-in theory is renormalizable too [20]. In the next section we will use the Feynman rules deduced from this simple construction of the generating functional to compute the in-in effective action in a concrete model.

3 Back-reaction in Inhomogeneous Cosmologies

In this section we apply Schwinger's in-in functional formalism to study back-reaction problems in expanding universes. We have already mentioned the possible physical relevance of the damping of anisotropies due to particle creation to explain the present observed properties of the Universe.

The dynamical evolution of a space-time caused by the interaction with some quantum field is ruled by the expectation value of the energy momentum tensor of the quantum field, see equation (1), and this includes the created particles. In order to obtain and analyze the semiclassical Einstein's equations we consider a non self-interacting conformally coupled massless scalar quantum field with action

$$S_m[\tilde{g}_{\mu\nu}, \phi] = -\frac{1}{2} \int d^n x \sqrt{-\tilde{g}} \left[\tilde{g}^{\mu\nu} \partial_\mu \phi \partial_\nu \phi + \frac{(n-2)}{4(n-1)} \tilde{R} \phi^2 \right] \ , \tag{17}$$

where the gravitational field is a smoothly perturbed spatially flat FRW space-time

$$\tilde{g}_{\mu\nu}(x) \equiv a^2(\eta)(\eta_{\mu\nu} + h_{\mu\nu}(x)) \ . \tag{18}$$

Here $h_{\mu\nu}(x)$ is a symmetric tensor representing small inhomogeneities, $a(x) = \exp(\omega(\eta))$ is the conformal factor depending on the conformal time $\eta \equiv \int dt/a$, and the flat metric $\eta_{\mu\nu}$ has signature $(-+\cdots+)$. Initially, we will work in n dimensions, as this will be useful to isolate the infinities of the theory by using dimensional regularization; but we will return to four dimensions when the infinities had been regularized.

The field action (17) can be expanded in powers of the inhomogeneity $h_{\mu\nu}$,

$$S_m[\tilde{g}_{\mu\nu}, \phi] = S_m^{(o)}[\phi] + \sum_{n=1}^{\infty} S_m^{(n)}[h_{\mu\nu}^{(n)}, \phi] \ . \tag{19}$$

Thus, the zero order term represents the action of a free scalar field in a flat FRW space-time and the higher perturbative terms carry all the information about the interaction with the gravitational inhomogeneities. Because of the assumed conformal coupling of the scalar field we can disregard the conformal factor $a(\eta)$ if we define a new matter field $\Phi(x) = \exp\left(\frac{n-2}{2}\omega(\eta)\right)\phi(x)$. Therefore, we can work with the quantum field Φ into a classical flat space-time geometry with inhomogeneities defined by $g_{\mu\nu} \equiv \eta_{\mu\nu} + h_{\mu\nu}(x)$.

Following the previous section the total in-in effective action in the semiclassical approximation is written [19] as

$$\Gamma[\omega^{\pm}, g_{\mu\nu}^{\pm}] = S_E[\omega^+, g_{\mu\nu}^+] - S_E[\omega^-, g_{\mu\nu}^-] + \Gamma_m[g_{\mu\nu}^+, g_{\mu\nu}^-] \ , \tag{20}$$

where $S_E[\omega, g_{\mu\nu}]$ is the usual Einstein action, where suitable quadratic terms in the curvature have been added in order to renormalize the infinities and fulfill the correct trace anomaly [13],

$$S_E[\omega, g_{\mu\nu}] \equiv S_E[\tilde{g}_{\mu\nu}] = \int d^n x (-g(x))^{1/2} \tag{21}$$

$$\times \left\{ \frac{1}{16\pi G}\tilde{R}(x) + \frac{\mu^{n-4}}{2880\pi^2(n-4)}\left[\tilde{R}_{\mu\nu\alpha\beta}(x)\tilde{R}^{\mu\nu\alpha\beta}(x) - \tilde{R}_{\mu\nu}(x)\tilde{R}^{\mu\nu}(x)\right] \right\} \ ,$$

and $\Gamma_m[g_{\mu\nu}^+, g_{\mu\nu}^-]$ is the remaining in-in effective action for the matter field which may be expressed in a path integral representation as,

$$\Gamma_m[g_{\mu\nu}^+, g_{\mu\nu}^-] = -i\ln \int \mathcal{D}[\Phi^+]\mathcal{D}[\Phi^-]\exp i\{S_m[g_{\mu\nu}^+, \Phi^+] - S_m[g_{\mu\nu}^-, \Phi^-]\} \ , \tag{22}$$

in the absence of classical matter sources. If we drop the tadpoles, because they are zero in dimensional regularization, isolate the independent terms in $g_{\mu\nu}^+$, and expand to second order in the perturbation, the above equation becomes

$$\Gamma_m[g_{\mu\nu}^+, g_{\mu\nu}^-] = \frac{i}{2} < (S_m^{(1)}[h_{\mu\nu}^+, \Phi^+])^2 > -i < (S_m^{(1)}[h_{\mu\nu}^+, \Phi^+]S_m^{(1)}[h_{\mu\nu}^-, \Phi^-]) >$$

$$+ O(h_{\mu\nu}^3) + \left(\begin{array}{c}\text{no contribution terms}\\ \text{to the variation } \delta g_{\mu\nu}^+\end{array}\right) \ , \tag{23}$$

where the bracket is defined as the expectation value with respect to the free term in (19),

$$< \mathcal{O}[\Phi(x)] > \equiv \frac{\int \mathcal{D}[\Phi]e^{iS_m^{(o)}[\Phi]}\mathcal{O}[\Phi(x)]}{\int \mathcal{D}[\Phi]e^{iS_m^{(o)}[\Phi]}} \ . \tag{24}$$

Then, we can use Wick's theorem in (23), and the Feynman rules determined by the generating functional in flat space calculated before,

$$< \Phi^+(x)\Phi^+(y) > = i\Delta_F(x-y) = -i\int \frac{d^np}{(2\pi)^n}\frac{e^{ip(x-y)}}{p^2 - i\epsilon}$$

$$< \Phi^-(x)\Phi^-(y) > = -i\Delta_D(x-y) = i\int \frac{d^np}{(2\pi)^n}\frac{e^{ip(x-y)}}{p^2 + i\epsilon}$$

$$< \Phi^+(x)\Phi^-(y) > = -i\Delta^+(x-y) = \int \frac{d^np}{(2\pi)^{n-1}}e^{ip(x-y)}\delta(p^2)\theta(-p^0) \ . \quad (25)$$

After a lenghtly calculation in which terms forming geometric invariants of the metric $g_{\mu\nu}$ have been grouped, and taking the limit to 4 dimensions, the finite in-in effective action to second order in $h_{\mu\nu}$ can be expressed as,

$$\Gamma_{\text{finite}}^{(2)}[\omega^\pm, g_{\mu\nu}^\pm] = \frac{1}{16\pi G}\int d^4x(-g^+(x))^{1/2}e^{2\omega^+}\left[R^+(x) + 6\omega_{;\mu}^+\omega^{+;\mu}\right]$$

$$+ \alpha\int d^4x(-g^+(x))^{1/2}\left[(R_{\mu\nu\alpha\beta}^+(x)R^{+\mu\nu\alpha\beta}(x) - R_{\mu\nu}^+(x)R^{+\mu\nu}(x))\ln(\mu e^{\omega^+})\right.$$

$$\left. -\frac{1}{12}R^+(x)R^+(x)\right]$$

$$+ \alpha\int d^4x(-g^+(x))^{1/2}\left[2R^{+\mu\nu}(x)\omega_{;\mu}^+\omega_{;\nu}^+ + R^+(x)\Box\,\omega^+ - 4\Box\,\omega^+(\omega_{;\nu}^+\omega^{+;\nu})\right.$$

$$\left. -3(\Box\,\omega^+)^2 - 2(\omega_{;\mu}^+\omega^{+;\mu})^2\right]$$

$$- \frac{\alpha}{4}\int d^4x d^4y(-g^+(x))^{1/2}(-g^+(y))^{1/2}$$

$$\times \left[3R_{\mu\nu\alpha\beta}^+(x)R^{+\mu\nu\alpha\beta}(y) - R^+(x)R^+(y)\right]K_1(x-y)$$

$$- \frac{\alpha}{4}\int d^4x d^4y(-g^+(x))^{1/2}(-g^-(y))^{1/2}$$

$$\times \left[3R_{\mu\nu\alpha\beta}^+(x)R^{-\mu\nu\alpha\beta}(y) - R^+(x)R^-(y)\right]K_2(x-y)$$

$$+ O(h_{\mu\nu}^3) + \left(\begin{array}{c}\text{No Contribution terms}\\\text{to the variation }\delta g_{\mu\nu}^+\end{array}\right) , \quad (26)$$

where $\alpha = (2880\pi^2)^{-1}$, $K_1(x-y)$ and $K_2(x-y)$ are the non-local pieces defined by the integrals

$$K_1(x-y) \equiv \frac{1}{2}\int \frac{d^4p}{(2\pi)^4}e^{ip(x-y)}\ln\left[\frac{(p^2 - i\epsilon)}{\mu_o^2}\right] ,$$

$$K_2(x-y) \equiv \int \frac{d^4p}{(2\pi)^4}e^{ip(x-y)}(-2i\pi)\theta(-p^0) , \quad (27)$$

μ is the renormalization parameter introduced in (21) and μ_o is a numerical factor.

By functional derivation of this effective action with respect to the classical space-time metric $g_{\mu\nu}^+(x)$ one obtains the semiclassical Einstein's equations (1) in an explicit form for our problem in hand. This action must be completed with a matter action for a classical isotropic fluid, otherwise it would not be consistent with an isotropic and homogeneous background; for its relevance in the early universe we should take a radiative perfect fluid.

4 Conclusions

In the last section we have derived, explicitly, the effective action to one loop order for a quantum scalar field coupled to small gravitational inhomogeneities in a flat FRW expanding universe. The computation has been done using Schwinger's in-in formalism also called the close-time-path functional formalism. This makes the action suitable for a direct evaluation of the so-called semiclassical Einstein's equations, i.e. Einstein's equations modified by the quantum effects produced by the interaction of the quantum field with the gravitational field. This constitutes the back-reaction problem; its solution is equivalent to the solution of the semiclassical equations. Of course, this is a highly non trivial problem, part of the problem; lies in the fact that the action is a higher derivative action (at this order it contains quadratic terms in the curvature) and this leads usually to runaway solutions. A consistent treatment of this problem within a perturbative scheme has been advocated by Simon [21], its aim is to get rid of the unphysical solutions that may otherwise arise [22]. This is especially important when there are solutions which indicate an instability of the classical background. Such a consistent perturbative framework has been applied, for instance, in the context of quantum cosmology [23].

Another source of difficulty is the appearence of the non local terms (27) in the action; the general treatment of these terms is not possible. Rather than trying to solve the general problem our aim for future work is to apply equation (26) and the corresponding semiclassical equations to some particular cosmological inhomogeneities. A special relevant case is that of inhomogeneities produced by cosmological topological defects. Topological defects such as cosmic strings are known to produce quite energetic particles during its formation and during its dynamical evolution [24], and the back-reaction of these on the string evolution may be significant. The back-reaction problem in cosmic strings has been considered by Hiscock [25], who uses the vacuum expectation value of the stress tensor of conformally coupled matter fields around a static, cylindrically symmetric cosmic string as a source in the linearized Einstein equations. The space-time metric resulting differs from that in which no back-reaction is taken into account, i.e. flat space with a deficit angle. In a similar way a static, spherically symmetric global monopole has been considered [26] by solving the linearized semiclassical Einstein's equations. It is found that the vacuum polarization effects may significantly alter the value of the monopole core mass when the symmetry breaking responsible for the monopole formation takes place near the Planck time. For a spherical monopole of mass 10^{16} GeV at a Compton wavelength the quantum correction to the mass is of the order of 10^3 GeV, which is large but not significant compared to the total mass. Nevertheless, semiclassical effects are important to decide the stability of such topological defects. In the approach developed here not only the static case, in which the vacuum energy is relevant, but also the dynamical case when the stress tensor expectation value contains also information on particle creation, can be estimated.

Notice that an important difference of the in-in formalism with respect to the in-out formalism is that the rate of particle creation cannot be read off directly from the effective action in the former case. In fact, in the in-out formalism the effective action is directly related to the transition amplitude from the "in" to the "out" vacua (vacuum persistence amplitude). When particles are produced the "in" and "out" vacua differ and thus the above amplitude gives a vacuum persistence probability which is less than unity. This means that the effective action contains an imaginary part; such imaginary part is directly related to the probability of particle creation [13]. In the in-in formalism, however, the effective action is not directly related to the vacuum persistence amplitude. The probability of particle creation, however, can be obtained in a simple way [19].

References

1.   S.W. Hawking, Phys. Rev. D46, 603 (1992).
2.   N.D. Birrell and P.C.W. Davies, *Quantum Fields in Curved Space*, Cambridge University Press, Cambridge, England (1982).
3.   W.G. Unruh, Phys. Rev. D14, 870 (1976); P. Candelas and D. Deutsch, Proc. R. Soc. London A354, 79 (1977).
4.   S.W. Hawking, Commun. Math. Phys. 43, 199 (1975).
5.   L. Parker, Phys. Rev. Lett. 21, 562 (1968); Phys. Rev. 183, 1057 (1969); Phys. Rev. D3, 346 (1971).
6.   A.A. Starobinsky and Y.B. Zel'dovich, Zk. Eksp. Teor. Fiz., 61, 2161 (1971) [Sov. Phys. JETP, 34, 1159 (1972)].
7.   J.A. Frieman, Phys. Rev. D39, 389 (1989); J. Céspedes and E. Verdaguer, Phys. Rev. D41, 1022 (1990); A. Campos and E. Verdaguer, Phys. Rev. D45, 4428 (1992).
8.   R.M. Wald, Commun. Math. Phys. 54, 1 (1977).
9.   V.N. Lukash and A.A. Starobinsky, Zk. Eksp. Teor. Fiz., 66, 1515 (1974) [Sov. Phys. JETP, 39, 742 (1974)]; V.N. Lukash, I.D. Novikov, A.A. Starobinsky and Y.B. Zel'dovich, Nuovo Cimento, 35B, 293 (1976).
10.   B.L. Hu and L. Parker, Phys. Rev. D17, 933 (1978).
11.   G.T. Horowitz, Phys. Rev. D21, 1445 (1980); G.T. Horowitz and R.M. Wald, Phys. Rev. D21, 1462 (1980); *ibid* D25, 3408 (1982).
12.   P.C.W. Davies and W.G. Unruh, Phys. Rev. D20, 388 (1979).
13.   J.B. Hartle, Phys. Rev. Lett. 39, 1373 (1977); J.B. Hartle and B.L. Hu, Phys. Rev. D20, 1772 (1979).
14.   B.L. Hu, *Quantum Field Theories in Relativistic Cosmologies*, in *Proceedings of the Second Marcel Grossmann Meeting*, edited by R. Ruffini (North Holland, 1982).
15.   J. Schwinger, J. Math. Phys. 2, 407 (1961).
16.   K. Chou, Z. Su, B. Hao and L. Yu, Phys. Rep. 118, 1 (1985).

17. R.D. Jordan, Phys. Rev. **D33**, 444 (1986).
18. R.D. Jordan, Phys. Rev. **D36**, 3604 (1987).
19. E. Calzetta and B.L. Hu, Phys. Rev. **D35**, 495 (1987).
20. P. Hajicek, *Time-loop formalism in Quantum Field Theory*, in *Proceedings of the Second Marcel Grossmann Meeting*, edited by R. Ruffini (North Holland, 1982).
21. J.Z. Simon, Phys. Rev. **D43**, 3308 (1991)
22. X. Jaén, J. Llosa and A. Molina, Phys. Rev. **D34**, 2302 (1986); J.Z. Simon, Phys. Rev. **D41**, 3720 (1990); *ibid* **D45**, 1953 (1992).
23. F.D. Mazzitelli, Phys. Rev. **D45**, 2814 (1992).
24. L. Parker, Phys. Rev. Lett., **59**, 1369 (1987); J. Garriga, D. Harari and E. Verdaguer, Nucl. Phys. **B339**, 560 (1990).
25. W.A. Hiscock, Phys. Lett. **B188**, 317 (1987).
26. W.A. Hiscock, Phys. Rev. **D37**, 2142 (1988); *ibid* Class. Quantum Grav. **7**, L235 (1990).

Quantization in a Colliding Plane Wave Spacetime

Miquel Dorca and Enric Verdaguer

Grup de Física Teòrica, Universitat Autònoma de Barcelona,
08193 Bellaterra (Barcelona), Spain

1 Introduction

Colliding wave spacetimes are exact solutions of Einstein equations which represent the head on collision of two exact gravitational plane waves and as such they are some of the simplest exact highly dynamical gravitational fields. The nonlinearity of General Relativity leads to some conspicuous features here [1], in our particular problem the collision gives a Killing-Cauchy horizon where a caustic is formed. When such gravitational fields interact with quantum matter one expects some important quantum effects such as vacuum polarization or the spontaneous creation of particles, it is the purpose of the present work to compute the creation of quantum particles. For this we quantize a massless scalar field in a colliding wave background and identify two physically meaningful vacua: the "in" and "out" vacua. They are, respectively, related to the flat "in" vacuum in the spacetime region before the collision and to a physically reasonable vacuum at the horizon. It is found that particles are spontaneously created with a spectrum inverse to the frequency; the spectrum is compatible in the long wavelenght limit with a thermal spectrum with a temperature which is inversely proportional to the focusing time of the waves. First, let us describe the physics of colliding waves.

2 Plane Waves

We start with a single gravitational plane wave. These are very simple solutions of Einstein's field equations with a group G_5 of isometries and an Abelian subgroup G_3 acting on null hypersurfaces [2]. Yet as it was first pointed out by Penrose [3] they have interesting global properties like the absence of spacelike Cauchy hypersurfaces as a consequence of their focusing properties. Exact plane waves focus null geodesics which collide with the wave. They are suposed to represent some physical situations in which strong gravitational waves may be produced such as in black hole collisions [4] or in some particular cases of travelling waves on cosmic strings [5] (in particular for strings with a deficit angle of π). The simplest way to represent plane waves is with the use of the so called *harmonic coordinates* (X, Y, U, V). If we restrict ourselves to one polarization,

$$ds^2 = dU\,dV - dX^2 - dY^2 + h(U)\left(X^2 - Y^2\right) dU^2 \ , \tag{1}$$

where $h(U)$ is an arbitrary positive bounded function such that when it is different from zero only in the interval $U \in (a, b)$, (a, b real parameters) leads to a sandwich wave. Note that for $U \notin (a, b)$ the spacetime is flat. The above coordinates do not display some of the symmetries of the spacetime. These are better seen with the use of the so called *group coordinates*, (x, y, u, v), which put (1) in the form: $ds^2 = dU\,dV - F^2(u)\,dX^2 - G^2(u)\,dY^2$. Here $U = u$, $V = v + x^2 FF' + y^2 GG'$, $X = xF(u)$, $Y = yG(u)$, where F and G are solutions to the equations $F''/F = h = -G''/G$. These coordinates are adapted to the Killing fields ∂_x and ∂_y. Using the weak energy condition one can see [6,7] that there is some $u = u_f$ (focusing surface) where the determinant of the transversal two dimensional metric vanishes, i.e. $F(u_f)G(u_f) = 0$. The group coordinates become singular at u_f although the spacetime is not. The focusing properties can be seen as follows. Assume, for instance, that $F(u_f) = 0$ and consider two particles following geodesics at some fixed coordinates (x_0, y_0) and $(x_0 + \Delta x_0, y_0)$, then at $u = u_f$ the particles will focus at the same time (we use u as an affine parameter).

The quantization of a field in this background is rather trivial [6] : no particle creation or vacuum polarization take place. For this quantization one takes a null hypersurface $u = const.$ (with $u < a$) as substitute Cauchy surface although null geodesics with fixed u do not register on that surface.

3 Colliding Plane Waves

Let us now consider the head on collision of two one polarization plane waves. Due to mutual focusing one expects rather dramatic consequences when two of such waves frontaly collide. In fact, the first solution representing the head on collision of two shock waves was obtained by Kahn and Penrose [8] and at the focusing surfaces of the waves true singularities of the spacetime are formed. Field quantization in that spacetime was considered by Yurtsever [9] but particle creation could be only computed approximately.

The collision of two plane waves does not always lead to a singularity; we shall consider a spacetime in which a Cauchy horizon, which may be interpreted as a *caustic* is formed at the focusing hypersurfaces of the plane waves. The spacetime may, in group coordinates, be separated into four regions, region IV ($u < 0$, $v < 0$) is flat (spacetime before the collision) i.e. $ds^2_{IV} = 4L_1 L_2\,du\,dv - dx^2 - dy^2$, the single plane wave regions: the u-wave (region II), ($0 \le u < \pi/2$, $v \le 0$), with metric,

$$ds^2{}_{II} = 4L_1L_2\left[1+\sin\left(u\right)\right]^2 dudv - \frac{1-\sin\left(u\right)}{1+\sin\left(u\right)}dx^2 - \left[1+\sin\left(u\right)\right]^2\cos^2\left(u\right)dy^2 \ ,$$

and the v-wave (region III), $(0 \leq v < \pi/2, \ u \leq 0)$, with the same metric as region II but changing u for v. And finally the interaction region (region I), is given by [10]

$$ds_I^2 = 4L_1L_2\left[1+\sin\left(u+v\right)\right]^2 dudv - \frac{1-\sin\left(u+v\right)}{1+\sin\left(u+v\right)}dx^2$$
$$- \left[1+\sin\left(u+v\right)\right]^2\cos^2\left(u-v\right)dy^2.$$

The parameters L_1 and L_2 have dimension of length and give the focusing times of the u-wave and v-wave respectively. Note that in the interaction region the metric has a group G_2 of isometries with ∂_x and ∂_y Killing fields along the transversal directions of the waves. The spacetime is regular everywhere but the coordinates become singular at the focusing surfaces $u = \pi/2$ and $v = \pi/2$ of the u and v waves respetively, and at the caustic $u + v = \pi/2$. The metric is continuous across the boundaries of the four regions $u = 0$ and $v = 0$, the first tangential derivatives are also continuous but the normal derivatives are not. Thus it satisfies the O'Brian-Synge conditions [11]; they guarantee that the Ricci tensor is zero everywhere (i.e. it is a vacuum solution) but the Weyl tensor acquires distributional values at the boundaries. The solution may be interpreted [12,13] as representing two shock gravitational waves followed by gravitational radiation.

To understand the global properties of that spacetime one must use appropriate coordinates in each region. In the flat region IV, the group coordinates are well adapted Lorentzian coordinates. But in the plane wave regions II and III we must use harmonic coordinates (X, Y, U, V). In these coordinates the metric, in region II, has the form (1) with $h(U) = 3/16L_1L_2\left(1+\sin u\right)^5$. It is useful to analyze geodesics in that region and one can see that only null geodesics with exact perpendicular incidence i.e. with momenta $p_x = p_y = 0$ will hit the surface $u = \pi/2$ and all others reach the interacion region. One sees in these harmonic coordinates that the boundary $v = 0$ is folded back in such a way that $u = \pi/2$ is the closing line of the boundary. This indicates that one may identify the "line" $u = \pi/2$ with the "point" $u = \pi/2$, $v = 0$. That point is a topological singularity called a *folding singularity* and is an accumulation point for the geodesics.

The interaction region is locally isometric to a region of the interior of the Schwarzschild spacetime with mass $M = \sqrt{L_1L_2}$. In that isometry the Schwarzshild coordinates are defined as: $t = x$, $r = M[1 + \sin(u + v)]$, $\phi = 1 + y/M$, $\theta = \pi/2 - (u - v)$. The caustic is mapped to the black hole horizon. Since appropriate coordinates to describe the horizon are the Kruskal-Szekeres coordinates, it is useful to introduce in region I a similar set of Kruskal-Szekeres like coordinates defined as follows. We first introduce time and space coordinates (ξ, η) by $\xi = u + v$, $\eta = v - u$ and define $\xi^* = 2M\ln\left[\left(1+\sin\xi\right)/2\cos^2\xi\right] - M\left(\sin\xi - 1\right)$. Then introduce null coordinates $\tilde{U} = \xi^* - x, \tilde{V} = \xi^* + x$ and finally $U' = -2M\exp\left(-\tilde{U}/4M\right) \leq 0$, $V' = -2M\exp\left(-\tilde{V}/4M\right) \leq 0$. The metric (2), in region I, reads

$$ds_I^2 = \frac{2\exp\left[(1 - \sin\xi)/2\right]}{(1 + \sin\xi)} dU'dV' - M^2(1 + \sin\xi)^2 d\eta^2 - (1 + \sin\xi)^2\cos^2\eta dy^2 \ ,$$

(2)

with $U'V' = 8M^2\cos^2\xi/(1 + \sin\xi)\exp\left[(\sin\xi - 1)/2\right]$, $U'/V' = \exp(x/2M)$. Notice that the metric looks manisfestly flat in these coordinates at the caustic $\xi = \pi/2$. This hypersurface is defined by $U' = 0$ and $V' = 0$ in the new coordinates and the vectors $\partial_{U'}$ and $\partial_{V'}$ become null Killing fields there.

4 Field Quantization

4.1 The "in" Vacuum

Let us now consider the quantization of a massless scalar field ϕ in the colliding wave spacetime. We have to solve the Klein-Gordon equation,

$$\Box_g\phi = 0 \ ,$$

(3)

in the different spacetime regions. Before the collision in region IV the spacetime is flat and the solutions of (3) may be expressed in terms of the ordinary flat space modes, which we define as "in" modes,

$$u_k^{in(IV)}(u, v, x, y) = \frac{1}{\sqrt{2k_-(2\pi)^3}} e^{-i2L_2k_-v - i2L_1k_+u + ik_xx + ik_yy} \ ,$$

(4)

where k_-, k_x and k_y are independent separation constants and $k_+k_- = (k_x^2 + k_y^2)/4$. These constitute a set of positive definite frequency modes which are well normalized on the boundary of region IV with regions II and III. The field operator ϕ can be expressed, as usual, in terms of the positive and negative frequency modes with creation and annihilation operators. These operators may be used to defined the "in" vacuum $|0\,\text{in}>$ and physically meaningful states for "in" particles. Although the boundary of region IV is not a Cauchy surface for the spacetime, since all modes propagate towards the interaction region they are well normalized on the boundary of the interaction region with the plane waves. That null surface is a Cauchy surface for the interaction region. The modes on such surface uniquely define the solutions in the interaction region. Since we will define the "out" modes on the caustic we will be interested in propagating the "in" modes up to the caustic. Now the Klein-Gordon equation cannot be solved exactly in region I; however the above task is greatly simplified because one can show that the "in" modes become blueshifted near the boundary between region I and II when $u \to \pi/2$ and between region I and III when $v \to \pi/2$. Thus the relevant part of the modes in the interaction region have high frequencies near the points $u = \pi/2$, $v = 0$ and $v = \pi/2$, $u = 0$ and consequently one may use the geometrical optics approximation to propagate them up to the caustic. The problem is thus reduced to compute the geodesics in region I near the above points and match them to the geodesics in region II (and III). It is worth to note that the geometrical optics approximation is exact in regions II and III. The final expression for the "in" modes near the caustic is [14]

$$u_k^{in(I)}\Big|_{\xi \lesssim \pi/2} = \frac{1}{\sqrt{(2\pi)2|k_x|}} \sum_l C_l^{(1)} Y_l^m \begin{cases} (-U'/2M)^{i4M|k_x|} & ; \ k_x \geq 0(V' = 0) \\ (-V'/2M)^{i4M|k_x|} & ; \ k_x \leq 0(U' = 0) \end{cases}$$

(5)

where $C_\alpha^{(1)}$ are coefficients and $Y_l^m (y/M, \pi/2 - \eta)$ are spherical harmonics. We have assumed for simplicity that the transversal coordinate y is a cyclic coodinate with associated momentum m. The coefficients $C_\alpha^{(1)}$ are restricted by mode normalization on the caustic.

4.2 The "out" Vacuum

Now we shall define the "out" vacuum $|0 \text{ out} >$. This can be easily done if we note that a natural set of positive frequency modes can be defined at the caustic using the null Killing fields $\partial_{U'}$ and $\partial_{V'}$. These modes are

$$u_k^{out} = \frac{1}{\sqrt{(2\pi)2\omega_\pm}} Y_l^m (y/M, \pi/2 - \eta) e^{-i\omega_+ U' - i\omega_- V'} .$$

(6)

with $\omega_\pm$ two separation constants satisfying the condition $16L_1L_2\omega_+\omega_- = l(l+1)$. It is easy to show that modes (6) are normalized at the caustic. The field operator ϕ can be writen in terms of these modes with the corresponding creation and annihilation "out" operators. Using such operators one may construct the "out" vacuum and the corresponding "out" particle states. This "out" vacuum is physically reasonable: a particle detector in free fall near the horizon will detect no particles if the quantum state is the "out" vacuum.

4.3 Particle Production

We can now calculate spontaneous particle creation by evaluating the "out" number operator N^{out} in the in vacuum, $< 0 \text{ in}|N^{out}|0 \text{ in} >$. This is obtained as usual [15] by means of the Bogoliubov coefficients $\beta_{k\bar{k}} = -(u_k^{in}, u_{\bar{k}}^{out*})$, i.e. $N_{\bar{k}}^{out} = \int d^3k |\beta_{k\bar{k}}|^2$ where $\int d^3k = (2\sqrt{L_1L_2})^{-1} \sum_m \int dk_- \int dk_x$. The final result is [14]

$$N_{\bar{k}}^{out} = \frac{\mathcal{G}}{8\pi\sqrt{L_1L_2}\,\omega_\pm} ,$$

(7)

where $\mathcal{G}$ is a geometrical factor independent of the frequency $\omega_\pm$. Equation (7) indicates that particles are produced with an spectrum which is inverse to the frequency. Note that in the long wavelength limit the spectrum is compatible with a thermal spectrum with a temperature $T = 1/8\pi\sqrt{L_1L_2}$, i.e. inversely proportional to the focusing time of the waves. This result is not unexpected since $1/4\sqrt{L_1L_2}$ is the surface gravity of the caustic.

References

1. J. B. Griffiths, *Colliding waves in general relativity* Clarendon Press, Oxford, (1991)
2. D. Kramer, H. Stephani, M. MacCallum and E. Herlt, *Exact solutions of Einstein's field equations* (Cambridge University Press, Cambridge, England, 1980).
3. R. Penrose, *Rev. Mod. Phys.* **37**, 215 (1965).
4. P. D. D'Eath, in *Sources of gravitational radiation*, edited by L. Smarr (Cambridge University Press, Cambridge, England, 1979).
5. D. Garfinkle, *Phys. Rev.* **D41**, 1112 (1989).
6. G. W. Gibbons, *Commun. Math. Phys.* **45** 191 (1975).
7. J. Garriga and E. Verdaguer, *Phys. Rev.* **D43**, 391 (1991).
8. K. Kahn and R. Penrose, *Nature (London)* **229**, 185 (1971).
9. U. Yurtsever, *Phys. Rev.* **D40**, 360 (1989).
10. S. Chandrasekhar and B. Xanthopoulos, *Proc. R. Soc. (London)* **A408**, 175 (1986).
11. S. O'Brian and J. L. Synge, *Commun. Dubl. Inst. Adv. Stud.* **A9**, (1952).
12. S. A. Hayward, *Class. Quantum Grav.* **6**, 1021 (1989).
13. U. Yurtsever, *Phys. Rev.* **D38**, 1706 (1988).
14. M. Dorca and E. Verdaguer. In preparation.
15. N. D. Birrell and P. C. W. Davies, *Quantum fields in curved space* (Cambridge University Press, Cambridge, England, 1982).

Coleman's Mechanism in
Jordan–Brans–Dicke Gravity

Luis J. Garay [1] and Juan García–Bellido [2]

[1]Instituto de Optica, C.S.I.C.,
Serrano 121, E–28006 Madrid, Spain
[2]Instituto de Estructura de la Materia, C.S.I.C.,
Serrano 123, E–28006 Madrid, Spain

Abstract: We consider the quantum gravity and cosmology of a Jordan–Brans–Dicke theory. Its constraint algebra is that of general relativity, as a consequence of the general covariance of scalar–tensor theories. We propose that boundary conditions must be imposed in the Jordan frame, in which particles satisfy the strong equivalence principle. We discuss both Hartle–Hawking and wormhole boundary conditions in the context of quantum cosmology. Wormholes may affect the constants of nature and, in particular, the Brans–Dicke parameter. Following Coleman's mechanism, we find a probability distribution for wormhole configurations which is strongly peaked at zero cosmological constant and infinite Brans–Dicke parameter. That is, we recover general relativity as the effective low energy theory of gravity.

Jordan–Brans–Dicke theory [1] is based on the idea of Mach that inertia arises from accelerations with respect to the general distribution of matter in the universe. Therefore, the inertial masses of elementary particles are not fundamental constants but represent the interaction of particles with a cosmic scalar field whose dynamics depend on the rest of the matter in the universe. Supposing that all matter particles have the same coupling to the scalar field, $m(\phi) = e^{\beta\phi}m$, the action in the so called Einstein frame can be written as [2]

$$\bar{S} = \frac{1}{16\pi G} \int d^4x \sqrt{-\bar{g}} \left(\bar{R} - \frac{1}{2}\bar{g}^{\alpha\beta}\bar{\nabla}_\alpha\phi\bar{\nabla}_\beta\phi \right) + \int e^{\beta\phi}m \, d\bar{s}. \qquad (1)$$

Physics must be invariant under conformal redefinitions of the metric since they correspond to an arbitrary choice of measuring units. We are thus free to choose the conformal frame in which we want to describe physics [1]. The most natural choice is the so called physical frame, in which observable particles have constant masses, since in this frame particles follow geodesics of the metric, thus satisfying the strong equivalence principle [3]. From the cosmological point of view, the comoving frame, in which a fundamental observer sees the universe as homogeneous and isotropic, is the physical frame since those observers will follow geodesics of the metric. In our case, the conformal redefinition of the metric that allows us to describe the theory of gravitation (1) in the physical frame is

$$\bar{g}_{\alpha\beta} = e^{-2\beta\phi} g_{\alpha\beta} \equiv \Phi g_{\alpha\beta}. \tag{2}$$

This is the so called Jordan frame, in which the action takes the JBD form

$$S = \frac{1}{16\pi G} \int d^4x \sqrt{-g} \left(\Phi R - \frac{\omega}{\Phi} g^{\alpha\beta} \nabla_\alpha \Phi \nabla_\beta \Phi \right) + S_m, \tag{3}$$

where Φ is a dimensionless scalar field and ω a constant given by $2\omega + 3 = (4\beta^2)^{-1}$. General relativity is recovered in the limit $\omega = \infty$, ($\beta = 0$). These theories are not ruled out by post–Newtonian experiments [4] nor by primordial nucleosynthesis bounds [5], and in fact have recently recovered great interest since they have been proposed as the arena for extended inflation [6], a new inflationary scenario which could solve some of the traditional problems of previous schemes [7]. Furthermore, as we have mentioned, JBD theory may well be the Planck scale theory of gravity, as predicted by string effective actions [8]. We analyze the Jordan–Brans–Dicke quantum cosmology [9] with special emphasis made on the effect of quantum wormholes on the low energy coupling constants [10].

The Hamiltonian constraints can be explicitly written in both the Jordan and Einstein frames. Both expressions are related through the canonical transformation (2)

$$\mathcal{H} = e^{-\beta\phi}\bar{\mathcal{H}}, \qquad \mathcal{H}_i = \bar{\mathcal{H}}_i. \tag{4}$$

The constraint algebra is the same in both frames and corresponds to that of general relativity, expressing the fact that JBD theory is generally covariant.

Boundary conditions must be imposed in the physical frame, where particles satisfy the strong equivalence principle. The boundary conditions in any other frame can be obtained by transforming those in the physical frame through the corresponding conformal redefinition. In JBD quantum cosmology we shall be concerned with boundary conditions both for the universe and for wormholes. The Hartle–Hawking wave function for the universe [11] is given by the Euclidean path integral over all compact four–metrics in the Jordan frame, over all possible BD field configurations and all matter fields. The saddle point configuration, which gives the dominant contribution to the path integral, will have a curvature scalar

$$R = \frac{2\omega}{2\omega + 3} \frac{4\Lambda}{\Phi} + \omega \left(\frac{\nabla\Phi}{\Phi} \right)^2, \tag{5}$$

and therefore will be compact when both the cosmological constant and the BD parameter are positive.

Non–perturbative quantum gravity effects due to non–trivial topologies, *e.g.* wormholes [12], will change the effective value of the physical parameters of the theory, in particular the Brans–Dicke parameter [10]. Wormholes may be interpreted, semiclassically, as throats joining two otherwise disconnected large regions of spacetime. In this picture one assumes the dilute wormhole approximation. JBD wormhole wave functions can be written as Euclidean path integrals over asymptotically flat spacetimes. One must also sum over all JBD field configurations whose Hamiltonian vanishes in the asymptotic region and similarly for the matter fields [13]. These boundary conditions have been imposed in the Jordan frame, although in the Einstein frame they take the same form, since the JBD field is constant at infinity.

Although Planck scale wormholes are not directly observable, they will produce effective interactions in the low energy physics that turn the coupling constants of nature into dynamical variables [14]. In particular, JBD wormholes will affect the kinetic term of the JBD field and therefore will modify the value of the low energy BD parameter.

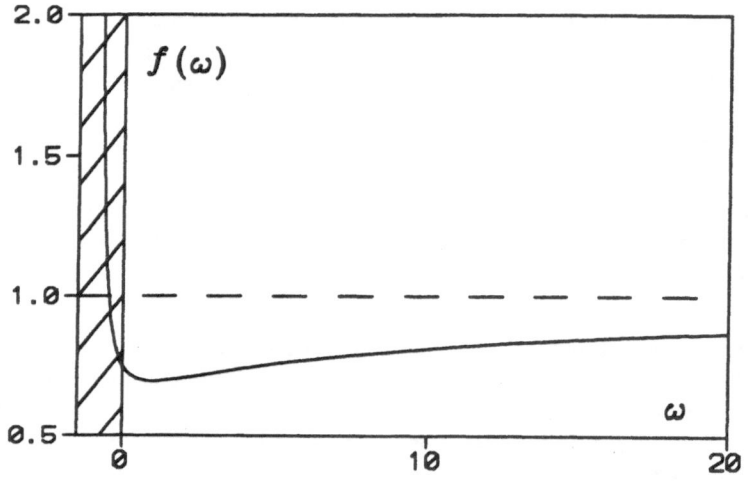

Fig. 1. Plot of the function $f(\omega)$. The $\omega < 0$ region has been excluded since it corresponds to non–compact saddle points. This function acquires its maximum at $\omega = \infty$, where $f(\infty) = 1$, which corresponds to the general relativity result.

Wormholes not only affect the constants of nature but also may provide a probability distribution for them. In general relativity this distribution is strongly peaked at zero cosmological constant. This is the so called Coleman's mechanism [15] for the vanishing of the cosmological constant. In JBD theory the probability distribution for the wormhole configurations α and thus for the coupling constants has the form [10]

$$\mathcal{Z}(\alpha) = \exp\left(\exp\frac{3\pi}{G_\alpha \Lambda_\alpha} f(\omega_\alpha)\right), \tag{6}$$

where $f(\omega_\alpha)$ is the function shown in fig.1, whose maximum value is obtained for $\omega_\alpha = \infty$, where $f(\infty) = 1$, for which we recover the general relativity result. $\mathcal{Z}(\alpha)$ is strongly peaked at $G_\alpha \Lambda_\alpha = 0$. Since this probability distribution is not normalizable, a suitable choice of the cutoff is needed in order to determine its maximum. In the context of general relativity, different cutoffs have been considered [15,16] that give different results. However, $G\Lambda$ seems to be the most natural choice for the cutoff since it is the adimensional vacuum energy, *i.e.* the cosmological constant in Planck units, which is observed to be less than 10^{-120}. With this choice for the cutoff, the probability distribution (6) acquires its maximum at $\omega_\alpha = \infty$. Thus we recover general relativity as the low energy effective theory of gravity, even though the fundamental high energy description may well be a scalar–tensor theory. This means that there can be no effective dynamical scalar field coupled to energy or matter. Note that this result can be generalized to any scalar–tensor theory with an arbitrary coupling $\omega(\Phi)$.

The usual interpretation [15] of this prediction is that Coleman's mechanism only ensures that the bottom line cosmological constant is zero. This does not exclude an inflationary universe with a non–zero false vacuum energy due to a certain phase transition. It just says that, whatever the effective potential for inflation is, the true minimum is at zero cosmological constant. However, it does preclude extended inflation since wormholes drive the BD parameter to infinity at all times and thus "freeze out" the evolution of the scalar field.

References

1. P. Jordan, *Nature* (London) **164** (1949) 637; *Z. Phys.* **157** (1959) 112;
    C. Brans and R. H. Dicke, *Phys. Rev.* **124** (1961) 925;
    R. H. Dicke, *Phys. Rev.* **125** (1962) 2163;
    C. Brans, *Phys. Rev.* **125** (1962) 2194;
    S. W. Hawking, *Comm. Math. Phys.* **25** (1972) 167.

2. See for instance, S. Weinberg, *Gravitation and Cosmology* (Wiley, New York, 1972).

3. J. A. Casas, J. García–Bellido and M. Quirós, *Class. Quant. Grav.* **9** (1992) 1371.

4. For a general review see C. M. Will, *Theory and Experiment in Gravitational Physics*, Cambridge University Press (1981); *Phys. Rep.* **113** (1984) 345.

5. J. A. Casas, J.García–Bellido and M. Quirós, *Mod. Phys. Lett.* **A 7** (1992) 447; *Phys. Lett.* **B 278** (1992) 94.

6. D. La and P. J. Steinhardt, *Phys. Rev. Lett.* **62** (1989) 376; *Phys. Lett.* **B 220** (1989) 375;
    E. J. Weinberg, *Phys. Rev.* **D 40** (1989) 3950;
    D. La, P. J. Steinhardt and E. W. Bertschinger, *Phys. Lett.* **B 231** (1989) 231;
    J. García–Bellido and M. Quirós, *Phys. Lett.* **B 243** (1990) 45; *Nucl. Phys.* **B 368** (1992) 463.

7.   A. H. Guth, *Phys. Rev.* **D 23** (1981) 347;
      A. D. Linde, *Phys. Lett.* **B 108** (1982) 389;
      A. Albrecht and P. J. Steinhardt, *Phys. Rev. Lett.* **48** (1982) 1220;
      A. H. Guth and E. J. Weinberg, *Nucl. Phys.* **B 212** (1983) 321.

8.   J. A. Casas, J. García–Bellido and M. Quirós, *Nucl. Phys.* **B 361** (1991) 713.

9.   S. del Campo and A. Vilenkin, *Phys. Rev.* **D 40** (1989) 688;
      A. K. Sanyal and B. Modak, *Int. J. Mod. Phys.* **A 7** (1992) 4039.

10.   L. J. Garay and J. García–Bellido, "Jordan–Brans–Dicke Quantum Wormholes and Coleman's Mechanism", preprint IEM–FT–59/92, submitted to Nuclear Physics B.

11.   J. B. Hartle and S. W. Hawking, *Phys. Rev.* **D 28** (1983) 2960.

12.   S. W. Hawking, *Mod. Phys. Lett.* **A 5** (1990) 145; *Mod. Phys. Lett.* **A 5** (1990) 453.

13.   S. W. Hawking and D. N. Page, *Phys. Rev.* **D 42** (1990) 2655;
      L. J. Garay, *Phys. Rev.* **D 44** (1991) 1059.

14.   S. W. Hawking, *Phys. Rev.* **D 37** (1988) 904;
      S. W. Hawking and R. Laflamme, *Phys. Lett.* **B 209** (1988) 39;
      S. Coleman, *Nucl. Phys.* **B 307** (1988) 864.

15.   S. Coleman, *Nucl. Phys.* **B 310** (1988) 643.

16.   J. Preskill, *Nucl. Phys.* **B 323** (1989) 141;
      B. Grinstein and M. B. Wise, *Phys. Lett.* **B 212** (1988) 407;
      B. Grinstein, *Nucl. Phys.* **B 321** (1989) 439;
      S. W. Hawking, *Nucl. Phys.* **B 335** (1990) 155.

No–boundary Condition in Multidimensional Gravity

G. A. Mena Marugán

Instituto de Optica, C.S.I.C., Serrano 121, E–28006 Madrid, Spain

Abstract: We consider D-dimensional Lovelock models whose matter action does not depend on the time derivatives of the metric, and which admit classical solutions in which the spacetime splits into a four-dimensional spacetime and an extra $(D-4)$-dimensional space, with matter fields independent of the extra $D-4$ coordinates. Freezing the extra degrees of freedom of the metric at the values they take in such classical solutions, we obtain an effective four-dimensional Einsteinian theory. The exponential action of the no-boundary D-dimensional classical solutions provides then a semiclassical approximation to the wave functions of this effective theory, and contains implicit information about the local behaviour of the semiclassical wave functions of the multidimensional model.

Consistency in unification theories generally requires that our universe possesses more than four dimensions [1]. For spacetimes of more than four dimensions, there exists a most natural topological generalization of Einstein gravity, which is Lovelock gravity [2]. The Lovelock Lagrangian is a linear combination of dimensionally continued Euler forms which contains the Hilbert-Einstein Lagrangian as a particular case [3]. In addition, the Lovelock dynamical equations are of the type of general relativity, inasmuch as they depend on the metric only up to its second spacetime derivatives [2].

In dealing with Lovelock theories of gravity, one has to adopt a perturbative formalism to obtain a singlevalued inversion of the gravitational momenta [4]. With this singlevalued inversion, one straightforwardly attains a singlevalued Hamiltonian for the system. In the quantum version of these theories, the physical states can be described by wave functions which are annihilated by the quantum operator associated to the corresponding Hamiltonian constraint. These wave functions are functionals in superspace, i.e., they solely depend on the values of the induced metric and the matter fields on the boundaries of the D-dimensional manifold, modulo $(D-1)$-diffeomorphisms [5].

Let us suppose now that our present universe can be described by an effective Einsteinian theory in which all the degrees of freedom of the multidimensional metric are frozen, except for those corresponding to a four-dimensional space-time. The physical states of such an effective theory will be represented by wave functions in a reduced superspace, that associated with the matter fields and induced metric of the four-dimensional spacetime. One could then expect that these wave functions should be obtainable from the wave functions of the D-dimensional theory, when restricting to those frozen values of the extra degrees of freedom which lead to the effective model in four dimensions. We will show that the relation that can be finally established between the wave functions of the D-dimensional and the effective four-dimensional models actually depends on the boundary conditions adopted in quantum cosmology.

The gravitational Lovelock action for a D-dimensional manifold $\mathcal{M}$ can be written [3]

$$
\begin{aligned}
S_G &= \int_{\mathcal{M}} \sum_{m=1}^{M} c_m L_m \\
&= \int_{\mathcal{M}} \sum_{m=1}^{M} \frac{c_m}{(D-2m)!} R^{a_1 a_2} ... R^{a_{2m-1} a_{2m}} e^{a_{2m+1}} ... e^{a_D} \epsilon_{a_1 ... a_D},
\end{aligned} \tag{1}
$$

where $M = \mathrm{int}\left(\frac{D-1}{2}\right)$, $\epsilon_{a_1 ... a_D}$ is the D-dimensional Levi-Civita tensor, the Lovelock coefficients c_m are real constants, $\{e^a 1$ is a D-tetrad for the manifold $\mathcal{M}$ and R^{ab} is its associated curvature two-form.

When the manifold $\mathcal{M}$ has a boundary $\partial\mathcal{M}$, the Lovelock action must be corrected with the surface terms [6]

$$
-\int_{\partial\mathcal{M}} \sum_{m=1}^{M} \frac{c_m m}{(D-2m)!} \int_0^1 ds \Theta^{a_1 a_2} R_s^{a_3 a_4} ... R_s^{a_{2m-1} a_{2m}} e^{a_{2m+1}} ... e^{a_D} \epsilon_{a_1 ... a_D}. \tag{2}
$$

Here, Θ^{ab} is the second fundamental form, and R_s^{ab} is the curvature two-form associated to the connection $\omega_s = \omega - s\Theta$, with ω the Levi-Civita connection.

We will restrict our discussion to Lovelock models whose matter Lagrangian is independent of the time derivatives of the induced metric. From Lovelock action, one can then obtain the gravitational momenta $\Pi^p_q = \delta S_G / \delta h_p{}^q$ ($p, q = 2,...,D$) [7]

$$
\Pi^p_q = h^{1/2} \sum_{m=1}^{M} c_m \sum_{s=0}^{m-1} \frac{(-4)^{m-s}}{2^{s+1}} \frac{m!(m-s)!}{s!(2m-2s)!} \delta^{p p_1 ... p_{2m-1}}_{q q_1 ... q_{2m-1}} \cdot
$$
$$
R^{q_1 q_2}{}_{p_1 p_2} ... R^{q_{2s-1} q_{2s}}{}_{p_{2s-1} p_{2s}} K^{q_{2s+1}}{}_{p_{2s+1}} ... K^{q_{2m-1}}{}_{p_{2m-1}}, \tag{3}
$$

with h the determinant of the induced metric, $\delta^{p ... p_{2m-1}}_{q ... q_{2m-1}}$ the generalized Kronecker delta and K^q_p the extrinsic curvature [8].

We will assume that the matter content of the system allows for the existence of a family of classical solutions for which: i) the spacetime is of the form $\mathcal{M}^4 \times \mathcal{M}^{D-4}$, with $\mathcal{M}^4$ a four-dimensional spacetime of compact constant time sections, and $\mathcal{M}^{D-4}$ a $(D-4)$-dimensional compact space; ii) the matter fields do not depend on the coordinates of the space $\mathcal{M}^{D-4}$; and iii) the inversion of the gravitational momenta can be single-valuedly and analytically defined, according to the perturbative formalism of Lovelock gravity [4], at least in the whole region of the gravitational phase space which is covered by the considered classical solutions. This last assumption should guarantee that the generalized Wheeler-DeWitt equation of the D-dimensional model is at least well-defined for wave functions which are strongly peaked about such classical solutions.

In studying these solutions, we can adopt the metric Ansatz

$$g_{ab} = \begin{pmatrix} g_{rs}(t, x^i) & 0 \\ 0 & g_{\alpha\beta}(y^\alpha) \end{pmatrix}, \tag{4}$$

where $r, s = 1,...,4$, $i = 2,3,4$, $\alpha, \beta = 5,...,D$, the x^i's are the spatial coordinates of the four-dimensional spacetime and the y^α's are the $D-4$ extra spatial coordinates. With this ansatz, the Lovelock Lagrangians L_m split in the following way [3]

$$L_m = {}^{(D-4)}L_m\,{}^{(4)}L_0 + m\,{}^{(D-4)}L_{m-1}\,{}^{(4)}L_1 + \frac{m(m-1)}{2}\,{}^{(D-4)}L_{m-2}\,{}^{(4)}L_2. \tag{5}$$

Here, the ${}^{(v)}L_m$'s denote evaluation of L_m on the manifold $\mathcal{M}^v$ ($v = 4$ or $D-4$), and L_m vanishes by convention if $m < 0$.

Integrating then the Lovelock action over all the coordinates y^α, and taking into account that L_2 is proportional to the Euler invariant in four dimensions, we obtain an effective Einsteinian gravitational action for the four-dimensional spacetime,

$$S_g^{eff} = c_1^{eff} \int_{\mathcal{M}^4} {}^{(4)}L_1 + \Lambda^{eff} \int_{\mathcal{M}^4} {}^{(4)}L_0 + \text{Surface Terms}. \tag{6}$$

We can likewise arrive at an effective matter action for the reduced model in four dimensions by integrating the action of the matter fields over y^α, provided these matter fields do not depend on the $D-4$ extra coordinates.

The quantization of the effective Einsteinian theory will lead, in particular, to the Wheeler-DeWitt equation in the reduced superspace corresponding to the four-dimensional spacetime. For each space $\mathcal{M}^{D-4}$, let $S_{cl}(h_{ij}, \phi_A)$ be the total action of the $\mathcal{M}^4 \times \mathcal{M}^{D-4}$ classical solution with induced three-metric h_{ij} and matter fields ϕ_A on the boundary of the four-dimensional spacetime. Then, $S_{cl}(h_{ij}, \phi_A)$ must be a solution to the Hamilton-Jacobi equation associated to the effective four-dimensional theory, and provided the semiclassical approximation is valid, the wave function $\Psi = e^{iS_{cl}(h_{ij}, \phi_A)}$ is a semiclassical solution to the Wheeler-DeWitt equation in the reduced superspace.

The action S_{cl} depends implicitly on the value $h_{\alpha\beta}^o(y^\alpha)$ taken by the metric of the extra $\mathcal{M}^{D-4}$ space, i. e., $S_{cl}(h_{ij}, \phi_A) \equiv S_{cl}(h_{ij}, \phi_A; h_{\alpha\beta}^o)$. One could expect that $\Psi = e^{iS_{cl}(h_{ij}, \phi_A; h_{\alpha\beta}^o)}$ should be given, in semiclassical approximation, by a wave function of the D-dimensional model, $\hat{\Psi}$, restricted to matter fields and induced metrics for which

$$h_{ij}(x^i, y^\alpha) = h_{ij}(x^i), \ h_{i\alpha}(x^i, y^\alpha) = 0,$$
$$h_{\alpha\beta}(x^i, y^\alpha) = h^o_{\alpha\beta}(y^\alpha), \ \phi_A(x^i, y^\alpha) = \phi_A(x^i). \tag{7}$$

The multidimensional wave function $\hat{\Psi}$ depends only on the values of the induced metric and matter fields on the boundary of the D-dimensional manifold $\mathcal{M}$. With our assumptions, this boundary must be the product of the compact manifold corresponding to $\mathcal{M}^{D-4}$ and the boundary of the four-dimensional manifold of the effective theory. Let us suppose that this last boundary is formed up by two disconnected three-surfaces of the same topology, associated, say, to an initial $t = t_o$ and a final $t = t_f$ constant time sections. Then, the restriction of the multidimensional wave functions to metrics with $h_{\alpha\beta}(x^i, y^\alpha) = h^o_{\alpha\beta}(y^\alpha)$ will present a dependence on $h^o_{\alpha\beta}$ coming from both disconnected boundaries of $\mathcal{M}$. This mixing in the dependence on $h^o_{\alpha\beta}$ prevent us to analyse in detail the implicit information about the superspace dependence of the multidimensional wave functions contained in $S_{cl}(h_{ij}, \phi_A; h^o_{\alpha\beta})$.

Nevertheless, we can overcome this problem by adopting a no-boundary condition [9], thus eliminating all the dependence of the wave functions on the initial boundary. We will hence suppose that the boundary of $\mathcal{M}^4$ is formed up by a sole connected three-surface: a section of constant time $t = t_f$. The D-dimensional manifold possesses then as its only boundary the product of that connected three-surface and the manifold $\mathcal{M}^{D-4}$.

When the manifold $\mathcal{M}$ presented a final and an initial boundaries, the Lovelock action had to be corrected with surface terms associated to both of these boundaries, obtaining in this way a total action $S_{cl}(h_{ij}, \phi_A; h^o_{\alpha\beta})$ (h_{ij} and ϕ_A referring now to the values of the induced three-metric and matter fields on the final boundary). If we now adopt a no-boundary condition, the initial boundary will not exist, and the admissible $\mathcal{M}^4 \times \mathcal{M}^{D-4}$ classical solutions should close smoothly in Euclidean time locally around the Wick rotated initial time t_o. The action for such no-boundary solutions can then be obtained from the action $S_{cl}(h_{ij}, \phi_A; h^o_{\alpha\beta})$, with initial boundary at $t = t_1$, by making t_1 tend to the initial time t_o and substracting at that limit the surface terms associated to the now non existing initial boundary. Using Eq.(2), one can show that the surface terms that must be added to S_{cl}, in order to obtain the action S^{NB}_{cl} of the no-boundary solutions, are [10]

$$\text{S.T.} = 2 \int d^3x d^{D-4}y \sum_{m=1}^{M} \sum_{r=0}^{1} \delta^{\beta_1 \dots \beta_{2(m-r)-2}}_{\alpha_1 \dots \alpha_{2(m-r)-2}} T^{\alpha_1 \dots \alpha_{2(m-r)-2}}_{\beta_1 \dots \beta_{2(m-r)-2}}(m, r) \Big|_{t=t_o}, \tag{8}$$

$$T^{\alpha_1 \dots \alpha_{2(m-r)-2}}_{\beta_1 \dots \beta_{2(m-r)-2}}(m, r) = -\frac{c_m m}{2^{m-1}} (^3h)^{1/2} (h^o)^{1/2} \bar{R}^{\alpha_1 \alpha_2}{}_{\beta_1 \beta_2} \cdots$$
$$\bar{R}^{\alpha_{2(m-r)-3} \alpha_{2(m-r)-2}}{}_{\beta_{2(m-r)-3} \beta_{2(m-r)-2}}.$$

$$\delta^{j_1 \dots j_{2r+1}}_{i_1 \dots i_{2r+1}} K^{i_1}{}_{j_1} \left[\delta^1_r (m-1) \left(\bar{R}^{i_2 r i_{2r+1}}{}_{j_{2r} j_{2r+1}} + \frac{2}{3} K^{i_{2r}}{}_{j_{2r}} K^{i_{2r+1}}{}_{j_{2r+1}} \right) + \delta^o_r \right], \tag{9}$$

^3h, h^o, $\bar{R}^{ij}_{kl}$ and $\bar{R}^{\alpha\beta}_{\gamma\delta}$ being the determinants and the Riemann tensors of the induced metrics h_{ij} and $h^o_{\alpha\beta}$, respectively.

The wave function $\Psi = e^{iS^{NB}_{cl}(h_{ij},\phi_A;h^o_{\alpha\beta})}$ provides a semiclassical solution to the Wheeler-DeWitt equation of the effective four-dimensional theory. The surface contribution (8) is a constant in the reduced superspace of that effective theory. Such a surface contribution depends, nevertheless, on the value $h^o_{\alpha\beta}$ taken by the metric of the $\mathcal{M}^{D-4}$ space. We will show that this dependence of the surface terms on $h^o_{\alpha\beta}$ allows the no-boundary action S^{NB}_{cl} to be a local solution for the Hamilton-Jacobi equation of the D-dimensional model.

Being $S^{NB}_{cl}(h_{ij}, \phi_A; h^o_{\alpha\beta})$ a solution to the Hamilton-Jacobi equation of the effective theory, one can see that such a no-boundary action will correspond to a solution for the D-dimensional Hamilton-Jacobi equation, evaluated at metrics and matter fields satisfying conditions (7), if it is fulfilled that

$$\frac{\delta S^{NB}_{cl}}{\delta h_{i\alpha}}(h_{ij},\phi_A;h^o_{\alpha\beta}) = \Pi^{i\alpha},$$

$$\frac{\delta S^{NB}_{cl}}{\delta h_{\alpha\beta}}(h_{ij},\phi_A;h^o_{\alpha\beta}) \equiv \frac{\delta S^{NB}_{cl}}{\delta h^o_{\alpha\beta}}(h_{ij},\phi_A;h^o_{\alpha\beta}) = \Pi^{\alpha\beta}, \qquad (10)$$

with $\Pi^{i\alpha}$ and $\Pi^{\alpha\beta}$ the classical momenta of the $\mathcal{M}^4 \times \mathcal{M}^{D-4}$ solutions at the final constant time surface, $t = t_f$.

One can first check from Eq.(3) that all the momenta $\Pi^{i\alpha} = h^{\alpha\beta}\Pi^i_{\ \beta}$ turn out to vanish in the considered classical solutions. Therefore, the first equation in (10) is fulfilled, since the no-boundary action presents no dependence on $h_{i\alpha}$.

One can also see [10] that, for the $\mathcal{M}^4 \times \mathcal{M}^{D-4}$ classical solutions,

$$\Pi^{\alpha\beta} = \sum_{m=1}^{M}\sum_{r=0}^{1} h^{\beta\gamma}\delta^{\alpha\beta_1\ldots\beta_{2(m-r)-2}}_{\gamma\alpha_1\ldots\alpha_{2(m-r)-2}}T^{\alpha_1\ldots\alpha_{2(m-r)-2}}_{\beta_1\ldots\beta_{2(m-r)-2}}(m,r), \qquad (11)$$

with $T^{\alpha_1\ldots\alpha_{2(m-r)-2}}_{\beta_1\ldots\beta_{2(m-r)-2}}(m,r)$ given by Eq.(9). The classical momenta $\Pi^{\alpha\beta}$ appearing in Eq.(10) can then be obtained by evaluating expression (11) on the final constant time surface. On the other hand, the functional derivative $\delta S^{NB}_{cl}/\delta h^o_{\alpha\beta}$ is the sum of the contributions coming from the no-boundary surface terms (8) and from the action $S_{cl}(h_{ij},\phi_A;h^o_{\alpha\beta})$. Since S_{cl} is the action of a classical solution for the variational problem with fixed $h_{\alpha\beta}(x^i,y^\alpha) = h^o_{\alpha\beta}(y^\alpha)$, both at initial and final times, we must have $\delta S_{cl}/\delta h^o_{\alpha\beta} = \Pi^{\alpha\beta}(t_f) - \Pi^{\alpha\beta}(t_o)$ (where the dependence on the spatial coordinates has not been explicitly displayed). As to the contributions of the surface terms, one can check from Eqs.(8,9,11) that $\delta(\text{S.T.})/\delta h^o_{\alpha\beta} = \Pi^{\alpha\beta}(t_o)$. Therefore, the action S^{NB}_{cl} satisfies as well the second equation appearing in (10).

We have thus shown that $S^{NB}_{cl}(h_{ij},\phi_A;h^o_{\alpha\beta})$ is a local solution to the D-dimensional Hamilton-Jacobi equation. Provided the semiclassical approximation is valid, the wave function $\Psi = e^{iS^{NB}_{cl}}$ is then a local semiclassical solution to the generalized Wheeler-DeWitt equation of the D-dimensional Lovelock model when evaluated at metrics and matter fields fulfilling restrictions (7). We can also assert now that not only the value of the no-boundary action $S_{cl}^{NB}(h_{ij}, \phi_A; h_{\alpha\beta}^o)$ coincides with that of a semiclassical action for the multidimensional model when restricted to metrics and matter fields satisfying such restrictions, but also that the values of the functional derivatives of this no-boundary action with respect to its implicit dependence on $h_{i\alpha}$ and $h_{\alpha\beta}$ coincide with those of the corresponding restricted functional derivatives of a semiclassical action in superspace. It is in this sense that the no-boundary wave functions of the effective four-dimensional model can be considered as relevant to the quantum theory of multidimensional gravity. Thus, the no-boundary action of the effective four-dimensional model contains more implicit information (in semiclassical approximation) about the superspace dependence of the multidimensional wave functions than what we initially expected.

References

1.   M. B. Green, J. H. Schwarz and E. Witten, *Superstring theory*, edited by P. V. Landshoff, D. W. Sciama and S. Weinberg (Cambridge University Press, Cambridge, 1987).
2.   D. Lovelock, Aequationes Math. 4, 127 (1970); J. Math. Phys. 12, 498 (1971).
3.   F. Müller-Hoissen, Class. Quantum Grav. 3, 665 (1986).
4.   G. A. Mena Marugán, Phys. Rev. D42, 2607 (1990); *Perturbative Formalism of Lovelock Gravity*, Phys. Rev. D, to appear.
5.   J. J. Halliwell, in *Proceedings of the Jerusalem Winter School on Quantum Cosmology and Baby Universes*, edited by T. Piran (Jerusalem, 1990).
6.   R. C. Myers, Phys. Rev. D36, 392 (1987)
7.   C. Teitelboim and J. Zanelli, Class. Quantum Grav. 4, L125 (1987).
8.   Our sign convention for the extrinsic curvature is that of R. M. Wald, *General Relativity* (Univ. of Chicago Press, Chicago, 1984).
9.   S. W. Hawking, Nucl. Phys. B239, 257 (1984).
10.   G. A. Mena Marugán, *Multidimensional Gravity and No-Boundary Condition*, Phys. Lett. B, to appear.

Distance of matter inside an Einstein-Strauss vacuole

M.Portilla [1] and J.J.Miralles [2]

[1]Universitat de València, Departament de Física Teòrica,
46100 Burjassot, València, Spain
[2]Universidad de Castilla-La Mancha, Departamento de Física Aplicada,
E.U.P.A Avda. España s/n, 02071 Albacete, Spain

In this note we study the distance-redshift relation for matter inside an Einstein vacuole. In a Friedmann model one has [1] $D(z) = D_F(z)$:

$$D_F(z) = \frac{zq_0 + (q_0 - 1)\left[-1 + \sqrt{1 + 2q_0 z}\right]}{H_0(1 + z)^2 q_0^2} \tag{1}$$

We expect some modifications to this law when considering matter inside a rich cluster, because the high density of the region strongly perturbs the general expansion of matter. The Einstein-Strauss vacuole, is certainly a highly idealized model of cluster, but it will give us a simple estimate of the goodness of (1) for regions of great density contrast. With this estimate, we shall consider, at the end of this note, the results of the method of determining the Hubble constant using X-ray and microwave observation of galaxy clusters.

So as to understand what follows we recall some of the fundamental points of the vacuole model [2]:

1) The space-time is a Friedmann Universe in the region I: $R > R_V$.

$$ds^2 = -dT^2 + a^2(T)dR^2 + a^2(T)\Sigma^2(R)\left(d\theta^2 + \sin^2\theta\ d\phi^2\right)$$

where $\Sigma(R)$ is $\sin R$, $\sinh R$, or R depending on the curvature of the hipersurfaces $T = constant$; and a Schwarzschild space-time of mass M in the region II: $R < R_V$.

2) It can be proved that the first and second fundamental forms are continuous through the temporal hipersurface $R = R_V$ if R_V verifies

$$\Sigma(R_V) = \left(\frac{2M}{H_0^2 a_0^3 \Omega_0}\right)^{\frac{1}{3}}, \tag{2}$$

where Ω_0 is the density parameter ($\Omega_0 = 2q_0$), $a_0 H_0\sqrt{|\Omega_0 - 1|} = 1$ for $\Omega_0 \neq 1$, and $a_0 H_0 = 2$ for $\Omega_0 = 1$.

3) There exists a coordinate system (Gaussian coordinates) such that the metric and its first derivatives are continuous.

4) The Schwarzschild radial coordinate of the vacuole is not stationary and evolves as $r(T) = a(T) \Sigma(R_V)$.

Let us consider a null congruence diverging from a point P in the Friedmann region subtending a solid angle $d\Omega$ with respect to the cosmological observer. Let dS be the transversal cross section of the congruence at some point Q inside the vacuole; and consider the cross section dS_V at the point V where the null congruence intersects the vacuole, The area distance is defined as $D^2 = \frac{dS}{d\Omega}$ [4].

Let us write it in the form

$$D_S = \sqrt{\frac{dS_V}{d\Omega}} \sqrt{\frac{dS}{dS_V}} = D_F(z_V) \sqrt{\frac{dS}{dS_V}} \tag{3}$$

The first factor is the area distance between P and V in a Friedmann universe. In order to get the ratio $\frac{dS}{dS_V}$, we need the vector field k^μ representing the congruence. Let us take the origin of coordinates at the point P in the Friedmann region. If $x^\mu(\lambda)$ is a null geodesic and λ an affine parameter, the tangent vector $k^\mu = \frac{dx^\mu}{d\lambda}$ satisfies $a(T)k^0 = C$, where C is an arbitrary constant. The value of this constant may be modified by scaling the affine parameter. We shall do the choice aftewards. A radial null curve is a geodesic in a Friedmann universe; then a vector field representing the congruence in region I is $k^\mu = (\frac{C}{a}, \frac{C}{a^2}, 0, 0)$. Let us compute its expansion $\theta = k^\mu_{;\mu}$ using the formula $k^\mu_{;\mu} = \frac{\partial_\mu(\sqrt{-g}\,k^\mu)}{\sqrt{-g}}$. The result is

$$\theta^I = \frac{2C}{a^2}\left(\frac{da}{dT} + \frac{d\ln\Sigma}{dR}\right). \tag{4}$$

Consider now, the part of the null congruence inside de vacuole. If we neglect the bending of light in this region, we can take Minkowski space as a good approximation to the Schwarzschild metric. Let us consider the observer at rest with respect the vacuole at instant V. It is straightforward to prove that for a radial congruence one has

$$\frac{dS}{dS_V} = \left(1 + \frac{1}{2}L\theta_V^{II}\right)^2 \tag{5}$$

where L is the area distance betwen events V and Q, and θ_V^{II} is the expansion of the null congruence using an affine parameter λ so that $d\lambda = $ distance in the same inertial frame. We can identify θ_V^{II} with θ_v^I if we choose C in (3) so that the affine parameter λ means distance for the cosmological observer too. Taking into account the relation $dl = |(k,u)| d\lambda$, one sees that this is achieved by choosing $C = a_V$. Equation (4) can be written then as

$$\theta_V^I = \frac{2}{a_V}\left(\dot{a}_V + (\ln\Sigma)_V'\right) \tag{6}$$

where $\dot{a} = \frac{da}{dT}$ and $(\ln\Sigma)' = \frac{d\ln\Sigma}{dR}$; and substituting into (3) we get, taking into account that $D_F(z_V) = a_V\Sigma_V$, and $\Sigma_V' = \sqrt{1-k\Sigma_V^2}$ can be written in the form

$$D = D_{FV} \left[1 + H_V L + \sqrt{1 - k\Sigma_V^2} \frac{L}{D_{FV}} \right] \tag{7}$$

where H_V is the Hubble constant when the null geodesics intersected the vacuole, which is related to H_0 by $H_V = H_0(1 + z_V)\sqrt{1 + 2q_0 z_V}$.

The distance L inside the vacuole may be here identified with the Schwarzschild radial coordinate of the point $V : L \simeq r_V$. So considering (2) we have

$$L = \frac{1}{1 + z} \left(\frac{2M}{H_0^2 \Omega_0} \right)^{\frac{1}{3}} . \tag{8}$$

Let us consider these relations at low redshift. The Friedmann area distance (1) is $D_F \simeq \frac{z}{H_0}$, and neglecting second order terms we have:

$$D = \frac{z}{H_0} + \left(\frac{2M}{H_0^2 \Omega_0} \right)^{\frac{1}{3}} . \tag{9}$$

Let us asume we have a direct estimation of the distance D of a cluster, for instance the one based in the combination of the microwave background decrement and the X–ray observations [3]. Let us call H_0^* the Hubble costant determined by assuming the Friedmaniann distance-redshift relation $D = \frac{z}{H_0^*}$. This will be related, according to this model, to the true Hubble constant by the expression

$$\frac{z}{H_0^*} = \frac{z}{H_0} + \left(\frac{2M}{H_0^2 \Omega_0} \right)^{\frac{1}{3}} . \tag{10}$$

It is convenient to introduce dimensionaless Hubble constant by the equations $H_0 = h H_{100}$, $H_0^* = h^* H_{100}$, with $H_{100} = 10^{-28} cm^{-1}$, then (10) turns out to be:

$$h^* = \frac{h}{1 + \alpha h^{\frac{1}{3}}}$$

$$\alpha = \frac{1}{z} (2M H_{100})^{\frac{1}{3}} \Omega_0^{-\frac{1}{3}} . \tag{11}$$

For an averange rich cluster one finds $M \simeq 10^{15}$ solar masses, so it will be useful to write α in the form $\alpha = \frac{M_{15}^{\frac{1}{3}}}{z} \times \left(\frac{0.03}{\Omega_0} \right)^{\frac{1}{3}} \times 10^{-2}$, where M_{15} is the cluster's mass in units of 10^{15} solar masses. Let us adopt $\Omega_0 = \Omega_{barionic} = 0.03$ according to the standard model for nucleosynthesis. The meaning of α is clear: h^* is a fraction of h when α is of order unity.

Birkinshaw [3], estimating the Hubble constant by this method obtained

$$0.05 \leq h_0^* \leq 0.5 \; for \; A576$$

$$0.1 \leq h_0^* \leq 0.7 \; for \; A2218 \; and \; A2319$$

This discrepancy may be interpreted using equations (11): Assuming $h = 0.8$, we have the next table

Cluster	z	M_c	α	h^*
A576	0.039	12.5	0.59	0.5
A2218	0.164	8.5	0.12	0.71
A2319	0.053	2.5	0.25	0.64

References

1.   Weinberg, S.: Gravitation and Cosmology. J. Wiley, New York (1972)
2.   Stephani, H.: General Relativity. Cambridge University Press, Cambridge (1982)
3.   Birkinshaw, M.: M.N.R.A.S. **187** (1979) 847
4.   Ellis, G.F.R.: Relativistic Cosmology (Enrico Fermi Summer School). Academic Press, New York (1971)

Conformally Stationary Cosmological Models

J.J. Ferrando, J.A. Morales, N. Portilla

Departament de Física Teòrica, Universitat de València,
46100-Burjasot (València), Spain

Abstract: The role played by conformally stationary space-times in Cosmology is discussed.

Let us consider a general property of the conformally stationary space-times due to Tauber-Weinberg [1] and Ehlers-Geren-Sachs [2]:

There exists an observer n measuring an isotropic distribution function solution of the Liouville equation if, and only if, the space-time admits a timelike conformal Killing vector ξ. Moreover, ξ and n are collinear.

This property is applicable to any relativistic gas. In particular, an isotropic cosmic microwave background radiation can exist only if the space-time is conformally stationary. And this radiation appears isotropic only with respect to the observer associated with the conformal Killing vector.

On the other hand, the redshift z in a conformally stationary space-time is given by[1]

$$1 + z = \sqrt{\frac{g_{00}(x_r)}{g_{00}(x_e)}} \tag{1}$$

where x_r and x_e are, respectively, the reception and emission events conected by a null geodesic. Substituting (1) in the following relation[2]

$$T_r = \frac{T_e}{1 + z} \tag{2}$$

where T is the effective radiation temperature, one has

---

[1] this expression is exactly the same as the redshift formula in a stationary space-time. Here $g_{00} = g(\xi, \xi)$, g being the space-time metric.

[2] Note that (2) is valid for a general state of radiation whose distribution function obeys the Liouville theorem.

$$T_r = \sqrt{\frac{g_{00}(x_e)}{g_{00}(x_r)}} T_e \tag{3}$$

If one admits the hypothesis that the background radiation was isotropic at decoupling time, the emission temperature is given by[3]

$$T_e = \frac{C}{\sqrt{-g_{00}(x_e)}}, \tag{4}$$

where C is a constant. Then (3) and (4) imply

$$T_r = \frac{C}{\sqrt{-g_{00}(x_r)}}, \tag{5}$$

that is to say, this radiation is isotropic now. Otherwise, *if the background radiation is anisotropic now with respect to the observer associated to the conformal Killing vector, then these anisotropies are primordial.*[4]

Therefore, we can have isotropic radiation in an inhomogeneous Friedmann perturbation, assuming that it is conformally stationary. This prompted us to reconsider an important result in Cosmology known as the Sachs-Wolfe effect [3]: inhomogeneities in the last scattering surface the Friedmann universe introducing a gravitational potential, which this one changes the energy of photons as a gravitational redshift. In this way one predicts microwave background anisotropies related to the variations of the gravitational potential over the last scattering surface. Recently [4], [5], we have proposed to revise this conclusion mainly because one can see that the model of universe used by Sachs-Wolfe (see below) is just a conformally space-time.

If we restrict ourselves to conformally static space-times, the metric may be written in the form

$$ds^2 = -\alpha^2(x^\mu)(dx^0)^2 + \Omega^2(x^\mu)\gamma_{ij}(x^k)dx^i dx^j \tag{6}$$

where $\frac{\alpha}{\Omega}$ is independent of x^0, the coordinate time adapted to the integrable conformal Killing vector, and $(x^\mu) = (x^0, x^i), i = 1, 2, 3$.

Elsewhere [6], we have obtained the expression of the energy-momentum tensor of the metric (6). Generically, it is inhomogeneous and has associated anisotropic pressures. Moreover, the energy density flux (with respect to n) is not zero. Then, this flux could be interpreted[5] as due to the relative motion of the matter with respect to the observer measuring isotropic radiation. Thus, an observer comoving with the matter will measure a dipolar anisotropy.

As an example, let us consider a conformally ststic potential perturbation to the Einstein-de Sitter universe,

$$ds^2 = a^2(\eta)[-(1 - 2\phi)d\eta^2 + (1 + 2\phi)\delta_{ij}dx^i dx^j] \tag{7}$$

---

[3] This is a consequence of the aforementioned Ehlers-Geren-Sachs results [2]
[4] i.e. these anisotropies were already present at decoupling time.
[5] Under the general algebraic conditions.

where the potential $\phi(x^i)$ is independent of the conformal time η. Let us assume the following approximation: ϕ and its first two derivatives are infinitesimal quantities of the same order. In this case, one can prove [7] that (modulo second order terms in ϕ) the metric (7) is equivalent to the pressure-less increasing scalar mode of the Sachs-Wolfe solution [3]. According to this result and the general properties of a conformally stationary space-time, the Sachs-Wolfe observer (i.e. the observer comoving with the matter) is not the observer associated with the conformal Killing vector. Therefore the Sachs-Wolfe observer cannot measure isotropic radiation at decoupling time. Hence, it is necessary to revise the Sachs-Wolfe effect [4], [5].

The above approximation just applies to regions where the density contrast is small. For domains with intermediate density contrast (i.e. of the order of unity) the terms containing $(d\phi)^2 \equiv \delta^{ij}\partial_i\phi\partial_j\phi$ cannot be neglected [8]. In this situation, two different contributions are present in the expression for the mean pressure. One of them is always positive and proportional to ϕ, being related to the mean gravitational energy of the matter. The other one is negative and proportional to $(d\phi)^2$. As the collective peculiar velocity of the matter results of the same order as $d\phi$, this term can be related to the dynamical pressure of a self-gravitating system of particles with a positive two point correlation function. Thus, when $(d\phi)^2$ is of the same order as ϕ, the metric (7) seems appropriated to deal with galaxy clusters. Therefore, the choice of a conformally stationary space-time seems quite good to describe the recent Universe after decoupling of matter radiation.

Before recombination matter and radiation were coupled. Balance kinetic equations describing their interaction have non-zero collision terms. The Liouville equation is not satisfied and, at that time, we cannot apply the Ehlers-Geren-Sachs result. Fortunately, this result has been generalized by Treciokas-Ellis [9] for an isotropic collision term. In this case the observer n measuring an isotropic distribution function is shear-free and conformally geodesic but, in general, n does not generate a conformal motion on the space-time. The metric (6) with α and Ω arbitrary functions belongs to this class of space-times. The question arises if a metric like (7), with $\phi(\eta, x^i)$, could describe a mixture of interacting matter and radiation. A preprint [10] containing our results in this direction is in progress.

References

1. Tauber, G. E., Weinberg, J. W.: Internal state of a gravitating gas. Phys. Rev. **122** (1961) 1342-1365
2. Ehlers, J., Geren, P., Sachs, R. K.: Isotropic solutions of Einstein-Liouville equations. J. Math. Phys **9** (1968) 1344-1349
3. Sachs, R. K., Wolfe, M.: Perturbations of a cosmological model and angular variations of the microwave background. Ap. J. **147** (1967) 73-90
4. Ferrando, J. J., Morales, J. A., Portilla, M.: Residual fluctuations in the microwave background at large angular scales: Revision of the Sachs-Wolfe effect. Phys. Rev. D (submitted)

5.   Ferrando, J. J., Morales, J. A., Portilla, M.: Revision of the Sachs-Wolfe effect. In Proceedings of the "Journées Relativistes" Amsterdam 1992. Class. Quantum Grav. (submitted).

6.   Ferrando, J. J., Morales, J. A., Portilla, M.: Inhomogeneous space-times admitting isotropic radiation: Vorticity-free case. Phys. Rev. D. **46** (1992) 578-584

7.   Moreno, J.: Distorsiones en la observación de la estructura a gran escala del Universo. Ph. D. Thesis, Unversitat de València (1992)

8.   Ferrando, J. J., Morales, J. A., Portilla, M.: The potential perturbation to Friedmann universes. Phys. Rev. D (submitted)

9.   Treciokas, R., Ellis, G. F. R.: Isotropic solutions of the Einstein-Boltzmann equations. Commun. Math. Phys. **23** (1971) 1-22.

10.   Ferrando, J. J., Morales, J. A., Portilla, M.: A model of inhomogeneous Universe with isotropic radiation (in preparation).

L-Rigidity in Newtonian Approximation

M. Barreda [1], J. Olivert [2]

[1]Departament de Matemàtiques i Informàtica, Universitat Jaume I,
12071-Castelló, Spain
[2]Departament de Física Teòrica, Universitat de València,
46100-Burjassot (València), Spain

Abstract: Newtonian limit of L-Rigidity is obtained. In this formalism, L-Rigidity is reduced to steady Newtonian rigid motions in a Newtonian frame of reference in which the observer is at rest.

1 Introduction

In a preceding paper [1], a definition of L-Rigidity was proposed with the purpose of relating the weak rigidity [2] and the dynamical rigidity [3], [4]. In the present work, we study the conditions of L-Rigidity by applying the Newtonian approximation in order to analyze how L-Rigidity particularizes in classical mechanics.

We take $(+ + + -)$ as the signature of the metric. Latin indices range from 1 to 3 and Greek ones have values from 1 to 4. We denote by $i_X g$ the inner product of the metric tensor g and a vector field X, and by $\mathcal{L}_X$ the Lie derivative with respect to X. Given a bitensor $I^{\lambda\alpha}(z, m)$, we adopt the convention of [3]: indices $\lambda, \mu, \ldots$ are always associated with the point z, and $\alpha, \beta, \ldots$ with the point m. Covariant derivatives may be carried out with respect to the coordinates of z or with respect to those of m. We symbolize these covariant derivatives by a dot and appropriate suffixes. We denote by $\sigma(z, m)$ the world-function and by $\bar{g}^\lambda{}_\alpha(z, m)$ the parallel propagator. In [5] the covariant derivatives of the world-function are obtained

$$\sigma^{.\lambda}(z, m) = -X^\lambda , \quad \sigma_{.\lambda\alpha} = -\bar{g}_{\lambda\alpha} - \int_0^1 u\, \bar{g}_\lambda{}^a \bar{g}^b{}_\alpha R_{\mathrm{apbq}} \frac{dx^\mathrm{p}}{du} \frac{dx^\mathrm{q}}{du} du + O_2 , \quad (1)$$

where $x(u) = \exp_z uX$ is the geodesic such that $x(1) = m$, whereas Latin indices have here values from 1 to 4 and denote indices at $x(u)$, when $0 < u < 1$.

We consider an almost-thermodynamic material scheme [6], an observer represented by a timelike curve L, a timelike unit vector field n^λ along L, and a tensor field $\Omega_{\lambda\mu}$ along L, antisymmetric and orthogonal to n^λ, which can represent the angular velocity of the body. We parametrize L as $z(s)$, s being an arbitrary parameter, so that $v^\lambda = dz^\lambda/ds$ is not unitary. We denote by $\hat{\Sigma}(s)$ the orthogonal hyperplane to $n^\lambda(s)$. Beginning from a transport law of [3], a derivative operator $\tilde{D}$ along L is defined by

$$\tilde{D}X^\lambda = \dot{X}^\lambda + \tilde{M}^\lambda{}_\mu X^\mu , \quad \tilde{M}^\lambda{}_\mu = \dot{n}^\lambda n_\mu - n^\lambda \dot{n}_\mu + \Omega^\lambda{}_\mu . \tag{2}$$

Here the dot means covariant differentiation along L. We denote by τ_s : $T_{z(0)}M \longrightarrow T_{z(s)}M$ the induced isomorphysm by the transport law of Dixon.

Hence it can be proved that there exist convex open sets $N_{z(s)} \subset T_{z(s)}M$ such that

$$N_{z(s)} = \tau_s \left(N_{z(0)} \right) , \tag{3}$$

$\exp_{z(s)}$ is a diffeomorphism of $N_{z(s)}$ into its image, and so that

$$\Sigma(s) = \exp_{z(s)} \left(N_{z(s)} \cap \hat{\Sigma}(s) \right) \tag{4}$$

are hypersurfaces formed by geodesics through $z(s)$ orthogonal to $n^\lambda(s)$.

Let $\Sigma = \bigcup_s \Sigma(s)$ be the universe tube containing L. Since the hypersurfaces $\Sigma(s)$ are disjoint, [7] defines a differentiable function χ on Σ given by

$$\chi(m) = s \quad \text{if} \quad m \in \Sigma(s) . \tag{5}$$

We denote by N^α the unitary vector field on Σ in the direction of $-n_\lambda \sigma^{\lambda\alpha}$, which coincides with the normal field on each hypersurface $\Sigma(s)$.

On the other hand, given a point $m \in \Sigma$ we consider the differentiable curve

$$\gamma_m(s) = \exp_{z(\chi(m)+s)} \left(\tau_{\chi(m)+s} \left(\tau_{\chi(m)}{}^{-1} \left(\exp_{z(\chi(m))} \right)^{-1} (m) \right) \right) , \tag{6}$$

which starts at m. From this, a vector field ω^α on Σ can be defined by

$$\omega^\alpha(m) = \left. \frac{dx^\alpha \circ \gamma_m}{ds} \right|_{s=0} . \tag{7}$$

Then from (3)–(6), we get

$$\omega(\chi) = 1 . \tag{8}$$

2 Some Expressions in the Newtonian Approximation

In this Section we use the notation and results presented in [8]; we denote by u^2 the magnitude of the coordinate velocity.

We consider a gravitationally bound system in which the quantities U (Newtonian potential), t_{ab} (components of stress tensor) and u^2 are all less than ϵ^2, ϵ^2 being the maximum value of the Newtonian potential. (The Newtonian potential at the center of the Sun is $\epsilon^2 \sim 10^{-5}$).

The tensor expressions used here are given in a comoving orthonormal frame with components (x_1, x_2, x_3, t).

We assume that the baryon "mass" density ρ_o is of order $O(\epsilon^0)$, and spatial partial derivatives preserve the original approximation orders while the time partial derivative increases these orders in one unit. We denote by d/dt the timelike derivative operator following the matter, given by $u_a \partial/\partial x_a + \partial/\partial t$.

The Newtonian expansion of the metric components is

$$g_{ab} = \delta_{ab} + O\left(\epsilon^2\right) , \quad g_{a4} = O\left(\epsilon^3\right) , \quad g_{44} = -1 + 2U + O\left(\epsilon^4\right) . \qquad (9)$$

So the Christoffel symbols are

$$\Gamma^a_{bc} = O\left(\epsilon^2\right) , \quad \Gamma^a_{b4} = \Gamma^4_{4b} = \Gamma^4_{44} = O\left(\epsilon^3\right) , \quad \Gamma^a_{44} = \Gamma^4_{a4} = -U_{,a} + O\left(\epsilon^4\right) . \qquad (10)$$

At Newtonian order the stress-energy tensor has the following components:

$$T^{ab} = \rho_o u_a u_b + t_{ab} + O\left(\epsilon^4\right) , \quad T^{a4} = \rho_o u_a + O\left(\epsilon^3\right) , \quad T^{44} = \rho_o + O\left(\epsilon^2\right) . \qquad (11)$$

3 Newtonian Equations of L-Rigidity. Consequences.

In [1] we defined an equivalent expression for the L-Rigidity, given by

$$\nabla_\alpha \omega^\alpha = 0 , \qquad (12)$$

$$\left(\tilde{D} T_m{}^{\lambda\mu}\right) (\chi(m)) = 0 , \quad \forall m \in \Sigma , \qquad (13)$$

$$g\left(\mathcal{L}_\omega N, N\right) = 0 , \qquad (14)$$

$$\tilde{D} v^\lambda = 0 , \qquad (15)$$

where

$$T_m{}^{\lambda\mu}(s) = \bar{g}^\lambda{}_\alpha \Big(z(s), \gamma_m\left(s - \chi(m)\right)\Big)$$
$$\times \bar{g}^\mu{}_\beta \Big(z(s), \gamma_m\left(s - \chi(m)\right)\Big) T^{\alpha\beta}\Big(\gamma_m\left(s - \chi(m)\right)\Big) .$$

It is easy to prove that the vector field ω^α transforms $\Sigma(s)$ into $\Sigma(s + ds)$, which indicates that ω^α determines the motion of the body. Moreover, L-Rigidity leads to a dynamical rigidity in weak fields.

Our purpose is to express (12)–(15) in the Newtonian approximation. To begin we take a timelike curve L parametrized with the coordinate time; then

$$v^i = O(\epsilon) , \quad v^4 = 1 , \qquad (16)$$

and a unit vector field n^λ along L without spatial components

$$n^i = 0 , \quad n^4 = 1 + U + O\left(\epsilon^4\right) . \qquad (17)$$

From (17) an antisymmetric tensor $\Omega_{\lambda\mu}$ along L and orthogonal to n^λ has the following components

$$\Omega_{ij} = O(\epsilon) , \quad \Omega_{i4} = 0 . \tag{18}$$

From this and taking into account (2), (9), (10), (17) and (18) we have:

$$\tilde{M}^i{}_j = \Omega_{ij} + O\left(\epsilon^3\right) , \quad \tilde{M}^i{}_4 = \tilde{M}^4{}_i = U_{,i} + O\left(\epsilon^4\right) , \quad \tilde{M}^4{}_4 = O\left(\epsilon^3\right). \tag{19}$$

To obtain the Newtonian approximation of (7) we are going to calculate $\sigma.^\lambda$ to this approximation order. Since this bivector generalizes the position vector – see (1) – and it is a Newtonian quantity, we calculate the Newtonian limit of $\sigma.^\lambda$ by treating space-time as flat. Thus

$$\sigma.^\lambda(z,m) = -r_\lambda(z,m) , \quad r_\lambda(z,m) = x_\lambda(m) - x_\lambda(z) . \tag{20}$$

By virtue of (2), (10), (19) and (20), the Newtonian expression for (7) is reduced to

$$w^a = v_a - \Omega_{ab}r_b + O\left(\epsilon^3\right) , \quad w^4 = 1 + O\left(\epsilon^4\right) . \tag{21}$$

Since the vector field w^α determines the motion of the body, (21) leads to

$$u_a = v_a - \Omega_{ab}r_b + O\left(\epsilon^3\right) ; \tag{22}$$

so, we obtain the Newtonian rigidity by choosing the motion of the body.

As we indicate above, the purpose of this work is to obtain the Newtonian limit of the L-Rigidity. We begin with the vanishing of the expansion of the vector field w^α.

Let $m \in \Sigma$; consider the hypersurface $\Sigma(\chi(m))$. Then, taking into account (4) and if we assume that $r_a\left(z(\chi(m)),m\right)$ is of order $O\left(\epsilon^0\right)$, it may be seen that

$$t(m) = \chi(m) + O\left(\epsilon^3\right) , \tag{23}$$

that is to say the submanifolds $\Sigma(t)$ are simultaneity hypersurfaces.

By virtue of (3), (4), (8), (20), (21) and (23), the partial derivatives of χ, are

$$\frac{\partial \chi}{\partial x_a} = 0 , \quad \frac{\partial \chi}{\partial t} = 1 + O\left(\epsilon^4\right) . \tag{24}$$

Thus, taking into account (10), (21), and (24) we have that, to the approximation order, the expansion of w^α vanishes.

With respect to the Newtonian approximation of (13), we must obtain the parallel propagator to this order. By differentiating (20) we get

$$\sigma.^\lambda{}_\alpha = -\delta^\lambda_\alpha , \tag{25}$$

and so taking into account (1), (9), (10) and (25) we obtain

$$\bar{g}^i{}_a = \delta^i_a + O\left(\epsilon^2\right) , \quad \bar{g}^4{}_a = \bar{g}^i{}_4 = O\left(\epsilon^3\right) , \quad \bar{g}^4{}_4 = 1 + O\left(\epsilon^4\right) . \tag{26}$$

Thus, from (2), (10), (11), (19), and (26), we get that the purely spatial part of (13) is verified to this order. Moreover, the mixed and timelike parts are reduced to

$$\frac{du_i}{dt} + \Omega_{ij}u_j = O\left(\epsilon^4\right) , \quad \frac{d\rho_o}{dt} = O\left(\epsilon^3\right) . \tag{27}$$

Below we analyze (14) in the Newtonian limit. By virtue of (9), (17), (21), (24) and (25), we have that, to this order, (14) is an identity.

To conclude, (15) can be written at this level of approximation as

$$\frac{dv_i}{dt} + \Omega_{ij} v_j = O\left(\epsilon^4\right) \ , \tag{28}$$

which follows from (2), (10), (16) and (19).

The last equation does not restrict the motion of the body, because this condition is verified by choosing a Newtonian frame of reference in which the observer is at rest [4].

If we analyze (22), (27) and (28) we get

$$\frac{d\Omega_{ij}}{dt} = O\left(\epsilon^4\right) \ , \tag{29}$$

which indicates that, to this order of approximation, the angular velocity is constant. Thus, from the conservation of rest mass, L-Rigidity leads to (22), (28) and (29), which are the expressions of a steady Newtonian rigid motion for an observer at rest.

The L-Rigidity must have this restriction, at this level of approximation, because L-Rigidity leads to a dynamical rigidity in weak fields, and the Newtonian limit of dynamical rigidity is consistent with the Newtonian rigidity only for steady motions [4].

References

1.   M. Barreda, J. Olivert: "Cases of Multipole Moments Conservation from L-Rigidity", in Proc. Encuentros Relativistas Españoles 91, ed. by J. Ibáñez (World Scientific, Singapore) to appear
2.   V. Del Olmo, J. Olivert: J. Math. Phys. 26 1311 (1985)
3.   W. G. Dixon: Proc. R. Soc. Lond. A 314 499 (1970)
4.   W. G. Dixon: "Extended Bodies in General Relativity: Their Description and Motion", in Proc. Isolated Gravitating Systems in General Relativity, ed. by J.Ehlers (North-Holland, Amsterdam, 1979) pp. 156–219
5.   J. L. Synge: "Relativity: The General Theory" (North-Holland, Amsterdam, 1972)
6.   J. J. Ferrando, J. Olivert: J. Math. Phys. 22 2223 (1981)
7.   W. G. Dixon: Phil. Trans. R. Soc. Lond. A 277 59 (1974)
8.   C. W. Misner, K. S. Thorne, J. A. Wheeler:"Gravitation" (Freeman, San Francisco, 1973)

Presymplectic Manifolds and Conservation Laws

V. Liern [1], J. Olivert [2]

[1]Departamento de Economía Financiera y Matemática, Universitat de València, 46010-València, Spain.
[2]Departament de Física Teòrica, Universitat de València, 46100-Burjassot (València), Spain

Abstract: In this paper we make use of a new structure called *seeded fibre bundle*. This allows us to combine the symplectic formalism and general relativity. A theorem of existence is obtained and some examples and properties are studied.

1 Introduction

In order to interpret the interaction of particles from a mathematical point of view, and within a non–quantum formalism, it is associated to a type of principal fibre bundles provided with connection. However, a particle which describes a specific movement is mathematically associated to a symplectic manifold on which a dynamic group acts. This association is an formalism independent of the observation, or better said, independent from the modification produced by the interaction that acts on the particle. However, this formalism can hardly be extended to the particles on which a given interaction acts, because in a symplectic manifold it can hardly be given a connection in a canonical way, and even less one that modifies the characteristic foliation of the symplectic manifold.

With the aim to be able to relate these ideas we need a new structure (the *seeded fibre bundle* one) which was introduced in the Ref. [7] . In that paper we proved that the results of elementary particles given by Souriau in Ref. [10] can be extended to general relativity. Furthermore, we showed that the class of the new fibre bundles was not empty. However, we did not give any theorem of existence.

In this work we prove the existence of these fibre bundles and we obtain, as a consequence of the theorem of existence, that a dynamic group in the symplectic manifold (the fibres) is also dynamic in the family of presimplectic manifolds of the seeded structure. Furthermore, it provides us with a method of construction that is used in the example given.

Finally, we study an example from the group $SU(2, \mathbb{C})$, and we apply Noether's Theorem in order to obtain some conservation laws. One of them is described by the following function:

$$f(\theta) = 4\sin^2 \frac{\theta}{2},$$

which we call *strangeness function*[1].

2 Seeded Fibre Bundles: Definition, Existence and Properties

Definition 2.1 Let $\lambda = (P, M, \pi, G)$ be a principal G–bundle provided with connection, (U, σ_U) a symplectic Hausdorff manifold left G–space. $\lambda[U] = (P_U, M, \pi_U, G)$, a fibre bundle with fibre type U associated to λ, is called *seeded fibre bundle*[2] if for every $x \in M$ exists $(V_x, \sigma_x) \subset P_U$ presymplectic regular manifold which satisfies:

(a) $\exists \Psi_x : V_x \longrightarrow \pi_U^{-1}(x)$ surjective submersion verifying

$$\pi_U^{-1}(x) = \frac{V_x}{\ker \Psi_{x_*}}.$$

(b) Given $x, y \in M$, if $V_x \cap V_y \neq \emptyset$, then $V_x = V_y$.
(c) $\ker \sigma_{xw} \subset Q_w$, Q being the horizontal distribution and $w \in V_x$.

Theorem 2.2 *Let $\lambda = (P, M, \pi, G)$ be a principal G–bundle provided with connection, (U, σ_U) a symplectic Hausdorff manifold. The fibre bundle associated to λ with fibre type U is a seeded fibre bundle if and only if a foliation S contained in the horizontal distribution exists.*

Proof. Let $\lambda[U] = (P_U, M, \pi_U, G)$ a seeded fibre bundle provided with connection (the induced by λ) and presymplectic manifolds family $\{(V_x, \sigma_x)\}_{x \in M}$. Then $E = \bigcup_{x \in M} V_x$ is a submanifold of P_U, as it is a disjoint union of submanifolds. Let us see that it is also presymplectic:

Let us take X, Y two vector fields tangent to E at $z \in E$; there is a unique V_x such that $z \in V_x$, and two vector fields X', Y' on V_x verifiying

$$i_{x_*} X'_z = X_z, \tag{1}$$

$$i_{x_*} Y'_z = Y_z, \tag{2}$$

---

[1] The greatest integer not larger than $f(\theta)$ take the values 0,1,2,3 which remind us the strangeness quantum–numbers.

[2] If we consider G the restricted Poincaré group, U an orbit of G in its coalgebra, and M the space–time manifold, it was seen in Ref. [7] that it is a seeded fibre bundle that allows to extend the results of the relativistic elementary particles of J. M. Souriau (see Ref. [10]) to general relativity.

where $i_x: V_x \longrightarrow E$ is the canonical immersion.

We can define in E the 2–form σ in the following way:

$$\sigma(X_z, Y_z) = \sigma_x(X'_z, Y'_z). \tag{3}$$

Given $v \in P_U$, one $x \in M$ such that $v \in V_x$ there exists. Then, from (c) of Definition (2.1), we have that

$$\ker \sigma_v = \ker \sigma_{xv} \subset Q_v. \tag{4}$$

Defining $S = \ker \sigma_\flat$ we obtain a foliation (notice that every $S_x = \ker \sigma_{\flat x}$ is one of them).

Conversely, let $S \subset Q$ be the foliation, let us take $\{H_p 1_{p \in P_U}$ the leaves of S, and we define

$$V_x = \pi_U^{-1}(x) \times H_p, \tag{5}$$

where $\pi_U(p) = x$. Let $p_1: \pi_U^{-1}(x) \times H_p \longrightarrow \pi_U^{-1}(x)$ be the canonical projection, then $\sigma_x = p_1^* \sigma_{\pi_U^{-1}(x)}$ is a presymplectic form defined in V_x. □

Corollary 2.3   *Under the assumptions in Theorem 2.2, if G (the structural group of the seeded fibre bundle) is dynamic in U, then G is dynamic in the presymplectic manifold family.*

Proof.   There is a dynamic action ϱ of G on U. Let H_p be a leaf of the foliation S. Let $V = U \times H_p$ (with the 2–form $\sigma_V = p_1^* \sigma_U$) we define the action $\varphi = \varrho \times \mathbb{1}_{H_p}$ that is dynamic. But, as each fibre is symplectomorphic to the fibre type U, we have $\varphi_x = \varphi|_{V_x}$ is dynamic. □

3 Symplectification in $SU(2, \mathbb{C})$

The properties of $SU(2, \mathbb{C})$ are well known. Its Lie algebra, $su(2)$, is the linear space of the traceless antihermitian matrices of the form iA (see Refs. [2],[3]), where

$$A = \begin{pmatrix} z & x + iy \\ x - iy & -z \end{pmatrix}, \qquad (x, y, z) \in \mathbb{R}^3. \tag{6}$$

Let σ_1, σ_2 and σ_3 be the Pauli matrices; $\{i\sigma_k\}_{k=1}^3$ form a basis for $su(2)$ that will be denoted by $\{\tilde{\sigma}_k\}_{k=1}^3$.

The *adjoint representation* of $SU(2, \mathbb{C})$ in its algebra will be given by

$$Ad(M, iA) = M \circ iA \circ M^{-1}. \tag{7}$$

In order to study the *coadjoint representation* we define the action of the elements in its coalgebra, $s^* u(2)$, on the algebra in this way:

$$\alpha(iA) = -\frac{1}{2} \text{tr}\, (\alpha \circ iA), \tag{8}$$

where $\alpha \in s^*u(2)$, and $iA \in su(2)$.

The matrices $\{\tilde{\sigma}_k\}_{k=1}^3$ are a basis of the coalgebra verifiying:

$$\tilde{\sigma}_i(\tilde{\sigma}_j) = \delta_{ij}, \quad \det \tilde{\sigma}_i = 1, \quad 1 \le i, j \le 3. \tag{9}$$

The *coadjoint representation* of $SU(2, \mathbb{C})$ in $s^*u(2)$ is given by

$$Ad^*(M, \alpha)(iA) = \alpha Ad(M, iA) = \alpha(M \circ iA \circ M^{-1}). \tag{10}$$

Taking into account (10) it is easy to prove the following Proposition

Proposition 3.1 *The isotopy group at* $\tilde{\sigma}_3 \in s^*u(2)$ *is a 1–dimensional connected Lie group, and its orbit* $\mathcal{O}$ *(for the action* Ad^**) is the unit sphere* S^2.

The elements of $\mathcal{O}$ can be written in polar cordinates

$$(\sin \theta \sin \varphi, \sin \theta \cos \varphi, \cos \theta), \quad 0 \le \theta, \varphi < 2\pi \tag{11}$$

and the Killing vector fields associated to $\{\tilde{\sigma}_k\}_{k=1}^3$ are

$$Z_{\tilde{\sigma}_1} = -2 \cos \varphi \frac{\partial}{\partial \theta} + \frac{2 \sin \varphi \cos \theta}{\sin \theta} \frac{\partial}{\partial \varphi}, \tag{12}$$

$$Z_{\tilde{\sigma}_2} = 2 \sin \varphi \frac{\partial}{\partial \theta} + \frac{2 \cos \varphi \cos \theta}{\sin \theta} \frac{\partial}{\partial \varphi}, \tag{13}$$

$$Z_{\tilde{\sigma}_3} = \frac{-2}{\sin \theta} \frac{\partial}{\partial \theta}. \tag{14}$$

The Lagrange 2–form of $\mathcal{O}$ is given by

$$\sigma_{\mathcal{O}} = \sigma \left(\frac{\partial}{\partial \theta}, \frac{\partial}{\partial \varphi} \right) = \frac{1}{2} \sin \theta d\theta \wedge d\varphi. \tag{15}$$

From (12)–(15) we obtain the following contractions

$$i_{Z_{\sigma_1}} \sigma_{\mathcal{O}} = \sin \varphi \cos \theta \, d\theta + \sin \theta \cos \varphi \, d\varphi, \tag{16}$$

$$i_{Z_{\sigma_2}} \sigma_{\mathcal{O}} = \cos \varphi \cos \theta \, d\theta + \sin \theta \sin \varphi \, d\varphi, \tag{17}$$

$$i_{Z_{\sigma_3}} \sigma_{\mathcal{O}} = - \sin \theta \, d\theta. \tag{18}$$

4. Example

Let M be the space–time manifold, and $W \subset M$ a 2–dimentional submanifold. We consider $V = S^2 \times W$ and we define the Lagrange form $\sigma_V = p_1^* \sigma_{S^2}$, where $p_1 : S^2 \times W \longrightarrow S^2$ is the canonical projection. Besides, we define an action of $SU(2, \mathbb{C})$ over V as

$$\varphi(g, (s, w)) = \left(Ad_g^*(s), w \right), \tag{19}$$

and from Corollay 2.3 it is dynamic.

Let $\lambda = (E, M, SU(2, \mathbb{C}), \pi)$ the principal fibre bundle with estructural group $SU(2, \mathbb{C})$ and basis M, we can construct the fibre bundle $\lambda[S^2] = (P, M, SU(2, \mathbb{C}), \pi_{S^2})$ associated to λ with fibre type S^2. If we ask for ker σ_V to be contained in the horizontal distribution of $\lambda[S^2]$, then from Theorem 2.2 we get that $\lambda[S^2]$ is a seeded fibre bundle.

Finally, let us study the Conservation Laws obtained from Noether's Theorem applied to V. From (16)–(18) and by the definition of dynamic group moment (see Ref. [1]), and making use of the fact that S^2 is connected, we obtain

$$J_1 = -\sin\theta \sin\varphi, \quad J_2 = \sin\theta \cos\varphi, \quad J_3 = \cos\theta - 1. \tag{20}$$

We call strageness function the following:

$$f(\theta) = J_1^2 + J_2^2 + J_3^2 = 4\sin^2\frac{\theta}{2} \tag{21}$$

which is obviously preserved.

References

1.   R. Abraham, J. E Marsden: " Foundations of Mechanics" (W.A. Benjamin, Cummings, Massachussets, 1978).

2.   F.J. Dyson: "Symmetry Groups in Nuclear and Particle Physics" (W. A. Benjamin, New York, Amsterdam, 1966).

3.   Y. Choquet Bruhat, C. de Witt, M. Dillard: "Analysis, Manifolds and Physics", (North-Holland, Amsterdam, 1982).

4.   D. Husemoller: "Fibre Bundles" (Springer-Verlag, New York, Heidelberg, Berlin, 1966).

5.   S. Kobayashi, K. Nomizu: "Foundations of Differential Geometry", Volume I (Interscience Publishers, New York, 1963).

6.   V. Liern: "Movimientos en Relatividad General" (Post-Graduate Thesis, Valencia, 1990).

7.   V. Liern, J. Olivert: "Extension of Geodesics Principle", in Proc. Encuentros Relativistas Españoles 91, ed. by J. Ibañez (World Scientific, Singapore), to appear.

8.   J. Olivert: J. Math. Phys. 21, 7 (1980).

9.   L. S. Pontrjagin: "Topological Groups" (Gordon and Breach, New York, London, Paris, 1977).

10.   J. M. Souriau: "Structure des Systèmes Dynamiques" (Dunod, Paris, 1970).

On a Project for a Repetition
of the Michelson – Morley Experiment

Ll. Bel [1], J. Martín [2]

[1]Laboratoire de Gravitation et Cosmologie Relativistes, CNRS/URA
769, Tour 22/12, 4, P. Jussieu, 75230 Paris, France
[2]Grupo de Física Teórica, Facultad de Ciencias, Universidad de
Salamanca, 37008 Salamanca, Spain

Abstract: Following the ideas of Ll. Bel expressed at ERE's 91 we give a definition of rigid congruences in both General and Special Relativity, and we try to make the definition plausible. To this end we recall Fermat's principle in General Relativity and we show that this principle allows us to reinterpret the "quotient metric" as the quadratic form which defines the optical length in a gravitational field. We apply the definition to the Earth–Sun system in the post–Newtonian approximation. Furthermore we compute the Fermat tensor and the corresponding relative variation of the speed of light in a Michelson–Morley-like experiment performed on the Earth's surface. According to all measurements to date, this quantity is extremely small (10^{-13}). Consistently with these results, negotiations regarding the repetition of the Michelson–Morley experiment, in its modern version (laser interferometry), are being achieved in order to test these results.

References

1. M. Born: Phys. Zeits. **10**, n.**22**, 814 (1909); Ann. der Phys. **30**, n. **11**, 1 (1909).
2. G. Herglotz: Ann. der Phys. **31** 393 (1910).
3. F. Noether: Ann. der Phys. **31** 919 (1910).
4. Recommandations I,II,III de la XXI Assemblée Générale de l'Union astronomique internationale.
5. Ll. Bel, in *Recent Developments in Gravitation* (eds. E. Verdaguer, J. Garriga, J. Cespedes). World Scientific, Singapore, 1990.
6. H. Weyl: Ann. der Phys. **54** 117 (1917).
7. T. Levi–Civita: Nuovo Cimento **16** 105 (1918).
8. V. Perlick: Class. Quantum Grav. **7** 1319 (1990).
9. I. Kovner: Astr. J. **351** 114 (1990).
10. M. G. Sagnac: C. R. Acad. Sc. Paris **157** 708, 1410 (1913); J. de Phys. **5**, t.**IV**, 177 (1914).

11.   F. Harres: Dissertation (Jena, 1911).
12.   B. Pogani: Ann. der Phys. **80** 217 (1926) and **85** 244 (1928); Naturwissenschaften **15** 177 (1927).
13.   A. A. Michelson, E. W. Morley: Am. J. Sci. **34** 333 (1887); G. Joss, Ann. Phys. **7** 385 (1930).
14.   A. A. Michelson: Astr. J. **LVI** 137 (1925); A. A. Michelson, H. G. Gale: Astr. J. **LVI** 140 (1925).
15.   R. J. Kennedy, E. Thorndike: Phys. Rev. **42** 400 (1932).
16.   A. Brillet, J.L. Hall: Phys. Rev. Let. **42** 549 (1979).
17.   J. M. Aguirregabiria, Ll. Bel, J. Martín, A. Molina: *Proceedings of the 1991 Relativity Meeting, Bilbo*. World Scientific.
18.   Ll. Bel, J. Martín: *Fermat's principle in general relativity*, submitted to Classical and Quantum Gravity.

Nonlinear Evolution of Cosmological Inhomogeneities

D. Sáez

Departamento de Física Teórica, Universidad de Valencia, Burjassot,
Valencia, Spain

Abstract: The nonlinear evolution of a cosmologically significant fluid is studied up to shell crossing. The magnetic part of the Weyl tensor, the pressure and the vorticity vanish. A suitable spatial grid is chosen. The relativistic Ellis equations are particularized on the world lines defined by the nodes of the grid and, then, the resulting equations are numerically solved. The integrations are performed in suitable Lagrangian inertial coordinates, in which the differential equations become ordinary. After the integration, a method to change from Lagrangian to Eulerian coordinates is applied. This approach has been outlined with the essential aim of studying the evolution of large scale cosmological structures. In this case, no important relativistic effects are expected, but a relativistic approach based on the Ellis formalism appears to be suitable due to two main reasons: it facilitates a rigorous choice of the initial conditions according to the gauge invariant Bardeen formalism, and it is not more involved than some nonrelativistic schemes. In order to test the method, its results are compared with the Tolman–Bondi solution in the spherically symmetric case. The comparison is very encouraging. In the limit of vanishing numerical errors, our method approaches an exact 3–dimension solution to the problem of structure evolution in the mildly nonlinear regime; hence, this solution improves on the Zel'dovich one, which is exact only in the 1–dimension case. Both solutions apply up to shell crossing. The extension of the proposed approach beyond caustic formation deserves attention.

The Great Attractor and the COBE Quadrupole

M. J. Fullana[1], D. P. Sáez[1] and J. V. Arnau[2]

[1]Departament de Física Teòrica, Universitat de València, Burjassot, València, Spain
[2]Departament de Matemàtica Aplicada i Astronomia, Universitat de València, Burjassot, València, Spain

Abstract: A nonlinear model for the Great Attractor is built. It is based on the Tolman-Bondi solution of the Einstein equations. The angular temperature distribution of the Cosmic Microwave Background produced by the Great Attractor is numerically obtained. Several realizations of the Great Attractor are studied. In all the cases, the distance from the Great Attractor to the Local Group is $\sim 43h^{-1}$ Mpc, the density contrast reduces to a half of the central value at a radius of $9h^{-1}\mathrm{Mpc} \leq R_c \leq 14h^{-1}\mathrm{Mpc}$, and the dipole due to the infall towards the inhomogeneity center is $1.33 \times 10^{-3} \leq D \leq 1.8 \times 10^{-3}$. A complete arbitrary background is assumed; the density parameter, Ω, and the reduced Hubble constant, h, ($H = 100h\,\mathrm{Km\ s^{-1}Mpc^{-1}}$) are $0.2 \leq \Omega \leq 1$ and $0.5 \leq h \leq 1$ respectively. The total quadrupole Q is split in two parts, the relativistic Doppler quadrupole, Q_D, and the reduced quadrupole, Q_r, produced by nonlinear gravity. The quadrupoles of the chosen realizations appear to satisfy the following inequalities $5 \times 10^{-7} \leq Q \leq 12.5 \times 10^{-7}$ and $-0.4 \times 10^{-7} < Q_r < 1.6 \times 10^{-7}$; this means that $|Q_r|$ ranges from 0.8 % to 3.2 % of the quadrupole Q_{rms} measured by COBE. Therefore, the subtraction of Q_r from Q_{rms} becomes irrelevant.

Springer-Verlag
and the Environment

We at Springer-Verlag firmly believe that an international science publisher has a special obligation to the environment, and our corporate policies consistently reflect this conviction.

We also expect our business partners – paper mills, printers, packaging manufacturers, etc. – to commit themselves to using environmentally friendly materials and production processes.

The paper in this book is made from low- or no-chlorine pulp and is acid free, in conformance with international standards for paper permanency.

Lecture Notes in Physics

For information about Vols. 1–384
please contact your bookseller or Springer-Verlag

New Series m: Monographs